From Fragmentation to Wholeness

Many of us wonder why, despite many sincere efforts towards positive social transformation, nothing seems to change. This wonderful book both answers that question and provides a roadmap for real change. Avoiding woolly thinking and a retreat into the fuzzy ways in which the notion of "spirituality" is so often used, the book provides a rigorous, comprehensive and philosophically sound model that both profoundly challenges received notions and provides, as few texts do, answers and a guide to practice. This is a foundational and fundamental book that has the potential to upend the ways in which we think the world and act within it.

—***John Robert Clammer*, Professor of Sociology,**
O.P. Jindal Global University, Former Professor,
United Nations University (Tokyo)

Author of *Cultural Rights and Justice: Sustainable Development, the Arts and the Body* (Palgrave MacMillan 2019)

> The 2030 Agenda was launched with great hope, yet has since been overshadowed by pandemic, conflict, and global uncertainty. Many have given up on the SDGs, but not Gupta and Singh. Rather than seeking external saviours, they show that the path to transformation lies within ourselves. This is not a self-help book, but a profound journey linking inner perception with external reality. The authors remind us that holding a mirror to oneself is essential, for individuals, societies, and nations.

—***Mr. Roy Wadia*, Regional Communications Chief,**
World Health Organization Western Pacific Region

We often think of ourselves as peace-loving, yet we inhabit a conflicted world. Gupta and Singh show why: the fragmentation of the world begins in us.

They argue that governance failures, including the shortcomings of the SDGs, will remain inevitable unless we transform how we perceive. Their book reveals how fear, identity, and memory distort perception, shaping the institutions we build and the conflicts we perpetuate.

This is not a guide to governance or reform. It is a profound examination of how inner disorder manifests as outer crises. Clear, challenging, and grounded in real-world examples, it offers fresh insights for readers interested in ecology, geopolitics, psychology, and personal transformation.

—**By *Pascal Laureyn*, freelance journalist**

From Fragmentation to Wholeness: Spirituality, Awareness and the Future Beyond the SDGs

BY

DEVESH GUPTA
Emerge, India

AND

NARESH SINGH
Jindal School of Government and Public Policy, India

emerald PUBLISHING

United Kingdom – North America – Japan – India – Malaysia – China

Emerald Publishing Limited
Emerald Publishing, Floor 5, Northspring, 21-23 Wellington Street, Leeds LS1 4DL

First edition 2026

Printed and bound by CPI Group (UK) Ltd, Croydon, CR0 4YY

British Library Cataloguing in Publication Data
A catalogue record for this book is available from the British Library

ISBN: 978-1-80686-678-6 (Print)
ISBN: 978-1-80686-677-9 (Online)
ISBN: 978-1-80686-679-3 (Epub)

Contents

About the Authors

Devesh Gupta is an educator, an author and a facilitator exploring the meeting point of consciousness, education and governance. He is the founder of Emerge, a space for re-examining learning, leadership and public policy from the root observing how fragmentation begins in the mind and how authentic action can arise from awareness rather than identity. His work examines how clarity of perception can give rise to ethical, intelligent and compassionate systems. Devesh serves as the Director of Education and Youth Initiatives at The Dais, an organisation working on youth engagement, education and systems change. He is also the Vice President of the Spirituality, Science, and Public Policy Network, a collaboration between Emerge and the Centre for Complexity Economics, Applied Spirituality and Public Policy (CEASP) at O.P. Jindal Global University, where he also teaches as a Visiting Faculty. Through Emerge and The Dais, he has facilitated programmes and dialogues with universities and public institutions including the University of Toronto, Queen's University of Charlotte, the University of Malaya, the University of South Africa (UNISA), Stanford University, the University of the South Pacific and the Indian Institute of Public Administration (IIPA) and has worked with the Indian Army and Air Force on ethics and leadership. His work through these platforms spans conflict and post-crisis regions like Afghanistan, to living in the Himalayas along the India–Nepal border for sustainable development and community engagement, to working with children on the coasts of Fiji and with students in rural and urban schools across India. Educational partnerships include Mayo College, Ajmer Mayo College Girls' School, Ajmer, Welham Boys' School, Dehradun The Daly College, Indore and Teach For India, with engagements supported by the US Embassy, UNDP, UNESCO, UNICEF and several state governments. Trained in international humanitarian and refugee law, he brings legal, psychological and philosophical insight into his work across education, leadership, peacebuilding and governance.

Prof. Dr. Naresh Singh is a Professor and the former Executive Dean, Jindal School of Government and Public Policy, O.P. Jindal Global University. He is the Director of the Centre for Complexity Economics, Applied Spirituality and Public Policy (CEASP), the Co-Director of the Centre for Legal Empowerment of the Poor, the Senior Fellow at the Jindal India Institute and the President of the Spirituality, Science and Policy Network. He has more than three decades of public policy experience in senior positions in the United Nations Development

Programme (where he was the Executive Director of the Commission on Legal Empowerment of the Poor) and the Federal Government of Canada. Professor Singh was a visiting scholar at several universities in North America, including Boston, Harvard, Fordham, McGill, Waterloo and Guelph. He has been a consultant to the United Nations Development Programme, the Commonwealth Secretariat, the Pan American Health Organization, the International Labour Organization and the Food and Agriculture Organization.

Preface: Seeing the Book as a Mirror

This book does not offer solutions in the conventional sense. It does not present methods, strategies or frameworks for implementation. It begins with a simple observation: that despite widespread efforts towards reform, the world remains caught in patterns of division, conflict and delay. Governance systems, educational institutions and even movements for sustainability seem unable to bring about lasting transformation.

The central argument of this book is that the disorder we face is not merely external or structural. It is a reflection of the inner condition of the human mind. The fragmentation seen in institutions is an extension of the fragmentation within thought, identity and perception. Unless this is clearly understood, even the best-designed policies will remain superficial.

When we say that 'unless this is clearly understood', it requires us to speak about what we mean by understanding, not knowing. To understand means to observe the nature and structure of something, its origin, what it does, its movement and its danger. And in seeing the danger clearly, not intellectually, there is an ending. We change when we see the problem clearly, not through thought, but when awareness perceives the danger directly. That perception is understanding, and in that understanding, there is ending, change and transformation.

Hence, the book starts as a movement of observation. This might be called meditation in a spiritual sense. We start with what is, and that observation of the outer brings awareness to the inner. In observing and understanding the movement of the self in relationship – what it does, its roots and consequences – there may be an ending. And that ending brings one to choiceless action.

While this is described linearly, it must be lived through direct perception, insight and change at every step. This is not a book of intellectual assembly but a movement that requires a different sense of reading and listening. To read rightly is to observe without the past. That is the essence of true perception.

So to proceed, one must realise that what obstructs perception in the world – the self – can also obstruct the very act of reading. In Part II, when the movement of the self is revealed, there may be resistance, agreement or disagreement. But all these are movements of the past. If that is seen, if the book can act as a mirror while you read, you may observe your own self, your reactions and identifications. To see that is to go beyond it. Let awareness do what it does. The book is merely a tool. When there is awareness, the book and its chapters fall into their right place. If you read from an aware mind, the awareness will negate what's false, and clear perception of false or truth is truth.

The book is structured as a movement. Part I (Chapters 1–3) begins with the outer world, the crises, systems and institutions we have created. This is not examined from a distance but as an invitation to observe how the world functions as a mirror. In understanding the world, we begin to understand ourselves in relationship.

Chapter 1 opens this inquiry by revealing how global disorder, from ecological collapse to institutional inertia, reflects the same fragmentation that operates within the individual. It is not merely a critique of external systems but an invitation to perceive the world as the projection of our inner state. Chapter 2 deepens this observation by showing how our core institutions – education, governance and health – are not neutral structures but embodiments of fear, control and psychological fragmentation. This chapter not only reveals that the global crisis is psychological in nature but also introduces the language and clarity of perception needed for the journey ahead. It acts as a shared foundation, so that when the reader encounters critiques of frameworks later, the insight is not theoretical but lived. Chapter 3 then extends this seeing to the SDGs and similar global frameworks. Even our most noble aspirations mirror the same movement of division, translating wholeness into targets, metrics and silos. Fragmentation hides in the language of unity.

Yet this is not where the inquiry ends; it is where it must turn inward. The crisis is not only structural. It is perceptual. Our responses are limited because they arise from the same identity and memory that created the problem. The outer and the inner are not two. To see this clearly is to prepare the ground for transformation.

Part II explores this inner movement, not thematically, but diagnostically and dissolutively. Each chapter focuses on a different facet of the self: fear, desire, time, knowledge, memory, relationship and identification. These are not truly separate, but we often experience them as such. What is revealed is their common root. Fear is desire in movement. Time is thought. Experience and image are both memory. Thought, desire, fear, time and identity are not different things. They are the self in movement. These chapters build progressively, emergently, yet, all show the movement of the self.

At first, this may appear repetitive. But the repetition is not of language. It is the repetition of life. Because the fragmented mind perceives these movements of fear, desire, thought, time as separate from each other, and also from its own self, we fail to see they are one movement: the illusion of separateness appearing as many. To see this, one must enter the mirror. The chapters do not merely describe the self. They allow us to observe its operation in real time. Each begins from 'what is': time, memory, knowledge, desire and reveals how these sustain fragmentation. In the seeing of 'what is', there is the ending of 'what divides'.

To understand, then, is to end. Not through resistance, practice or delay but through clear perception of what is. Each chapter in Part II follows this arc: What is it? How is it sustained? What does it do? And is there freedom in seeing it wholly?

One may explore Advaita Vedanta or Mahayana Buddhism, study Freud, Heidegger or Foucault, engage with ontology, epistemology or systems theory.

Yet none of these, however profound, can end the self. As long as there is a knower who stands apart from the known, the self continues. Insight does not lie in accumulation. It arises when there is direct perception, and that perception is only possible when inquiry is not analysis, not comparison, not escape, but a total seeing of the self in action. This movement of seeing unfolds most clearly when one begins to observe their own fragmentation, not abstractly, but as it expresses itself through identity, time, image and becoming. That is the arc of Part II. If the earlier sections opened a door, it is here that the mirror is held up most directly. In that mirror, the self may be seen, not explained. And when it is seen, not as a theory but as fact, its movement ends. Then, intelligence acts, not the individual.

Part III turns from perception to action. Not action as decision or choice but as the natural flowering of intelligence. When the self ends, what remains is wholeness. From that wholeness, action arises. Such action has no centre, no reward, no ideology. It is love in motion where perception is action.

The intention behind this book is not to scale impact in the conventional sense. It does not seek followers, frameworks or movements. That would only repeat the very patterns it exposes. What matters is not reach but depth. The content of consciousness is shared. When there is total transformation in one human being, it changes the field for all. This change happens in physical time but arises from a different ground, beyond psychological time.

This is not the isolated breakthrough of a heroic individual but the emergence of a single intelligence: a whole mind beyond time, acting through any brain that is in communion and presence. Such a mind sees without division. It is not personal. It is not shaped by memory. It has no centre. In that centreless awareness, perception is no longer filtered through fear, desire or image. What acts is not the individual but intelligence itself which is wholeness, love and clarity.

The intention, then, is to create a mirror and a space for those who arrive here, to help them inquire, not to influence. A mind in inquiry often travels far and wide to find order, from gods to religions, from monasteries to psychedelics, from knowledge to control. But eventually, it must return to the seeing of itself. Until that happens, no method or movement can help.

This book intends to be that pointer: a friend, a companion, a quiet guide, so that the reader does not lose themselves in psychological time, repeating the cycle of effort, action and suffering. In a world on fire, urgency demands clarity. Right education and love can cut through conditioning. When one is ready to be aware, this journey need not be one of endless pain. Seeing clearly ends the delay.

That is the work this book invites: not the application of ideas but the ending of the 'me', so that what remains is a space through which awareness and love may act. You are not being led through this book. You are being invited to observe, and in that observation, to meet the whole of life.

Introduction

SDGs at the 2030 Horizon: A Time for Reflection

As the finish line for the 2030 Agenda on the Sustainable Development Goals (SDGs) emerges on the horizon, the time for deep reflection, assessment of strategies and policies and evaluation of progress against targets and indicators has arrived. With 10 of the 15 years (2015–2030) now behind us, this book proposes to take a deep dive into the lessons learned and to pay attention to what we have come to see as the fundamental missing dimension. We recognise that what we propose may appear far removed from typical approaches to sustainable development to date and might even be considered soft or impractical by many. At the same time, we understand that others will find immense value in what we propose, though they may need time to digest, reflect on, investigate, enquire and understand before actively considering them as part of what succeeds the SDGs post-2030. Hence the timeliness of this book.

The SDGs have represented a milestone in international cooperation, mobilising governments, civil society, the private sector and millions of citizens towards a common vision of peace, prosperity, people, planet and partnership (United Nations, 2015). In the decade since their adoption, progress has been achieved in many areas. Extreme poverty has been further reduced, millions have gained access to renewable energy, maternal and child mortality rates have declined and new commitments to gender equality and digital inclusion have emerged (UNDP, 2022). Efforts to address climate change and biodiversity loss, though incomplete, have spurred unprecedented levels of investment in clean energy and ecosystem restoration (Rockström et al., 2023).

Yet progress has been uneven and insufficient. And progression always doesn't mean change. Despite notable successes, projections confirm that only a fraction of the Goals will be fully achieved by 2030. Some of the most urgent challenges including rising inequality, violent conflict, biodiversity loss, unsustainable consumption and global health vulnerabilities remain unresolved (United Nations, 2025).

From Fragmentation to Wholeness, 1–7

doi:10.1108/978-1-80686-677-920261001

Contradictions and Structural Limits

The unfinished business of the SDGs reflects not only gaps in resources, political will and coordination, but also deeper structural and human factors. Outer reforms alone like technological solutions, financial flows and institutional agreements are not sufficient to address problems that are complex, interconnected and deeply rooted in human behaviour. Greed, fear, exclusion and short-term competition for power and resources continue to undermine cooperation. These patterns of thought and behaviour have driven conflict, ecological destruction and persistent inequality. The global community has thus reached a paradoxical point. Even when/if scientific and political consensus exists on urgent action, climate change being a prime example, progress and change remains fragmented. The gap lies not in knowledge or resources, but in perception and sensitivity. Can we identify, see and understand the source of our greed, fear and competition for power and resources? Because without understanding the problem, there is no real solution. This book seeks to answer precisely that question.

Four Decades of Sustainable Development Thinking

Much has been learned about sustainable development over the last four decades, beginning with the World Commission on Environment and Development's landmark report 'Our Common Future' (WCED, 1987). The international community and national entities have designed projects, programmes and policies with a multiplicity of outputs, outcomes and indicators. Some relied on outmoded linear approaches; others experimented with systems thinking and, more recently, complex adaptive systems perspectives. Financial innovations, such as blended finance, which uses public funds to de-risk larger pools of private investment have attempted to mobilise resources at scale (UNEP, 2016). Parallel to this, scholars and practitioners have critiqued the limitations, and even harms, of neo classical economics, developing alternatives such as circular economy models (Geissdoerfer et al., 2017), doughnut economics (Raworth, 2017), degrowth (Kallis, 2011), post-growth economics (Jackson, 2017) and capital approaches that seek to maintain or enhance human, social and natural capital bases while meeting the needs of present and future generations (Costanza et al., 2014). Significant progress has also been achieved in green technologies and renewable energy transitions.

Academically, there has been a marked expansion in sustainability science and the study of sustainable development as a practice (Clark & Harley, 2020). Sub-disciplines like behavioural economics, complexity economics and ecological economics have gained recognition for their relevance to sustainability. At the policy level, terms such as polycrisis, metacrisis and super-wicked problems have entered the discourse (Levin et al., 2012). Tools such as scenario building, design thinking, strategic foresight, network science and complexity-informed policy instruments are increasingly in vogue (Singh & Bhatnagar, 2023).

The Missing Dimension: Inner Systems

Amidst this proliferation of approaches, one must not lose sight of the fundamental nature of sustainable development: wholeness. As Clark and Harley (2020) emphasise, fragmentation into 'environmental sustainability', 'economic sustainability' and 'social sustainability' is counterproductive. Sustainability requires recognition of the dynamic interplay of economic, social and ecological subsystems as a single whole system. Yet, as valuable as this perspective is, it remains an *outer systems* view. The fundamental source of fragmentation does not lie only in systems but in the human psyche itself, what may be called the *inner dimension*. Work on this inner dimension remains limited and insufficiently deep. The existing literature describes these dimensions in terms of psychology, spirituality and consciousness, pointing to individual and collective mindsets, values, beliefs, worldviews and associated cognitive, emotional and relational capacities (Ives et al., 2023; Wamsler, 2020; Wamsler & Bristow, 2022).

Inner transformation refers to a profound shift in perspective towards a relational paradigm, emphasising interdependence and connectedness with others and the world, while having a deeply caring and compassionate quality of relationship (Walsh et al., 2020). This view resonates with complexity science's relational ontology (Meadows, 2008) but expands it by integrating intra-being relationality, including one's relationship to oneself, body, emotions and perceptions (Nhat Hanh, 1987).

Ives et al. (2023) conclude that the quality of our relationships, with ourselves, with others, and with the biosphere, ultimately creates sustainable or unsustainable cultures and structures. Importantly, inner transformation is not an introspective substitute for practical action but a necessary ground for activating latent human potential at individual, collective and systemic levels. Building on this, Singh and Bhatnagar (2023) argue that the illusion of separation among humans and between humans and nature is the root of unsustainability. The Inner Development Goals (IDGs, 2025) similarly highlight the bridging role of five inner dimensions, being, thinking, relating, collaborating and acting, across 23 identified skills. Reviews by Wamsler (2021, 2022) reinforce the importance of integrating inner and outer transformation for sustainability and climate action. Despite these promising developments, progress remains inadequate. Most approaches still attempt to integrate fragments rather than challenge the very fragmentation of thought. Robert Pirsig's insight is instructive here:

> The true system, the real system, is our present construction of systematic thought itself, rationality itself. If a factory is torn down but the rationality which produced it is left standing, then that rationality will simply produce another factory. If a revolution destroys a government, but the systematic patterns of thought that produced that government are left intact, then those patterns will repeat themselves in the succeeding government. There's so much talk about the system. And so little understanding. (Pirsig, 1974, p. 93)

This book takes that insight seriously. It argues that unless the rationality of fragmentation in human consciousness is directly seen, any reform, no matter how innovative, will reproduce disorder.

From Outer Systems to Inner Perception

The recognition of inner dimensions raises a deeper question: why do we fragment reality at all? Complexity science helps us see interconnections, yet even integration remains within the logic of fragmentation, seeking to assemble wholes from parts. As Gupta (2024) argues, wholeness is not the sum of fragments but the ground from which all perception and action must arise, and argues that inner and outer is just one movement. Similarly, institutions mirror the consciousness of those who create them. When the self, driven by fear, desire and identity, becomes the basis of governance or reform, disorder is perpetuated under the guise of progress. This perspective resonates with Jiddu Krishnamurti's (1983) insight that 'to transform the world, we must begin with ourselves; and what is important in beginning with ourselves is the intention.' The fragmentation of thought, dividing observer and observed, self and other, inner and outer, generates conflict at every scale. Bohm (1980) similarly emphasised that fragmentation is not only a mistake in thinking but a lived reality that structures societies, institutions and even scientific paradigms. Hence, the challenge of sustainable development is not merely to design better frameworks but to see directly how the conditioned patterns of thought with its the pursuit of certainty, the grasping at identity, the flight from fear replicate themselves in policy, economics and education. Awareness of this movement, without choice or control, opens the space for intelligence to act.

The Arc of Inquiry

The movement of this book follows a simple but radical trajectory:

- *Diagnosis*: Our crises are not separate. Climate breakdown, inequality and political disorder are outward expressions of the same fragmentation within human consciousness. Reforms fail because they mirror this inner division.
- *Inquiry*: The root of disorder lies in the self, in memory, image, knowledge and psychological time. By enquiring into these foundations, we see how violence, fear and conflict are created and perpetuated.
- Response: When fragmentation ends in awareness, action arises from wholeness. This opens new possibilities for governance, institutions, education and human relationship, not as frameworks to be implemented but as living expressions of perception.

Chapter Roadmap

Part I: Diagnosis – The Outer World Mirrors Inner Division

- *Chapter 1*: *Polycrisis as a Reflection of Inner Fragmentation* Shows how ecological, political, technological and social crises are interconnected reflections of inward disorder.
- *Chapter 2*: *Human Condition-Psychological Roots of Institutional Breakdown* Demonstrates how governance, education and health systems reproduce the same fear, control and competition embedded in the psyche.
- *Chapter 3*: *Beyond Goals and Frameworks: Seeing the Root of Global Crisis* Analyses contradictions within the SDGs, revealing that without a shift in perception, frameworks recycle disorder.

Part II: Inward Inquiry – Understanding the Source of Fragmentation

- *Chapter 4: The Self as the Source and Effect of Fragmentation, Conflict and Violence* Explores how the self generates division, conflict and violence, while also being shaped by the very fragmentation it creates.
- *Chapter 5: The Trap of Knowledge and Belief* Shows how accumulated knowledge and ideology distort perception and prevent renewal.
- *Chapter 6: Relationship, Memory and Image* Examines how relationships are shaped by images and memory rather than direct encounter.
- *Chapter 7: Psychological Time and Inertia* Reveals how projection into the future sustains disorder and prevents transformation.
- *Chapter 8: Love Without Image* Describes how love and compassion naturally emerge when the self and its images end.

Part III: Holistic Response

- *Chapter 9: Choiceless Action and Intelligence* Proposes action arising from awareness itself rather than from choice, ideology or identity.
- *Chapter 10: Conscious Governance and Leadership* Perceives leadership and governance as reflections of wholeness, not instruments of power.
- *Chapter 11: Policy, Education, Systems – Acting from Coherence* Shows how coherence can inform policy and institutions when action flows from perception.
- *Chapter 12: Living Institutions: Mirrors, Spaces and Emergence* Explores institutions as living spaces for awareness and relationship, not control and conformity.
- *Chapter 13: Grand Conclusions and Summary – Awareness Is the Change* Concludes by showing that transformation does not lie in reform or integration but in awareness ending fragmentation.

Invitation

This book is not a framework to be implemented, nor a set of prescriptions to be adopted. It is an inquiry into perception itself. The crises of our age: climate change, inequality, political polarisation, institutional fragility are mirrors of our inner fragmentation. Without understanding the self, reforms will remain partial, reactive and contradictory.

- *For policymakers*, the book illuminates why reforms stall and how genuine transformation requires coherence of perception, not only new designs.
- *For scholars*, it bridges complexity science with explorations of consciousness, providing a foundation for perceiving inner and outer dimensions as one.
- *For educators*, it reveals how institutions mirror conditioning and how awareness can open a different possibility of learning.
- *For citizens*, it emphasises that responsibility begins in seeing that the global crisis is not only external but also internal.

What follows is not a vision to be preserved or a goal to be achieved. It is an invitation to see directly. And in that seeing which is choiceless, unconditioned and whole, lies the possibility of a new action, one not rooted in fear or fragmentation but in awareness itself.

References

Bohm, D. (1980). *Wholeness and the implicate order*. Routledge.

Clark, W. C., & Harley, A. G. (2020). Sustainability science: Toward a synthesis. *Annual Review of Environment and Resources*, *45*(1), 331–386. https://doi.org/10.1146/annurev-environ-012420-043621

Costanza, R., Daly, H., Fioramonti, L., Giovannini, E., Kubiszewski, I., Mortensen, L. F., Pickett, K. E., Ragnarsdóttir, K. V., de Vogli, R., & Wilkinson, R. (2014). Development: Time to leave GDP behind. *Nature*, *505*(7483), 283–285. https://doi.org/10.1038/505283a

Geissdoerfer, M., Savaget, P., Bocken, N. M. P., & Hultink, E. J. (2017). The circular economy – a new sustainability paradigm? *Journal of Cleaner Production*, *143*, 757–768. https://doi.org/10.1016/j.jclepro.2016.12.048

Gupta, D. (2024). *Eternal movement*. Emerge Publications. https://doi.org/10.65169/q7d2z6

Inner, Development Goals. (2025). *The Inner Development Goals framework: Five dimensions and 23 skills for sustainable development and leadership*. Inner Development Goals Initiative. https://www.innerdevelopmentgoals.org

Ives, C. D., Schäpke, N., Woiwode, C., & Wamsler, C. (2023). Imagine sustainability: Integrated inner–outer transformation in research, education and practice. *Sustainability Science*, *18*, 2777–2786. https://doi.org/10.1007/s11625-023-01368-3

Jackson, T. (2017). *Prosperity without growth: Foundations for the economy of tomorrow* (2nd ed.). Routledge.

Kallis, G. (2011). In defence of degrowth. *Ecological Economics*, *70*(5), 873–880. https://doi.org/10.1016/j.ecolecon.2010.12.007

Krishnamurti, J. (1983). *The future of humanity*. Harper & Row.
Levin, K., Cashore, B., Bernstein, S., & Auld, G. (2012). Overcoming the tragedy of super wicked problems: Constraining our future selves to ameliorate global climate change. *Policy Sciences*, *45*(2), 123–152. https://doi.org/10.1007/s11077-012-9151-0
Meadows, D. H. (2008). *Thinking in systems: A primer*. Chelsea Green Publishing.
Nhat Hanh, T. (1987). *Interbeing: Fourteen guidelines for engaged Buddhism*. Parallax Press.
Pirsig, R. M. (1974). *Zen and the art of motorcycle maintenance*. William Morrow.
Raworth, K. (2017). *Doughnut economics: Seven ways to think like a 21st-century economist*. Chelsea Green Publishing.
Rockström, J., Gupta, J., Qin, D., Lade, S. J., Abrams, J. F., Jacobson, L., Rocha, J. C., Zimm, C., Narain, A., Steffen, W., Lenton, T. M., Bai, X., Schlüter, M., Managi, S., Broadgate, W., Gaffney, O., Potočnik, J., & Bhaduri, A. (2023). Safe and just Earth system boundaries. *Nature*, *619*(7968), 102–111. https://doi.org/10.1038/s41586-023-06083-8
Singh, N., & Bhatnagar, D. (2023). *Applied spirituality and sustainable development policy*. Emerald Publishing.
United Nations. (2015). *Transforming our world: The 2030 agenda for sustainable development*. United Nations.
United Nations. (2025). *Sustainable development goals report 2025*. United Nations. https://unstats.un.org/sdgs/report/2025/
United Nations Development Programme. (2022). *Human development report 2021/2022: Uncertain times, unsettled lives*. UNDP.
United Nations Environment Programme. (2016). *Unlocking the sustainable potential of land resources: Evaluation systems, strategies and tools*. UNEP.
Walsh, Z., Böhme, J., & Wamsler, C. (2020). Towards a relational paradigm in sustainability research, practice, and education. *Ambio*, *49*(12), 350–360. https://doi.org/10.1007/s13280-020-01322-y
Wamsler, C. (2020). Education for sustainability: Fostering a more conscious society and transformation towards sustainability. *Sustainability Science*, *15*(1), 1–16. https://doi.org/10.1108/IJSHE-04-2019-0152
Wamsler, C. (2021). Climate change, mindfulness, and inner transformation: The need for a new sustainability paradigm. *Global Environmental Change*, *71*, 102373. https://doi.org/10.1016/j.gloenvcha.2021.102373
Wamsler, C., & Bristow, J. (2022). The role of inner capacities in transformative climate adaptation: A framework for policy and practice. *Environmental Science & Policy*, *133*, 144–157. https://doi.org/10.1016/j.envsci.2022.03.020
World Commission on Environment and Development. (1987). *Our common future*. Oxford University Press.

Part I

Diagnosis: The Outer World Mirrors Inner Division

Humanity stands at a threshold. The 21st century has been described as an age of the polycrisis, a convergence of ecological, economic, political and psychological disruptions that intensify one another. Climate change drives displacement, inequality breeds unrest and technological acceleration deepens both connection and control. What appear as separate emergencies are facets of a single movement: disorder extending through every domain of life.

The Sustainable Development Goals were conceived as a universal roadmap for peace and prosperity, yet progress remains uneven and contradictions multiply. Emissions rise even as we pledge to reduce them, inequality grows despite commitments to inclusion. These are not merely failures of implementation but reflections of how humanity thinks and acts. The very consciousness that fragments life into sectors : environment, economy, education, governance, also fragments action. In seeking to integrate the parts, we rarely question the perception that divided them.

Reform, therefore, often reproduces the crisis it aims to solve. Policies and technologies rearrange symptoms while the root confusion endures: the belief that change lies in the future, in better systems, rather than in the awakening of perception now. Conflict among nations mirrors conflict within the self; competition for power echoes the insecurity of individuals seeking continuity through identity and possession. The world, in its disorder, is a projection of the human mind.

The chapters in this part trace this recognition – from the landscape of global crisis, through the contradictions of reform, to the realisation that the outer world mirrors the inner movement of thought. To understand and transform society, we must look not outward for new frameworks but inward, into the consciousness that creates them. The outer world is the self writ large, and in seeing that fact lies the beginning of a different order.

Chapter 1

Polycrisis as a Reflection of Inner Fragmentation

Abstract

The opening chapter establishes the ground of the inquiry: that the global polycrisis is not a collection of unrelated events but a reflection of inner fragmentation. It examines how ecological collapse, institutional fatigue and widening inequality arise from the same movement of thought that divides observer and observed, inner and outer. The chapter traces how fear, desire and the pursuit of control generate external systems of domination, while reform attempts though sincere, merely rearrange the surface of disorder. Governance, education, health and technology each become mirrors revealing the same pattern: action without understanding. Even complexity and systems theories, while aspiring towards wholeness, remain partial when they arise from fragmented perception. The chapter concludes that the real crisis lies in the actor, the conditioned mind itself. To respond coherently, one must begin not from reaction or reform but from perception that sees division as it arises. Such seeing is the beginning of transformation.

Keywords: Polycrisis; inner fragmentation; systemic reflection; reactive governance; complexity of mind; perceptual shift; outer and inner; wholeness of action

The multiple ecological, political, technological and social crises we face today are not separate phenomena. They are interconnected reflections of a deeper fragmentation, not only in systems but in consciousness itself. What we call the polycrisis is not just the convergence of multiple global challenges; it is the outward projection of inward disorder.

This chapter invites a shift in perception. It asks whether the crises of the world: climate collapse, rising inequality, digital disruption, institutional inertia, are symptoms of something more fundamental: a fractured way of seeing, organising

From Fragmentation to Wholeness, 11–40

doi:10.1108/978-1-80686-677-920261002

and being. When the self is divided, its actions fragment the world. When thought operates in silos: as fear, desire, identity and control, institutions mirror that fragmentation in policy, structure and purpose.

We begin not by proposing solutions, but by questioning the foundations of the problem itself. Only when the perception and the mirror is clear can action be whole.

1.1 Introduction: The World in Crisis

The world stands at a threshold. Rising temperatures, mass extinction, political disintegration, economic disparity and technological overreach no longer appear as isolated crises, but as facets of a single, interconnected emergency. The Intergovernmental Panel on Climate Change (IPCC, 2023) warns of irreversible ecological tipping points. Over one million species face extinction due to habitat loss, pollution and global warming (IPBES, 2019). The Global Footprint Network reports that humanity consumes resources 1.7 times faster than the Earth can regenerate them (Global Footprint Network, 2022). Global inequality continues to rise (UNDP, 2023), youth unemployment exceeds 13% worldwide (ILO, 2023) and mental health disorders are now amongst the leading causes of disability across all age groups (WHO, 2022). At the same time, democratic institutions are eroding and authoritarianism is gaining ground in many parts of the world.

Even the tools meant to address these issues like international agreements, sustainable development frameworks and multilateral institutions are increasingly seen as inadequate, outpaced by the scale and depth of disruption. According to the 2023 Sustainable Development Goals (SDGs) Progress Report, only 12% of the goals are on track to be achieved by 2030, while nearly half are showing minimal or moderate progress (United Nations, 2023). Climate action, biodiversity preservation, food security and peacebuilding as goals lie at the heart of sustainability but remain far from realisation. Development policies frequently operate in contradiction: agricultural expansion is subsidised while forest conservation is promoted, and military spending soars even as peacebuilding is discussed. Governments pledge to phase out fossil fuels while continuing to provide over a trillion dollars in subsidies to oil, gas and coal industries (IEA, 2023). They encourage sustainable agriculture and biodiversity restoration while simultaneously subsidising monocultures such as corn, soy and palm oil that degrade soil and accelerate deforestation (Clapp, 2021; FAO, 2022). Healthy diets are promoted in public health campaigns even as sugar, processed foods and industrial meat production are supported by policy. Massive investments are directed towards highways and airports while commitments are made to reduce transport emissions and electric vehicle subsidies expand even though electricity grids remain largely fossil fuel-dependent (IEA, 2023). States emphasise inclusive growth while tax systems disproportionately favour corporations and the wealthy, deepening inequality (Piketty, 2020; UNDP, 2023). They warn against digital surveillance and emphasise ethical Artificial Intelligence (AI), even as they expand investments in facial recognition and algorithmic policing (Zuboff, 2019). Peace agreements are celebrated while global military expenditure reaches record levels and biodiversity targets are announced while infrastructure corridors, mines and dams consume the very ecosystems those targets are meant to protect (SIPRI, 2023; UNEP, 2023).

The 2025 SDG report, coming 10 years after implementation started, acknowledges some degree of progress:

> More than half the world's population now benefits from some form of social protection, up by 10 percentage points compared to a decade ago. Child marriage and maternal and child mortality rates have fallen, and more young people, especially girls, complete school. Women now hold 27 per cent of parliamentary seats worldwide, up from 22 per cent. Access to electricity and clean cooking has expanded. Internet connectivity has increased by 70 per cent, opening new horizons. Around the world, young people, communities, civil society and local leaders are stepping up their action to deliver on the promise of the SDGs. (United Nations, 2025)

However, it hastens to add that:

> we face a global development emergency. Over 800 million people are trapped in extreme poverty. Carbon dioxide levels are at the highest in over two million years, and 2024 was the hottest year on record, surpassing the 1.5°C threshold. Peace and security have worsened, with over 120 million people forced from their homes, more than double the number in 2015.

The 2025 report then goes on to call for more of what United Nations (UN) reports have always emphasised, including massive investment, removing barriers to the full participation of women in economies and public life, ending gender-based violence and ensuring equal access to education, employment and leadership. The call includes tackling poverty and inequality head-on, closing the digital divide and ensuring that technology, including AI, is harnessed inclusively and responsibly. With five years now left for the final reckoning of the 2030 agenda, the time has come for the international community to go deeper than it has done so far. Yet this progress is double-edged. Women's participation may be rising, and internet connectivity expanding, but what use is such progress when the same systems are employed to intensify division, fuel nationalism and wage wars, whether knowingly or unknowingly? The very tools that are celebrated as markers of advancement can also reinforce fragmentation when they are captured by fear, identity and power.

This convergence of crises has come to be known as the 'polycrisis', a term used to describe not just multiple simultaneous emergencies, but the ways in which those emergencies interact, reinforce and amplify each other. Adam Tooze (2022) describes it as a 'tangled web' where economic, ecological, political and health crises feed off one another in unpredictable ways. The World Economic Forum (WEF, 2023) defines *polycrisis* as a condition in which 'multiple global risks interact in ways that significantly exceed the sum of their parts', generating cascading failures across human and planetary systems. In response, policymakers and institutions worldwide are calling for resilience, innovation and new governance models.

But when we examine these crises, a further question arises: who is calling them 'poly', and who is looking at them in parts? Is the very perception of polycrisis itself fragmented, leading to fragmented actions? Perhaps division is not only the effect but also the cause. Separation is itself both cause and consequence, perpetuating disorder. This suggests that the inquiry cannot remain at the level of policy or governance, it must reach deeper. We must ask not only about the root cause of conflict and fragmentation in society, but also why our understanding, our responses and our lives are themselves so fragmented. From there, the inquiry must go further: are we truly separate from the world we describe?

Beneath the urgency of institutional appeals lies a more fundamental question, one that is rarely asked: Who is the actor behind action? Who is responding to this crisis? This book begins with the proposition that the most urgent crisis of our time is not only systemic or ecological, but perceptual. The way we see the world, ourselves, and others is shaped by fragmentation. Our thought is divided. Our knowledge is compartmentalised. Our action is rooted in identity, fear and control. We have inherited institutions, ideologies and beliefs that reflect this inner division, and now, as those institutions falter, we continue to search for solutions without questioning the actor who is acting.

In that sense, the polycrisis is not just 'out there'. It is 'in here'. It is us. Climate change, for instance, is not merely the result of bad policy or industrial excess. It is the outcome of a consciousness that sees nature as separate from itself, a resource to be extracted, controlled and commodified. Conflict is not only about weapons or borders; it is born from identification, fear and the construction of the 'other'. Inequality is not just a structural issue; it reflects a mode of living driven by comparison, ambition and psychological insecurity. In each case, the crisis expresses something deeper: the architecture of human consciousness itself.

This insight is not new. Jiddu Krishnamurti once said, 'You are the world, and the world is you' (Krishnamurti, 1972). The physicist Bohm (1992) warned that the greatest danger to humanity is the illusion that the world's problems are separate from the consciousness that created them. Ulrich Beck (1992), writing on the 'risk society', showed how modern institutions increasingly produce the very crises they seek to manage. McGilchrist (2019) argues that attention shapes perception, and when attention is fragmented, the world itself appears divided. Yet despite such insights, policy discourse remains largely externalised, still attempting to treat crises as separate entities, still seeking to act upon them as though the actor stood apart. Even complexity theory, which rightly challenges linear and reductionist thinking, often fails to examine the condition of the observer. Frameworks like the SDGs emphasise integration and partnership, but they do not question the psychological ground from which action emerges.

To act from fragmentation, even with the best intentions, is to replicate fragmentation. A divided self cannot build a whole world. This chapter sets the stage for the journey ahead. It examines how the polycrisis has been diagnosed by global institutions, systems theorists and development practitioners, and then turns that diagnosis inside out. It explores the fallacy of treating crises in isolation, the limits of systems thinking without self-awareness, and the mirror-like nature of outer and inner disorder. It proposes that before we innovate, reform or redesign, we must perceive. We must see clearly how the self acts in the world, and how the world reflects the self and shapes the self. Once the mind has actually seen and inquired into this, not intellectually, but directly, there is the possibility of arriving at a space where fragmentation is not, and from there, perhaps, a perception and action that is whole. Without understanding the self, there is no understanding. And understanding itself is that which is not conditioned, biased or fragmented.

For without seeing the fragmentation within, no action, however strategic or sustainable, can be truly whole, good, non-divisive or unifying.

1.2 The Fallacy of Disconnection

The modern world is wired, networked and globalised. We speak of interdependence and interconnectedness across borders, value chains and ecosystems. Yet, when it comes to responding to crisis, our approach is almost always fragmented. Climate change is tackled through carbon targets, inequality through financial inclusion, conflict through security frameworks, education through curriculum reform and mental health through individual coping mechanisms. Each problem is framed, studied and responded to as if it were separate.

This tendency is not accidental; rather, it is embedded in how we know, how we govern and how we act. For example, governments organise themselves into ministries: of environment, of finance, of education, of health (IPCC, 2021; UNDP, 2020). Universities, experts and specialists divide the world into disciplines: economics, psychology, biology, political science. Even within multilateral development efforts like the SDGs, where 'integration' is a buzzword, the practical implementation remains compartmentalised, each goal owned by a different agency, each project measured through isolated indicators. What emerges is not a coherent response, but a patchwork of efforts that often contradict one another. Agricultural expansion is funded even as forests are protected. Growth is incentivised even as sustainability is promoted (UNEP, 2023; UNDP, 2023).

This is not merely a logistical failure. It is a failure of perception. The fallacy of disconnection assumes that the world can be understood and changed by breaking it into parts. The whole contains the fragment, but the fragments cannot make up the whole. In that sense, when we try to integrate fragments, that itself is an action of fragmentation because the integrator itself is a fragment. Critical systems theorists have noted this risk: when integration is pursued without reflecting on the standpoint of the observer, it becomes 'managerial holism', a reproduction of the same divisions under the guise of unity (Jackson, 2019; Midgley, 2000). This is the legacy of reductionist science and industrial rationality: a way of thinking that treats 'wholes' as nothing more than the sum of their components. But in the real world, there are no isolated events. Environmental degradation cannot be divorced from economic structures, which are shaped by political interests, which are influenced by cultural narratives, which are rooted in psychological and historical conditioning. Every action reverberates across domains. Every crisis is entangled with every other.

The systems theorist Meadows (1999) warned of this decades ago: 'You can't just tinker with one part of a system and expect the rest to remain unchanged.' Yet most interventions do exactly that, they apply technical fixes to isolated symptoms without understanding the movement of the whole. Capra (1996) likewise emphasised that systems thinking must be rooted in a deep perception of interconnectedness, not just modelled through data but understood as a lived reality. More recently, scholars in governance and complexity have pointed out that even sophisticated systems approaches often carry forward the assumptions of control and separation if they fail to question the consciousness of the actors themselves (Geyer & Cairney, 2015).

At the same time, when we attempt to fix the system without inquiring into whether the system itself is a product of the fragmented self, we risk perpetuating the very disorder we seek to overcome. The vision, purpose and structures of most systems arise from a mind shaped by fear, competition and division. Any attempt to repair such systems is still being undertaken by the fragmented mind within the context of a fragmented system. What appears as reform is often continuity in disguise, because the source of the action remains unchanged.

This becomes clear when we look across domains.

Governance

When governments face crises such as inequality or social unrest, they typically respond with policies designed to stabilise the system: financial redistribution, security measures or institutional reforms. While these interventions may alleviate certain pressures, they rarely confront the fact that governance itself has been structured around control, domination and competition. These qualities reflect the fragmented mind, which fears disorder and therefore seeks to impose order externally. Without addressing this root, reform sustains the same power dynamics under new appearances.

Therefore, we have not been able to address the root insecurity in human beings and in relationships between human beings, because we have always assumed that by changing the environment around the individual, we could transform the individual within. From Hobbes' *Leviathan*, which grounded political order in fear of violent death (Hobbes, 1651/2012), to modern disciplinary societies analysed by Foucault, where power operates through surveillance, normalisation and control (Foucault, 1977), governance has relied on fear and desire as mechanisms of regulation. Societies have attempted to shape behaviour through punishment and reward, embedding these logics within capitalism, communism and other systems. Yet as long as fear and desire are used as tools to end fear and desire, the cycle only continues.

At moments of immediate urgency, temporary order may indeed require enforcement, when time does not allow for deeper transformation. Even thinkers like Giddens (1984) have noted how structures both constrain and reproduce social action, making institutional fixes self-reinforcing. Ostrom (1990), in her work on collective action and the commons, demonstrated how governance innovations can sustain cooperation, but they still operate within the same fragmented assumptions of self-interest and rational choice. Without inner transformation, the patterns of conflict and insecurity persist.

Given the present urgency, whether planetary boundaries, wars or the collapse of ecosystems, what is needed is not temporary modification for adjustment, guided by self-interest and fragmented identification with memory, ideas, thoughts or objects, but a fundamental and permanent change. Only then can governance cease to be an extension of division and become a reflection of wholeness.

Education

In education, reforms often target testing systems, curricula or teacher training. While these may generate improvements in efficiency or measurable outcomes, they do not question the deeper structure in which students are conditioned to compare themselves, to compete and to pursue knowledge as a means to secure advantage. This conditioning mirrors the divisions in society at large and reproduces them through each new generation. Unless there is awareness of this conditioning, reforms remain trapped in the same cycle: a fragmented system shaping fragmented individuals to sustain a fragmented society.

When we educate children to be efficient and competitive above all else, we create machines with no compassion, human beings coded to divide and separate. Greater emphasis is placed on skills, degrees and external validations than on understanding. Freire (1970) described this as the 'banking model' of education, where students are treated as passive recipients of information rather than active participants in awareness and transformation. Technical precision is cultivated, but when there is no love and compassion in the heart, that precision can be turned towards destruction: the killing of civilians, children or those labelled as 'non-civilians'. Krishnamurti (1953/2009) likewise warned that education which prioritises conformity and ambition produces conflict-ridden individuals, incapable of peace either inwardly or outwardly. The division between 'terrorists' and 'civilians', and the intellectual justification of the state or law to kill at convenience by calling themselves 'state', are not separate from this wrong kind of education. They reflect an education that has prized knowledge without wisdom, efficiency without empathy and achievement without awareness. As Nussbaum (2010) has argued, when compassion and the capacity for critical moral imagination are excluded, education risks creating skilled professionals who are emotionally stunted and socially dangerous.

This becomes clearer when we contrast mainstream systems with alternative visions. Mainstream systems like standardised testing regimes, science, technology, engineering and mathematics (STEM)-first policies, global university rankings reinforce comparison and hierarchy. They measure intelligence in narrow ways, rewarding speed, memorisation and productivity while ignoring the qualities of perception, empathy and capacity to inquire freely and relate. They prepare students to adapt to economic competition rather than to question its premises. In contrast, alternative visions of education, from holistic approaches in progressive schools to traditions of contemplative pedagogy place the emphasis on awareness, relationship and the flowering of the whole human being. Where mainstream systems produce fragmented specialists, alternative visions seek holistic individuals who can live with clarity and compassion, staying true to the meaning of the word individual, which means indivisible.

Yet even many 'alternative' approaches risk being absorbed into the same logic if they are framed as methods to enhance performance or productivity. True transformation requires not another reform of content or technique, but an awakening of perception: seeing how conditioning shapes thought, comparison and ambition. Without such awareness, education continues to replicate the disorder of the world within the minds of the young. With it, there arises the possibility of an education that is not about becoming someone, but about understanding oneself, and from that understanding, living in harmony with others and with the world.

Health and Caregiving

In the field of health, efforts to expand access to care or introduce new technologies are important but frequently miss the deeper reality of fragmentation. Health systems often treat the body in isolation, focusing narrowly on symptoms, organs or conditions without attending to the whole person in relationship with mind, society and environment. The biomedical model, dominant across much of the modern world, excels at technical intervention but frequently neglects the psychological, social and spiritual dimensions of well-being (Engel, 1977).

The rise of geriatric care, for example, brings questions not only about medical access but also about how societies value or marginalise the elderly. Are older persons regarded merely as patients, or as human beings with wisdom, relationship and dignity? Gender disparities in care, too, reveal that health systems mirror broader patterns of neglect and inequality. Women across many societies continue to face under-diagnosis of pain, limited reproductive rights and unequal access to essential services (WHO, 2023). These inequities are not only medical failures; they are reflections of social conditioning that undervalues certain lives over others.

When health is treated only as a technical problem, the living reality of relationship, compassion and dignity is overlooked. Ivan Illich (1976) argued decades ago that medical systems risk 'medicalising' life itself, creating dependence on institutions while neglecting the inherent capacities of individuals and communities to care for one another. More recent thinkers on caregiving remind us that health is not only about curing disease but about the quality of presence and relationship between caregiver and patient (Tronto, 1993). Without compassion, even the most advanced technology can become dehumanising. With compassion, even the simplest act of care, holding a hand, listening deeply, can restore dignity and healing.

Climate and Environment

Responses to the climate crisis frequently focus on technological fixes: renewable energy, carbon markets or geoengineering. These may reduce certain harms, yet the deeper disorder remains. The system generating ecological collapse is rooted in the fragmented mindset of exploitation, consumption and control over nature. Without perceiving this, new fixes risk repeating the same logic under different names: a green economy that still treats the Earth as a resource to be managed, rather than recognising the inseparability of human life and natural intelligence.

The danger lies not only in the technologies themselves, but in the mindset that drives them. Renewable energy projects, for example, are essential, yet many large-scale solar, wind and hydropower projects displace communities and damage ecosystems, reproducing the same extractive logic as fossil fuels. Carbon markets and offset schemes often allow wealthy nations and corporations to purchase the right to pollute, outsourcing ecological burdens to poorer communities in the Global South (Böhm et al., 2012). Even geoengineering proposals, framed as solutions to planetary-scale crises, embody the same assumption that humanity can dominate and manipulate the climate system without unforeseen consequences (Hamilton, 2013).

Such approaches obscure the root issue: the way human beings perceive their relationship with nature. As environmental philosophers have argued, ecological collapse cannot be separated from a worldview that treats the Earth as an object for human use (Naess, 1989). Capra (1996) likewise emphasised that sustainability requires a fundamental shift in perception from viewing the Earth as a collection of resources to be managed, towards understanding it as a living system in which human beings are embedded. Indigenous thinkers have long articulated this truth. Robin Kimmerer (2013), writing from Potawatomi traditions, describes the Earth as a gift and teacher, calling for reciprocity rather than extraction. Shiva (2005) critiques industrial agriculture and globalisation as forms of ecological violence, reminding us that local, community-based practices embody a relationship of care with the land. Unless such a shift in consciousness takes place, rooted in gratitude, reciprocity and interdependence, the climate crisis will be met with reforms that only extend the same paradigm of exploitation in greener forms.

Yet even reciprocity and interdependence can imply separation, as if humans and the Earth were distinct actors in negotiation. When you stand outside the Earth, separated, you perish. Life itself breaks what is cut off: water becomes stagnant, a leaf decays when detached and so too do the mind and society when divided. This recognition should not be misused as a justification for inaction. It is thought that seeks to perpetuate itself, clinging to continuity even when it is destructive. What is aligned with life, however, does not require justification, it naturally breaks away from patterns of division and exploitation, and in doing so points towards a different way of living.

True ecological intelligence is not about managing resources but about realising the indivisibility of life, seeing that human well-being and planetary well-being are not two separate projects. Without this perception, climate action remains fragmented: a series of fixes on the surface of a much deeper disorder. Ecological intelligence, in this sense, is not a separate faculty added to human capabilities but inseparable from intelligence itself, the capacity to see, live and act as one movement without division.

Technology

Technology is the accumulation of memory as experience and knowledge, fragments of human discovery combined into tools. As Krishnamurti warned, the danger is not in knowledge itself but in its use without compassion: a person who learns how to split the atom, but whose heart is untouched by love, becomes a monster. Knowledge without the ground of intelligence is destructive, because it carries within it the seed of division.

In technology, this pattern of fragmented fixing is striking. When digital systems generate problems such as addiction, surveillance or misinformation, the response is usually to regulate or redesign features. Yet the underlying drive that produces these technologies – profit, power and control, is rarely questioned. As Zuboff (2019) shows, surveillance capitalism monetises every aspect of life, converting personal experience into data for extraction. Surface reforms may soften the harms, but the deeper orientation towards commodification remains intact.

AI illustrates this tendency. It is promoted as a tool for solving human problems, yet its development is shaped by competition and exploitation. Most large systems are built by monopolies seeking dominance, not human or ecological well-being (Crawford, 2021). Ethical guidelines are written, but they operate within the same logic that sustains inequality and ecological destruction. The result is that fixes within this paradigm reinforce the very disorders they claim to solve.

But the challenge extends beyond corporate or state systems. Consider dependence on generative tools like Generative Pre-trained Transformer. For those capable of inquiry and able to discern the true from the false, such tools can serve exploration rather than exploitation. Their use then reflects goodness not as an imposed ideal, but as the natural outcome of clear perception. In this way, one can give knowledge its rightful place without clinging to it or mistaking it for wisdom. But for those who approach such tools without awareness, the machine becomes another layer of conditioning. They may memorise what is generated without understanding its meaning, quote endlessly without coherence and rely on patterns without perception. Without wholeness, technology multiplies fragmentation: offering answers without comprehension, information without insight and knowledge without wisdom.

Technology also refines the means of violence. Lethal autonomous weapons, faster fighter aircraft, guided munitions and long-range missiles are not neutral advances; they are more precise ways of killing that distance the actor from the consequences of violence. The development and display of such instruments paraded on national days, celebrated in recruitment drives, or enumerated in media with range and payload as badges of prestige, produce a troubling psychology of power. The declared reach of a missile can become a surrogate for security, pride or greatness for populations that otherwise endure insecurity and material deprivation. That sense of grandeur, 'how far it can reach' offers a fragile psychological safety, a symbolic victory that can mask hunger, social need and moral cost. In this way technology becomes not only an instrument of domination but a balm for collective insecurity, encouraging a false dream of strength while the underlying human conditions remain unaddressed.

The consequences are already visible. In Palestine, the Israeli military has employed an AI-assisted system known as *'Lavender'* to generate target lists for airstrikes in Gaza, identifying up to 37,000 potential targets based on large datasets, often with minimal human oversight and with significant risk to civilians (Human Rights Watch, 2024; McKernan & Davies, 2024). The Israeli military also uses digital tools and AI in Gaza to help determine who or what to attack, tools whose use may exacerbate risk to civilians (Human Rights Watch, 2024).

In governance, China's surveillance architecture in Xinjiang exemplifies how digital systems enforce domination: biometric data, facial recognition, mobile phone monitoring and predictive policing algorithms are deployed ubiquitously to monitor Uyghurs and other minorities (Human Rights Watch, 2019; OHCHR, 2022). The state aggregates multiple data streams in platforms like the Integrated Joint Operations Platform (IJOP) to flag subjects deemed 'risky' and to manage daily life through automated policing (Human Rights Watch, 2019).

In each case, technology does not simply 'fix' problems but reproduces fear, conformity and control. The deeper question, then, is not how to make technology 'better', but how to perceive the consciousness that produces it. When technology arises from a mind rooted in fear, ambition and separation, it inevitably extends fragmentation in digital form. As Lanier (2018) observes, digital systems shape not only markets but also relationships and attention, amplifying division rather than communion. Unless there is a transformation in perception, a seeing that intelligence and compassion are inseparable, technological innovation will remain another mask of disorder.

People are creating what sells, and what sells is what people are. At present, what people are is fragmented, and so what they produce and consume reflects that fragmentation. This is not only the problem of technology or economy; it is the problem of thought and consciousness itself. Bohm (1992) emphasised that thought is a system: fragmented thought inevitably generates fragmented outcomes, whether in politics, economics or technology. Without a radical change in consciousness, every new tool or system merely becomes another expression of the same fragmentation.

Beyond Reform: The Crisis of Perception

Across all these domains, the movement is the same. The fragmented mind creates fragmented systems, and those systems generate crises. In turn, attempts to fix those crises are undertaken from within the same fragmentation, and the cycle repeats. What is required is not merely reform but a radical shift in perception: to see the whole, to question the self that divides and to allow action to arise from a place of intelligence beyond fragmentation.

Most human beings do not have a direct perception of reality but depend upon tools, frameworks and accumulated knowledge. While technology, expertise and experience are useful, without direct perception they inevitably lead to separation, illusion, division and conflict.

Even complexity theory, for all its emphasis on interconnection, often remains confined within a scientific or managerial mindset. It models feedback loops, emergent behaviour and adaptive systems, but rarely asks: *Who is observing? What is the state of the mind that is intervening in the system?* Without this, even sophisticated models become instruments of a fragmented consciousness trying to manage what it cannot perceive in totality.

For this reason, Morin (2008) spoke of complex thought as something deeper than complexity theory. He argued that it was not enough to build models of chaos, agent-based interactions or nonlinear dynamics. What was required was a reorientation of how we think and know. His principle of dialogic emphasised the ability to hold together opposing concepts like order and disorder, unity and multiplicity, reason and emotion, without reducing one to the other. Complex thinking sought to integrate science with philosophy, poetry and wisdom traditions, making it a mode of thought rather than a technical methodology. In this, Morin pointed to an important truth: observation is not neutral. It is shaped by worldview, conditioning and language.

This was a profound shift. Morin invited science, policy and education to recognise their own limits and to reform themselves in light of those limits. He called for a way of thinking that would be self-reflective, relational and open to uncertainty. In doing so, he laid important groundwork for seeing knowledge itself as a living process rather than a fixed structure.

Yet the movement of inquiry does not stop there. Morin's complex thought takes us to the threshold where we begin to ask not only how we think but also what thought itself is. Here, thinkers like David Bohm push the question further, showing that fragmentation lies not only in disciplines and institutions but in the very nature, structure and operation as movement of thought.

Bohm (1996), in his work on thought and wholeness, put it sharply: 'Fragmentation is not just a symptom. It is the way thought operates.' Bohm's insight reframes the problem entirely. It is not just that we see a fragmented world; it is that our very way of seeing is fragmented. The fragmentation lies not in the world, but in thought itself, in how it separates observer from observed, self from other, idea from reality. When this way of thinking becomes the foundation of science, policy, education and economics, we inherit systems that reflect the same fragmentation. The disconnect is not only across ministries and disciplines; it is in the actor who is trying to connect them.

Consider the SDGs, often celebrated as a landmark framework for integrated development. With 17 goals, 169 targets and over 230 indicators, the SDGs aim to address poverty, health, gender, climate, education and governance in a unified agenda. Yet in practice, implementation efforts remain siloed. Ministries responsible for Goal 13 (climate action) often operate independently from those focused on Goal 8 (economic growth) or Goal 2 (agriculture and food security). The interlinkages are acknowledged rhetorically, but rarely acted upon from a perceptual ground of wholeness (United Nations, 2023). This is despite the availability of frameworks like planetary boundaries, which map the limits within which humanity can operate safely (Rockström et al., 2023), providing a scientific context for the kind of integrated action the SDGs were intended to embody. The result is institutional confusion, contradictory objectives and superficial impact assessments.

This is not a critique of intent. It is a critique of consciousness. Whether action is guided by intent and knowledge or arises without them, it can still be mechanical because the quality of action depends not on its content but on the state of the observer. Even the most integrated frameworks become reactive when they are applied by an observer who is not whole, who is not aware from within.

The neuroscientist McGilchrist (2019) offers a related insight in his study of brain lateralisation. In *The Master and His Emissary*, he argues that the modern world is dominated by the left hemisphere's mode of attention: analytic, fragmentary, obsessed with categorisation and control. The right hemisphere, which sees relationships, context and the living whole, has been subordinated. As a result, we have created a culture that 'sees the parts but not the whole, the map but not the territory, the system but not the life within it'. McGilchrist's diagnosis is biological, but the implication is civilisational. The way we attend to the world shapes the world we co-create.

Yet it is important not to reduce his insight into another scheme of labelling by calling some people 'left-brained' and others 'right-brained'. McGilchrist himself warns against such simplifications. The point is not to prefer one hemisphere over the other, but to see how imbalance in modes of attention mirrors imbalance in culture. To turn this into another framework for identity, comparison, or becoming would only repeat the very fragmentation we are trying to understand. Wholeness is not found in privileging one part, but in ending division altogether.

Even if McGilchrist's account is accurate, the left hemisphere cannot, by effort or desire, transform itself into the right. The very attempt to do so arises from the left's own drive for control, improvement and becoming. This is why the solution cannot lie in preferring one mode or training the brain into balance through technique. It lies in seeing directly how the structure of thought perpetuates its own disorder and imbalance.

This same misperception drives the global emphasis on resilience. In the face of multiple crises, resilience has become a dominant buzzword. Institutions must be resilient. Communities must be resilient. The climate system must be made resilient through adaptation and geoengineering. Yet rarely is the resilience of perception examined. What is the quality of attention with which we see crisis? Are we reacting from fear and habit, or observing with clarity and depth?

The fragmented observer, when faced with systemic breakdown, instinctively reaches for more management: data, strategy, targets, measurement. But the system is not just breaking down. It is reflecting the incoherence and fragmentation of the actor. More data will not resolve a crisis created by the mind that measures. More control will not create order if the actor is driven by fear. The very effort to solve from fragmentation reproduces fragmentation. The very desire for psychological security creates both physical and psychological insecurity, because that desire itself is the action of an insecure mind.

This is why, despite decades of global summits, treaties and sustainable development plans, the crisis persists and deepens. The tools of reform are built within the same architecture of thought that caused the problem and the need for reform. Every institution created by the fragmented mind inherits its logic. Every such reform effort carries its seeds of fragmentation and continues its own disorder.

We imagine that change can be achieved through systems, legislation or revolutions. But what is the use of all this if inwardly we remain the same? The insecurity between nations is not separate from the insecurity within the human mind. Inward disorder cannot give rise to outward harmony. Yet modern governance largely assumes the opposite that improved design, inclusive processes or smarter technologies can overcome the incoherence at the root.

This is the fallacy. Not just a methodological error but a perceptual one.

When the actor remains fragmented, the action, no matter how visionary, remains partial. When thought operates in fragments, systems built from thought will fragment further. The very perception of the false, seen not intellectually but directly, is already the perception of truth. This is not pessimism. It is a call to turn inward, to look at the ground of perception itself, and to ask: *Is it possible to act from a mind that sees the whole?*

This question forms the basis for the next section.

1.3 The World as Mirror

If fragmentation drives reform and governance, it is equally visible in the world's crises, which mirror back the state of our own consciousness. In moments of global emergency, we instinctively look outward. We analyse data, assess systems and compare countries. We ask: What's wrong with the world? But rarely do we ask a more intimate question: What is the world showing us about ourselves? And is the world separate from us?

These are not just metaphors. It is a precise description of how human consciousness and the world it inhabits are not two separate realms. Every institution, every policy, every ecological collapse is a manifestation of human thought, desire, memory and fear. The climate crisis is not just an external phenomenon caused by emissions and consumption; it is a mirror of the psychological division between human beings and nature and the cause and effect of human action in pursuance of desire. War is not only the failure of diplomacy; it is the materialisation of the inner violence we live with daily: comparison, competition, resentment, insecurity. Inequality is not merely a systemic flaw; it reflects the inner drive to dominate, to become, to be more than another.

The world is not separate from the self. The self projects itself into the world, and what we call 'crisis' is the externalisation of that projection. This insight runs through the work of Jiddu Krishnamurti, who observed, *'The world is you, and you are the world'* (Krishnamurti, 1972). It is not a poetic sentiment but a call to perceive crisis as a mirror, not a target. The disorder in our systems is not only around us, it is within us. The fragmentation in politics is not separate from the fragmentation in relationship. The endless pursuit of growth and control is not different from the inner movement of ambition and fear. We have made the world in our own image, and now we find ourselves trapped in it, trying to escape without seeing that we are its architects. We think there is a society out there, but society is nothing but the relationship of people.

This does not mean the world is only psychological. Material forces, historical injustices and structural inequities are real. But to stop there is to stop short. For behind every structure lies a set of values, assumptions and perceptions that gave rise to it. Behind every extractive economy is a worldview that sees the Earth as separate. Behind every arms race is a mind that feels threatened. Behind every broken education system is a culture that prioritises achievement over understanding. And in seeing this, one comes upon a deeper fact: the world is not separate from the mind that created it. This moment is the only moment in which the entire past meets, because consciousness is not personal. A human being is the history of humankind.

This mirroring is visible in some of the most high-profile crises of our time. Climate policy is often framed in the language of control: carbon targets, net-zero deadlines, geoengineering milestones, treating the Earth as an object to be managed rather than a living whole of which we are a part (Capra, 1996). The same division is evident in geopolitics. Nations operate primarily on the basis of foreign policy self-interest, even when claiming moral grounds. In the ongoing Israel–Palestine (including Gaza) conflict, Israeli military action is justified under the banner of self-defence, while extensive evidence from UN agencies, human rights organisations and independent observers documents large-scale civilian casualties, mass displacement and the destruction of essential infrastructure (Human Rights Watch, 2024; UNOCHA, 2024). Some international legal experts, drawing on reports by organisations including Al-Haq (2024), have argued that these conditions meet the threshold for genocide. The long-standing structural conditions that perpetuate violence, occupation, blockade and systemic discrimination, remain unaddressed.

Even the act of naming this violence, whether as a *war*, *conflict*, *international armed conflict*, *ethnic cleansing* or *genocide*, is conditioned by political self-interest. Language itself becomes an instrument of conflict, words that should clarify reality instead conceal it. Each label carries different legal and moral consequences, yet these consequences rarely translate into coherent global action. International law operates within the same fragmented logic it seeks to correct. Mechanisms of accountability are invoked selectively, and implementation seldom follows recognition. What is visible to all remains unanswered in practice.

The question of representation itself becomes part of this confusion. Does a militant organisation like Hamas, operating within layers of occupation and retaliation, truly speak for the people of Palestine? When bombs fall on Gaza, what is being targeted, a government, a group or millions of civilians trapped within a closed geography? The very framing of justification and response continues to mirror the mind that divides, identifies and retaliates, unable to see the shared humanity it destroys.

Seen more deeply, this is the movement of thought itself, lost in the illusion of separateness, identifying and dividing, creating and using violence. Technology, as the accumulation of past knowledge, is drawn into this same movement, serving preservation, expansion and hatred, which are all part of its own nature.

A similar mirroring can be seen in the paralysis of the United Nations Security Council (UNSC) during major humanitarian crises. While the UN Charter frames the Council as the primary body for maintaining international peace and security, its decision-making is constrained by the veto power of its five permanent members. In practice, this means that geopolitical alliances and national interests often override urgent humanitarian imperatives. In multiple recent votes on ceasefire resolutions concerning Gaza, permanent members, including the United States, have exercised their veto to block measures supported by the majority of the international community (UN Digital Library, 2023–2024). This has occurred despite repeated warnings from UN agencies of famine, mass displacement and systemic destruction (UNOCHA, 2024; WFP, 2024). The UNSC's inability to act decisively reflects not only a procedural flaw but the same underlying fragmentation: the prioritisation of identity, alliance and perceived strategic advantage over the recognition of shared humanity.

In the mirror, this is the same pattern that appears in personal life when loyalty to one's image, group or belief overrides compassion or truth. The institutional structure amplifies the stakes, but the root is the same: a mind that sees through division inevitably acts in ways that protect the part at the expense of the whole.

When we observe carefully, we see that every crisis carries within it the structure of the self: the movement of comparison, the longing for permanence, the drive for control, the escape from insecurity. These are not abstractions. They are present in daily life, in how we speak, relate, consume, work and govern.

This insight, that the world reflects the self, is not an invitation to withdraw into inner life or ignore material injustice. On the contrary, it is a radical broadening of responsibility. If the world is shaped by our perception, then responsibility begins not with action alone but with seeing. Seeing how we hold on to images, pursue outcomes for security and reduce others to labels and roles. These invisible threads become institutions. What begins as thought becomes a norm; what begins as fear becomes a border; what begins as comparison becomes inequality.

Even social reform, when enacted from a fragmented self, becomes another extension of fragmentation. We attempt to fix what we do not fully understand. We organise compassion into programmes while retaining the ego that separates the helper from the helped. We seek justice through structures that reproduce control. We campaign for inclusion while operating from inherited ideas of power and competition. The result is a repetition of patterns, well-meaning but fundamentally reactive (Freire, 1970). Insights from neuroscience also remind us that the sense of self is not a fixed entity but a construction of memory and narrative (Damasio, 2010).

To break this pattern, we must observe the movement of thought itself. Thought, when not seen for what it is, becomes the hidden architect of the world. It divides time into past and future, constructs identities and sustains conflict through memory. It seeks certainty in belief, security in achievement and connection through image, all of which ultimately prevent direct perception of life as it is.

David Bohm, in his dialogues with Krishnamurti, emphasised this point repeatedly: *'What we take to be reality is in fact shaped by our thought. But thought does not know this, it projects and then forgets its projection'* (Bohm & Krishnamurti, 1999). This forgetfulness is not trivial. It allows us to act as if the world's disorder is separate from our own inner lives. It sustains the illusion that the crisis is 'out there' – climate, poverty, politics, rather than seeing that every crisis is a symptom of deeper perceptual conditioning.

In this sense, the call to 'change the world' becomes not just a matter of innovation or activism but of insight. Not the insight of information, but the insight of seeing. Seeing that the self, as we commonly understand it, is a psychological construction made of time, memory and image. Seeing that the continuity of this self is the continuity of the same world we wish to transform. When the self operates as the centre of action, even the most well-meaning act becomes an extension of fear, control or identity. But when action arises from perception, when the mind is still and attentive, there is a different quality of movement. That movement does not reproduce division. It does not seek security through systems. It acts because it sees, not because it believes.

This is not philosophy. It is the foundation of clarity, non-division and sustainability. For no system will endure whether ecological, economic or social, if it is based on a consciousness of fragmentation. To sustain life, one must live from a sense of wholeness, not as an ideal but as a fact of perception. The Earth is whole. Life is whole. The mind that sees this does not divide its response into psychological ambitions, goals and projects. It acts from coherence, not from fear.

This chapter does not deny the importance of policy, structure or collective will. But it asks: Can these emerge from a different source? A source not rooted in self-image, fragmented self-interest, ideology or reactive judgement, but in direct seeing, moment to moment. Such seeing is rare, but it is possible. And without it, no plan, however strategic, can bring about lasting transformation.

The mirror of the world is not meant to condemn. It is meant to reveal. When we see clearly that our systems are our own projection, then we are no longer trapped by blame or paralysis. We are invited into responsibility, not only for what we do, but for how we see and perceive. The next section will examine this further by exploring how complexity and systems thinking, while valuable, often remain insufficient when the observer as the self is not transformed.

1.4 Complexity and Systems Thinking: A Partial View

In response to the failures of fragmented approaches, many scholars, practitioners and institutions have turned to complexity science and systems thinking. These frameworks mark a clear evolution from linear, mechanistic models of change. They emphasise feedback loops, emergence, interdependence and the unpredictability of dynamic systems. They challenge the illusion of simple causality and highlight that interventions in one part of a system reverberate across others. This shift in thinking is both necessary and important.

Yet even here, a subtle danger remains. Complexity can become another framework applied by a fragmented self. The model changes, but the actor remains the same. Without transformation in the observer, systems thinking risks becoming an intellectual tool rather than a perceptual shift.

The language of systems is powerful: whole systems, nonlinearity, fractals, co-evolution, resilience. But these terms are often held as concepts, not lived as realities. Interdependence may be accepted in theory but not embodied in relationship. Emergence may be referenced in governance models but not perceived in the psyche of the policymaker. What begins as a challenge to reductionism is quietly reabsorbed into the very paradigm it sought to transcend.

This becomes evident in practice. Governments hire complexity consultants, fund scenario-building workshops or insert 'systems language' into strategy documents. Multilateral organisations speak of 'whole-of-government' approaches and 'interconnected challenges'. Yet interventions continue to emerge from a centralised, managerial mindset, seeking control through mapping, prediction and integration plans. The system becomes something to manage rather than something to see from within (Morin, 2008).

Wicked policy problems: climate breakdown, biodiversity loss, geopolitical instability, structural inequality resist solution precisely because they are sustained by the very divisions that responses often reinforce. Traditional governance frameworks, built on linear and compartmentalised methodologies, try to balance competing interests, but this balancing act still takes place within the same fragmented paradigm. As Ostrom (2010) showed in her work on polycentric governance, diverse and interconnected systems require adaptive, self-organising approaches. Yet these only function when actors are capable of cooperation beyond narrow self-interest.

The distinction here is fundamental. Complexity can be held as knowledge, as a framework to analyse or it can be seen as life itself. The former operates from the outside, dissecting the system into parts and seeking leverage points. The latter dissolves the boundary between observer and observed. It recognises that the state of the observer determines the quality of intervention. This is rarely acknowledged in policy discourse. Even when complexity thinking invokes humility, uncertainty or emergence, these qualities are often reduced to management principles. The actor is trained to 'embrace ambiguity', but continues to operate from identity, role, ambition or ideology. Action arises from the intellect, guided by past frameworks, and therefore remains tethered to the past. In that attachment lies separation. The actor is no longer in direct touch with reality, but with an image of it, projecting what should be onto a fragmented perception of what is. Illusion then carries forward the very problems it seeks to end. As articulated in *Governance Without the Self* (Gupta, 2025): *'The system does not need new tools. It needs a new actor, one who sees without separation.'* That seeing is not a method. It is not another framework. It is a quality of perception that emerges when the mind is free of self-image, accumulated knowledge and psychological time. Only such a mind can act without division and therefore without perpetuating the crisis it seeks to resolve.

When stripped of this perceptual depth, complexity thinking begins to mirror the very fragmentation it critiques. It becomes another branch of knowledge fused with data science, policy design and systems mapping, producing new solutions from an old mind. The frameworks evolve, but the consciousness remains unchanged. The result is often more sophisticated processes, but not transformation (Bateson, 2002; Capra, 1996; Senge, 1990).

This limitation is not abstract. It carries real consequences. Attempts to systematise compassion, design for emergence or map relationships can easily become mechanisms of control when enacted by a mind that still seeks certainty. The paradox of trying to manage complexity while remaining psychologically identified with knowledge, time or ideology lies at the heart of many reform failures.

The proliferation of 'complexity-informed' initiatives illustrates this. Public policy labs, systems change incubators and adaptive leadership programmes often begin with profound questions but culminate in refined processes rather than deeper perception. Stakeholder maps are drawn, leverage points identified, scenarios modelled. But who is doing this? With what quality of attention? Are actors aware of the subtle motivations, fear of irrelevance, attachment to expertise, desire for impact or recognition that shape their work? Without such awareness, the tools become masks for the same actor: the self, fragmented and striving.

Leading systems thinkers themselves have warned against this drift. Morin (2008) called for a 'reform of thought' that could hold paradox and wholeness. Capra (1996) argued that a true systems worldview must rest on ecological consciousness, not just analytic interconnection. Senge (1990) emphasised that systems intelligence requires a 'shift of mind' rather than new strategies. Yet in practice, these deeper invitations are often sidelined in favour of frameworks that can be applied without inner transformation.

This is not to dismiss the value of complexity. On the contrary, its greatest potential lies precisely in the invitation it carries: to perceive reality directly, as a living whole. Such perception is not conceptual. It is available here and now, to any mind capable of observing without the interference of ego, role or knowledge. For this to happen, complexity thinking must cease to be a professional toolkit and become a mirror to the self.

In many ways, systems thinking has opened an important doorway. It has brought the language of interdependence into the mainstream. It has questioned linear causality and dissolved the illusion of isolated agency. But the opportunity is missed when the framework replaces the fact, when the map replaces the territory, when process replaces perception. We speak of systems but continue to act from the same fragmented centre.

The deepest implication of systems thinking, then, is not about systems at all. It is about the actor who sees the system. Without clarity in the observer, reform becomes complexity layered upon confusion. The map grows more detailed, but the traveller remains lost.

This calls for a profound shift, not in strategy but in seeing. It is not enough to know about complexity. One must see from wholeness. And to see from wholeness is to question the fragmentation within.

This leads directly into the next movement of inquiry. If the real crisis is not complexity but the observer, then we must ask: Who is this observer? What is the structure of the self that acts, relates and seeks to transform? Is that structure itself the root of the problem?

Having seen the limits of frameworks, we arrive at the more fundamental question: the actor behind all action. For no matter how advanced the tools, how integrated the models or how visionary the policies, it is the human being who applies them. And if that actor is fragmented, the action will carry the same fragmentation. This is where the real crisis begins.

1.5 The Real Crisis Is the Actor

When a house is on fire, it is natural to focus on the flames. We bring in tools, technologies and teams to extinguish the fire, rebuild the walls and install alarms for the future. But what if the fire is being lit, over and over again, by the same person who brings the tools? What if the one holding the hose is also holding the match?

This is the situation humanity finds itself in. Every reform, innovation or policy is enacted by a human being, a person with motivations, fears, assumptions, knowledge and identity. And yet, the central question is rarely asked: Who is this actor? What is the structure of the one who is acting?

Modern governance, development and sustainability discourse tend to focus on systems, structures and outcomes. The actor is either taken for granted or abstracted into roles: the policymaker, the entrepreneur, the civil society leader. We analyse what they do, measure how they perform and evaluate what impact they produce. But the inner condition of the actor, how they perceive, relate and identify is left unexamined. It is assumed to be neutral or irrelevant.

This is a dangerous omission. For the world is not changed by systems alone, it is changed by the quality of perception within those who participate in them.

The self that acts in the world is not a neutral instrument. It is shaped by psychological patterns: fear, memory, desire, image, comparison, ambition. These patterns are not personal flaws, they are the architecture of the self as it is commonly experienced. When this structure is not seen clearly, it drives action in ways that often contradict stated intentions. A development practitioner may speak of equity, while operating from a subtle superiority. A reformer may demand justice, while driven by resentment or attachment to ideology. A policymaker may promote cooperation, while protecting identity and power. These contradictions are not moral failings. They are consequences of acting from a fragmented self.

Fragmentation is visible not only in individual relationships but in the functioning of global institutions. The UNSC, for example, is mandated to maintain international peace, yet the veto power of its five permanent members has repeatedly blocked action in crises where mass atrocities have occurred. In Syria, multiple resolutions aimed at halting the use of chemical weapons and the targeting of civilians were vetoed (Security Council Report, 2023). In Palestine, vetoes have prevented ceasefire resolutions even as humanitarian law violations were documented, reflecting how national alliances and political identity override the stated commitment to human rights.

The Responsibility to Protect (RtoP) doctrine was invoked in Libya in 2011, leading to swift international military intervention, but has been conspicuously absent in Myanmar despite overwhelming evidence of crimes against humanity against the Rohingya (Bellamy, 2015; UNHRC, 2018). These selective applications reveal that territorial claims, strategic alliances and ideological attachments often take precedence over the well-being of human beings.

This selectivity is mirrored elsewhere. The United States and NATO fund Ukraine in the name of protecting its children from Russian forces, while some of the same nations fund Israel as it deploys the Israeli Defence Force in operations that have killed children. Here, symbol and identity are prioritised over life itself. Flags, nations and ideas as constructs of thought have become more important than human survival. We are willing to take life for what is not true and has only been made real for us by our thoughts.

The paralysis around Gaza illustrates this problem with painful clarity. The doctrine of RtoP, adopted unanimously at the 2005 World Summit, was intended as a safeguard against mass atrocity. It obligates states to protect their populations, commits the international community to assist, and, when a state is 'manifestly failing', mandates collective action through the Security Council. Yet in Gaza, as in Syria and Myanmar before it, this responsibility has collapsed at the altar of political alliances. Across 2023–2025, resolutions calling for an immediate ceasefire and unimpeded humanitarian access were supported by the overwhelming majority of Council members, yet repeatedly blocked by a single veto. The result is paralysis in the very body tasked with maintaining peace.

The General Assembly has stepped in under its 'Uniting for Peace' procedure, passing resolutions demanding an 'immediate, unconditional and permanent ceasefire'. These reveal the conscience of the majority but carry no enforcement power. The International Court of Justice (ICJ), in *South Africa v. Israel*, has ordered provisional measures under the Genocide Convention, directing that humanitarian assistance be allowed and that acts of genocide be prevented. The International Criminal Court prosecutor has even sought arrest warrants for leaders of both Hamas and Israel. But here too, enforcement depends on state cooperation. Law without compliance power is not protection; it is paperwork.

Why, then, does RtoP not move in Gaza? The answer is not technological, we have the evidence, the satellites, the real-time reporting. Nor is it normative, we already have a dense body of international law. The Geneva Conventions (1949) and their Additional Protocols codify the laws of armed conflict, prohibiting deliberate attacks on civilians, starvation as a method of warfare and collective punishment. The Genocide Convention (1948) obligates states not only to punish but also to prevent acts committed with the intent to destroy, in whole or in part, a national, ethnic, racial or religious group. The UN Charter (1945) prohibits aggressive war, affirms the sovereign equality of states and vests primary responsibility for peace and security in the Security Council. The United Nations General Assembly (1948), though formally non-binding, has become the foundation for a web of binding treaties and customary norms. Jus cogens norms represent peremptory principles of international law, such as prohibitions on genocide, slavery and torture, from which no derogation is permitted. And the principle of *pacta sunt servanda*, enshrined in the Vienna Convention on the Law of Treaties (May 23, 1969), holds that treaties in force are binding and must be performed in good faith.

On paper, these are among the strongest legal instruments humanity has produced: universal in scope, binding in law and often described as non-derogable. Yet none of them enforces itself. They are frameworks, agreements and words inscribed by human beings, and their application is mediated through states, governments and leaders. Here the deeper problem emerges. One state invokes the Genocide Convention in an ICJ case; another denies that its threshold has been met. One government cites the Geneva Conventions to condemn its adversary; another claims the same text as justification for its own actions. The law remains constant, but its interpretation and enforcement fracture along political alliances and identities.

Thus, even the most sacred norms become another mirror of the self. They are applied or withheld not according to universal principle, but according to the fragmented psychology of the actor: the self conditioned by fear, memory, ambition and identity. Law becomes another fragment of thought, invoked when convenient, ignored when inconvenient and endlessly contested in its application. The problem, therefore, is not only institutional paralysis but perceptual limitation. Without transformation in the actor, even jus cogens norms remain subject to the same selectivity and contradiction. And so, while the fire rages, we keep layering norms upon norms, frameworks upon frameworks, without seeing that it is the one holding the hose who is also holding the match.

At the surface, fragmentation appears as role-based separation: we divide ourselves into sectors, functions and hierarchies. At a deeper level, it operates through identity: we see ourselves as Indian or American, Hindu or Christian, activist or academic. At its core, fragmentation arises from thought identifying with image: the image of who I am, who I should be, what I must achieve or how I must be seen (Bohm, 1992).

When action arises from this centre, it inevitably creates conflict, resistance and control. It may do so subtly, wrapped in the language of reform and progress. But the underlying movement remains the same: the self, trying to secure itself through action. The world becomes an object to shape, a threat to manage or a stage to perform upon. Even the most beautiful vision of sustainability, equity and peace becomes another extension of this centre. And then, the common insecurity which has been created leads to maintaining that insecurity. The demand of this fragmented world is to keep its fragments intact, which only leads to more physical and psychological insecurity.

This is why many sincere efforts fail to transform the world. They do not fail in intent, or even in design, but in the absence of self-awareness. The actor has not seen and has not been seen. And when the actor cannot see and is not seen and understood, its movement remains mechanical and repeats the very patterns it seeks to transcend.

This insight is not cynical. It is liberating. For it points to a different possibility: action without a centre. Action that is not driven by image, ambition or identification, but by perception, intelligence and care. Such action cannot be planned in advance. It arises naturally when the mind is quiet, when it sees clearly, without distortion. From that seeing, a response flows, not as a result of thought, but as a movement of intelligence, an optimum response to the whole that has been clearly seen and understood (Bohm, 1996; Krishnamurti, 1954).

This is not mystical. It is the most practical thing possible. When we see clearly, without the filter of the self, what needs to be done becomes obvious and happens on its own without a separate doer making an effort or an action of will. We listen without defence. We observe without projection. We speak without fear. From this clarity, right action emerges not as a strategy, but as a direct expression of coherence.

This capacity is within all of us. But it is rarely lived because our systems of education, leadership and governance do not ask us to look inward. They ask us to perform, to specialise, to manage. We are trained to apply knowledge, but not to observe the knower. We are encouraged to innovate, but not to question the psychological motives that drive innovation. As a result, we keep seeking better frameworks without ever examining the framework of the self: the assumptions, images and fears that shape perception.

This is the real crisis. Not the breakdown of systems, but the absence of inward clarity in those who build and maintain them.

When we examine closely, we see that all collective crises: polarisation, environmental degradation, economic inequality, violence, are maintained by this absence. Not just among elites or leaders, but in each of us. We carry the same impulses: the drive to assert identity, the fear of insignificance, the need to control. These impulses shape not only policy, but daily life. They distort our relationships, our sense of responsibility and our capacity to see, listen, perceive, learn, relate, respond, understand and act and make it all conditional, conditioned, limited and fragmented.

So the question is not just institutional or global. It is deeply individual and collective, but each one must see for themselves. Can we observe the movement of the self in ourselves? Can we see how desire, fear and memory shape our consciousness and what we perceive from it? Can we see, without judgement, the impulse to be right, to be known, to be secure? And can that seeing bring about an end to that movement, not through suppression or discipline, but through understanding?

This is the beginning of transformation, not in society, but in the actor. And when the actor changes, the action changes. When the observer is no longer caught in knowledge, be it psychological or technical, identity, comparison or time, the response to crisis is no longer reactive. It is no longer an extension of fragmentation. It is whole.

Change in the consciousness of one person changes the consciousness of everyone, in time, beyond time. And in this movement, when the actions of many actors and observers change, there is an emergence of transformation in society, of which we are already a part when we are ready to see.

Gupta (2024) described as the ground of true responsibility: 'To be responsible is not to carry the weight of the world, but to be free of the division that created it.' Responsibility begins with awareness. And awareness begins when the self is seen clearly, not as a problem to be fixed but as a movement to be observed. That observation is the turning point. With it comes the summation of energy for right action to emerge. What was previously dissipated in the fragmentation of thought, replicating itself psychologically and manifesting as a divided society, becomes available as an undivided capacity to respond. This undivided energy is awareness and responsibility itself, the ground from which coherent and compassionate action flows. Awareness itself is responsible.

From that point, everything changes. Leadership is no longer role-based, it is perceptual. Governance is no longer driven by ideology, it is rooted in insight. Development is no longer organised around outcomes, it flows from relationship. The world does not need more experts acting from fragmentation. It needs human beings who can see.

As we transition into the final section of this chapter, this insight will be deepened further. We have looked at the failure of systems (1.1), the mirroring of self in world (1.2), the limitation of frameworks (1.3) and the structure of the actor (1.4). Now we must ask: What does this mean for action itself? If action from a fragmented self perpetuates the crisis, is there an action that arises from wholeness? Can such action exist in policy, leadership and everyday life? What is its nature?

These questions will lead us into the final section: a reflection on perception, awareness and the possibility of acting without the continuity of the self.

1.6 Conclusion – From Reaction to Perception: A Different Ground for Action

In the face of mounting crisis, the instinct is to act. We are taught that to be responsible is to intervene, to fix, to respond with urgency. And indeed, there are moments when action is essential: when lives must be protected, ecosystems defended, injustices opposed. But beneath the rush to act lies a deeper question: From where does this action arise?

Most action in the world today arises from reaction. That is, from psychological movement driven by fear, hope, memory, identity or ideology. We react to crisis with urgency but also with conditioning. We act not only because we care, but because we are anxious, ambitious, ashamed or attached. We identify with causes, roles and missions. We want to win, to be seen, to fix the world. Even our noblest efforts are often fuelled by the self-image of the 'I', which acts to become something or to protect something.

Such action, though it may produce temporary results, cannot bring lasting coherence. For it arises from the same ground of fragmentation that created the crisis in the first place. The identity may change from aggressor to activist, from politician to reformer but the centre remains the same: a divided self acting upon a divided world. The map has changed, but the traveller is still lost (Bohm, 1992).

This is why, as seen repeatedly in governance and development, laws and moral frameworks often fail to translate into coherent action in a fragmented society. Efforts to change outer structures, whether through treaties, reforms or institutional redesign, without addressing inner fragmentation tend to replicate the same patterns of conflict and exploitation (Morin, 2008; Ostrom, 2010). For example, while international humanitarian law prohibits indiscriminate attacks on civilians, military spending continues to dwarf humanitarian funding, with global defence expenditure reaching $2.4 trillion in 2023 compared to less than $60 billion for humanitarian response (SIPRI, 2024; UNOCHA, 2023).

To see this is not to retreat into passivity. It is to question the very ground of action. Is there an action that arises not from reaction, but from perception? From seeing directly, without the interference of the self?

Such action does not begin with a plan. It begins with observation. Not of others, but of oneself. Observation of how thought moves, how identity forms, how fear distorts, how ambition conceals. This observation is not introspection. It is not analysis. It is the choiceless awareness of fact without interpretation, without justification. In that awareness, the movement of the self is seen clearly. And in that clarity, the movement loses its momentum. There is a stillness.

From that stillness, action may arise. But now, it is no longer a projection of the self. It is not 'my' action, in the usual sense. It is action without centre, ethical, fluid, precise, responding to what is without carrying the weight of the past. Bohm (1992) described this as action that emerges from insight, not from accumulated knowledge or memory. It is a movement of intelligence, not of ego.

This kind of action cannot be codified. It cannot be designed into systems. But it can be lived. It can inform how we listen, how we lead, how we speak, how we design policy, how we walk through institutions. It is not a method. It is a different quality of being, one that sees relationship not as structure, but as movement.

Gupta (2025) describes this as the core shift in governance: not new models, but new perception. The suggestion was not to abandon reform, but to ground it in a different soil. A soil not of knowledge but of awareness. Not of role, but of presence. From that ground, even the most complex institutions can be renewed, not because we imposed better frameworks, but because we acted from wholeness.

Wholeness is not an ideal. It is not something to achieve. It is the nature of reality when not divided by thought. It is the field in which perception becomes action, in which clarity becomes compassion. In that field, decision-making changes. Relationships change. Systems may still be needed, but they are no longer substitutes for understanding.

This does not mean that action becomes perfect, or that life becomes simple. Complexity remains. Uncertainty remains. But the actor is no longer caught in becoming. There is no agenda, no role to fulfil, no self to protect. From such a place, response is both humble and clear. It does not seek control. It does not fear uncertainty. It meets the moment with intelligence that is not inherited, but living.

This is the foundation from which a truly sustainable world may emerge. Not because we engineered it, but because we acted without separation. And that begins not in the system, but in the self, or rather, in the ending of the self as the centre of action.

When the self no longer dominates perception, a different movement unfolds, one not bound to outcome, and therefore not constrained by fear or ambition. It acts because it sees, not because it seeks. This seeing is not emotional or conceptual; it is direct, unmediated and whole.

Urgency, then, does not demand faster reaction, but deeper perception. With the climate crisis and humanity approaching critical tipping points, we must act instantly. Transformation cannot take time. Yet because transformation cannot come from force, punishment or fear, this urgency itself demands compassion. It demands creating spaces where those who cannot yet see may see, where they feel safe, free and loved. Urgency calls for action that embodies freedom and compassion, not as ideals but as living realities.

The invitation, then, is not to reject frameworks or institutions, but to go deeper than them. To see the self that uses them. And to question whether that self can ever bring wholeness to a world it sees through division.

In the next chapter, we begin this inquiry. What is this self? How is it formed? How does it sustain fragmentation even in its efforts to unify? Only by understanding the self, clearly, directly, can we understand the world. For the two are not separate. They are one movement, seen from different sides.

References

Adam Tooze. (2022). *Chartbook #131: Welcome to the world of the polycrisis*. Chartbook.

Al-Haq. (2024). *Statements and legal analyses on Gaza/occupied Palestinian territory, 2024*. Al-Haq.

Bateson, G. (2002). *Mind and nature: A necessary unity. Advances in systems theory, complexity, and the human sciences*. Hampton Press. (Original work published 1979).

Beck, U. (1992). *Risk society: Towards a new modernity*. Sage Publications.

Bellamy, R. (2015). Climate engineering and the governance of geoengineering: A social science perspective. *Wiley Interdisciplinary Reviews: Climate Change*, *6*(6), 597–615.

Bohm, D. (1992). *Thought as a system*. Routledge.

Bohm, D. (1996). *On dialogue*. Routledge.

Bohm, D., & Krishnamurti, J. (1999). *The limits of thought: Discussions between J. Krishnamurti and David Bohm*. Routledge.

Böhm, S., Misoczky, M. C., & De Carvalho, J. L. F. (2012). Greening capitalism? A Marxist critique of carbon markets. *Organization Studies*, *33*(11), 1617–1638.

Capra, F. (1996). *The web of life: A new scientific understanding of living systems*. Anchor Books.

Clapp, J. (2021). *Food* (3rd ed.). Polity.

Crawford, K. (2021). *Atlas of AI: Power, politics, and the planetary costs of artificial intelligence*. Yale University Press.

Damasio, A. (2010). *Self comes to mind: Constructing the conscious brain*. Pantheon.

Engel, G. L. (1977). The need for a new medical model: A challenge for biomedicine. *Science*, *196*(4286), 129–136. https://doi.org/10.1126/science.847460

Food and Agriculture Organization of the United Nations. (2022). *The state of food and agriculture 2022: Repurposing food and agricultural policies to make healthy diets more affordable*. FAO.

Foucault, M. (1977). *Discipline and punish: The birth of the prison* A. Sheridan, Trans.. Pantheon.

Freire, P. (1970). *Pedagogy of the oppressed*. Continuum.

Geneva Conventions, Aug 12, 1949, 75 U.N.T.S. 31.

Geyer, R. & Cairney, P. (Eds.). (2015). *Handbook on complexity and public policy*. Edward Elgar.

Giddens, A. (1984). *The constitution of society: Outline of the theory of structuration*. University of California Press.

Global Footprint Network. (2022). *Earth Overshoot Day 2022: 1.7 Earths explainer*. Global Footprint Network.

Gupta, D. (2024). *Eternal movement*. Emerge Publications. https://doi.org/10.65169/q7d2z6

Gupta, D. (2025). *Governance without the self: Consciousness and the end of fragmentation, systems, and complexity*. Emerge Publications. https://doi.org/10.65169/a9m3p4

Hamilton, C. (2013). *Earthmasters: The dawn of the age of climate engineering*. Yale University Press.

Hobbes, T. (2012). In N. Malcolm (Ed.), *Leviathan*. Oxford University Press. (Original work published 1651).

Human Rights Watch. (2019). *China's algorithms of repression: Reverse engineering a Xinjiang police mass surveillance app*. HRW.

Human Rights Watch. (2024). *Gaza: Israel's apparent use of AI in targeting raises risk to civilians*. HRW.

Intergovernmental Panel on Climate Change. (2021). *Climate change 2021: The physical science basis (AR6 WG I)*. Cambridge University Press.

Intergovernmental Panel on Climate Change. (2023). *AR6 synthesis report: Climate change 2023*. IPCC.

Intergovernmental Science-Policy Platform on Biodiversity and Ecosystem Services. (2019). *Global assessment report on biodiversity and ecosystem services*. IPBES.

International Energy Agency. (2023). *Fossil fuel consumption subsidies 2022–2023: A record high*. IEA.

International Labour Organization. (2023). *Global employment trends for youth 2023: Investing in transforming futures for young people*. ILO.

Jackson, M. C. (2019). *Critical systems thinking and the management of complexity*. Wiley.

Kimmerer, R. W. (2013). *Braiding sweetgrass: Indigenous wisdom, scientific knowledge, and the teachings of plants*. Milkweed Editions.

Krishnamurti, J. (1953/2009). *Education and the significance of life*. HarperCollins.

Krishnamurti, J. (1954). *The first and last freedom*. Victor Gollancz.

Krishnamurti, J. (1972). *The awakening of intelligence*. Victor Gollancz.

Lanier, J. (2018). *Ten arguments for deleting your social media accounts right now*. Henry Holt.

McGilchrist, I. (2019). *The master and his emissary: The divided brain and the making of the Western world* (2nd ed.). Yale University Press.

McKernan, B., & Davies, H. (2024, April 3). 'The machine did it coldly': Israel used AI to identify 37,000 Hamas targets. *The Guardian*. https://www.theguardian.com/world/2024/apr/03/israel-gaza-ai-database-hamas-airstrikes

Meadows, D. H. (1999). *Leverage points: Places to intervene in a system*. Sustainability Institute.

Midgley, G. (2000). *Systemic intervention: Philosophy, methodology, and practice*. Kluwer Academic/Plenum.

Morin, E. (2008). *On complexity*. Hampton Press.

Naess, A. (1989). *Ecology, community and lifestyle*. Cambridge University Press.

Nussbaum, M. C. (2010). *Not for profit: Why democracy needs the humanities*. Princeton University Press.

Office of the United Nations High Commissioner for Human Rights. (2022). *Assessment of human rights concerns in the Xinjiang Uyghur autonomous region, People's Republic of China*. OHCHR.

Ostrom, E. (1990). *Governing the commons: The evolution of institutions for collective action*. Cambridge University Press.

Ostrom, E. (2010). Polycentric systems for coping with collective action and global environmental change. *Global Environmental Change*, *20*(4), 550–557. https://doi.org/10.1016/j.gloenvcha.2010.07.004

Piketty, T. (2020). *Capital and ideology*. Harvard University Press.

Rockström, J., Gupta, J., Qin, D., Lade, S. J., Abrams, J. F., Andersen, L. S., Armstrong McKay, D. I., Bai, X., Bala, G., Bunn, S. E., Ciobanu, D., DeClerck, F., Ebi, K., Gifford, L., Gordon, C., Hasan, S., Kanie, N., Lenton, T. M., Loriani, S., … Zhang, X. (2023). Safe and just Earth system boundaries. *Nature*, *619*(7968), 521–532. https://doi.org/10.1038/s41586-023-06083-8

Security Council Report. (2023). *The veto: Trends and explanations*. SCR.

Senge, P. M. (1990). *The fifth discipline: The art and practice of the learning organization*. Doubleday.

Shiva, V. (2005). *Earth democracy: Justice, sustainability and peace*. South End Press.

Stockholm International Peace Research Institute. (2023). *SIPRI yearbook 2023: Armaments, disarmament and international security*. SIPRI.

Stockholm International Peace Research Institute. (2024). *Trends in world military expenditure 2024 (Fact Sheet)*. SIPRI.

Tronto, J. C. (1993). *Moral boundaries: A political argument for an ethic of care*. Routledge.

United Nations. (1945). *Charter of the United Nations*.

United Nations. (1948). *Universal declaration of human rights*.

United Nations. (2015). *Transforming our world: The 2030 agenda for sustainable development*. United Nations.

United Nations. (2023). *The sustainable development goals report 2023*. United Nations.

United Nations. (2025). *The sustainable development goals report 2025*. United Nations.

United Nations Development Programme. (2020). *Human development report 2020: The next frontier—Human development and the anthropocene*. UNDP.

United Nations Digital Library. (2023–2024). *Security council draft resolutions and voting records on Gaza ceasefire*. UNDL.

United Nations Environment Programme. (2023). *State of finance for nature 2023*. UNEP.

United Nations General Assembly. (1948). *Universal declaration of human rights*. https://www.un.org/en/about-us/universal-declaration-of-human-rights

United Nations Human Rights Council. (2018). *Report of the special rapporteur on the issue of human rights obligations relating to the enjoyment of a safe, clean, healthy and sustainable environment (A/73/188)*. United Nations. https://digitallibrary.un.org/record/1474985/files/A_HRC_37_59-EN.pdf

United Nations Office for the Coordination of Humanitarian Affairs. (2023). *Global humanitarian overview 2023*. UNOCHA.

United Nations Office for the Coordination of Humanitarian Affairs. (2024). *Occupied Palestinian territory: Situation reports and humanitarian updates (Gaza)*. UNOCHA.

Vienna Convention on the Law of Treaties, May 23, 1969, 1155 U.N.T.S. 331.

World Economic Forum. (2023). *The global risks report 2023* (18th ed.). WEF.

World Food Programme. (2024). *Gaza and West Bank: Emergency food security assessments and IPC updates*. WFP.

World Health Organization. (2022). *World mental health report: Transforming mental health for all.* WHO.

World Health Organization. (2023). *World health statistics 2023: Monitoring health for the SDGs*. WHO.

Zuboff, S. (2019). *The age of surveillance capitalism: The fight for a human future at the new frontier of power*. PublicAffairs.

Chapter 2

Human Condition: Psychological Roots of Institutional Breakdown

Abstract

Following the opening enquiry into how the polycrisis mirrors inner fragmentation, this chapter turns inward to uncover the psychological architecture from which the outer world arises. It serves as a moment of shared reflection, a pause to establish the language and clarity needed before the inquiry proceeds deeper. This chapter explores what is meant by consciousness and examines how thought, language and memory construct the self, an image sustained by identification, comparison and continuity. Institutions mirror this same architecture, organising life through fear, control and measurement. The analysis reveals how specialisation, the observer–observed division and psychological time perpetuate fragmentation, giving rise to inequality, competition and institutional inertia. Even frameworks like the Sustainable Development Goals, though envisioned as transformative, extend the same movement of postponement and control. By understanding the illusion of an independent actor, the reader begins to see how the self is both the creator and the product of disorder. Transformation, this chapter concludes, does not arise through reform or design but through perception itself. This foundation prepares the ground for the next chapter's examination of why global frameworks, despite good intention, cannot heal the very division from which they are born.

Keywords: Consciousness; psychological structures; institutional mirroring; identification; observer-illusion; temporal conditioning; systemic inertia; division; self-understanding

The human condition, marked by fear, desire and division, is not a private or isolated experience. It expresses itself outwardly through the institutions we build and the systems we sustain. Governance, education and health mirror the same

From Fragmentation to Wholeness, 41–71

Published under exclusive licence by Emerald Publishing Limited
doi:10.1108/978-1-80686-677-920261003

psychological movement: the search for order through control, security through exclusion and achievement through competition.

Understanding this movement is essential before going further. Without perception at the very beginning, the observer risks falling back into the past – interpreting what follows as knowledge rather than insight. But transformation and communion are not goals at the end of a path; they exist only in the immediacy of seeing. This chapter is written to create such a space of perception and communion, so that as we proceed towards the critique of frameworks in Chapter 3 and the deeper inquiry into consciousness in Part II, the reading itself is not mechanical but alive.

Though its themes may appear close to those of Part II, the difference is crucial. Chapter 2 lays the ground for freedom, not as a result, but as a presence that must be there from the start.

2.1 Introduction

Chapter 1 concluded with a radical proposition: the polycrisis is not only a matter of failing systems, but of the fragmented perception of the actor who creates, builds and manages them. Climate breakdown, armed conflict, economic inequality and institutional paralysis are not simply the result of external forces; they are movements (Gupta, 2024), expressions of an inner architecture shaped by fear, identity and control (Bohm, 1992; Krishnamurti, 1983; McGilchrist, 2019). This architecture sustains itself, repeating endlessly, because it is not being brought to an end.

This chapter turns the inquiry inward. If the world reflects the self, then we must understand the nature of this self, not as a philosophical abstraction, but as a living psychological structure present in each of us. We will explore how consciousness is conditioned, how it becomes fragmented and how this fragmentation creates and sustains global disorder. In doing so, the discussion links directly to the structural challenges faced in achieving the Sustainable Development Goals (SDGs), where institutional and technical reforms often falter because they are implemented from the same divided perception that generated the crises (United Nations, 2023).

Here, the inquiry into the 'human condition' begins with the self as a psychological construct. The self is the accumulation of memories, both physical records and their emotional residues of fear, desire and attachment, as well as the ideas, images and experiences that are registered or created. By 'understanding,' we do not mean intellectual agreement or theoretical explanation. We mean insight that transforms human beings at every step of inquiry, not at its conclusion. This is learning that engages both the emotional and intellectual dimensions and requires a state of direct presence and awareness. To observe in this way is to look without attachment to one's identities or to the already fragmented, identified self. Such observation is not a tool towards an end; it is both the means and the end of this book.

By 'movement of the self,' we refer to how the self comes into being, how it operates and whether its continuity can come to an end. In this exploration, the term 'human condition' does not refer to an abstract universalism that erases difference, nor does it deny the realities of structural injustice and historical inequality. Rather, it points to the common architecture of thought and identity that underlies human action across cultures, ideologies and institutions (Tajfel & Turner, 1986; Varela et al., 1991). This architecture shapes how we see, how we relate and how we act. It is the unseen constant beneath the variability of political systems, governance models and development frameworks.

By examining this architecture, we can begin to understand why outer reforms often fail to produce coherence and why even the most well-intentioned efforts can reproduce the very crises they seek to overcome. The inquiry ahead is not about how to better manage complexity, but whether the mind of the actor can perceive without fragmentation. If such perception is possible, it could transform the foundations of governance, policy and sustainability.

2.2 The Content of Consciousness

Consciousness, as understood here, is the total field of our inner life: thought, memory, emotion and sensation. Within this field lies what is often called the 'self,' a centre we experience as 'I' or 'me.' This self is not an isolated, immutable entity but a construction assembled over time through personal experience, cultural conditioning, language and collective memory (Damasio, 2010; Varela et al., 1991). From birth, human beings are immersed in relationships and social structures that shape identity. Family narratives, national history, religious traditions, economic systems and educational institutions provide the frameworks through which we interpret the world. Through repeated reinforcement, these interpretations solidify into the sense of a stable 'self' who acts, chooses and decides. Social identity theory (Tajfel & Turner, 1986) further explains how belonging to groups becomes a core part of this construction, creating both internal cohesion and external division. Awareness, in contrast, is not part of this field of consciousness but the capacity to perceive it in its entirety without being shaped by its content.

Consciousness as Content

Krishnamurti (1991) described this field of inner life as the 'content of consciousness': the totality of beliefs, experiences, knowledge and conditioning. This content is not separate from consciousness; it is consciousness as we ordinarily know it. Thought, which is always rooted in past experience, organises this content into a narrative of self: who I am, who I should be, what I believe and what I fear. Bohm (1992) emphasised that this process is inherently fragmentary. Thought divides the field of life into categories as self and other, us and them, past and future and then acts as though these divisions are inherent in reality. The self becomes a centre around which perception is organised, filtering all experience through the lens of identity and memory. Damasio's (2010) neuroscientific work supports this by showing how the brain constructs a continuously updated 'autobiographical self' that integrates bodily states, emotional experiences and social cues into a narrative of continuity that masks its underlying fragmentation.

The Filtered Nature of Perception

This filtering is not neutral. Contemporary neuroscience shows that perception is shaped by priors like beliefs, expectations and predictive models that constrain what we perceive as 'objective' reality (Friston, 2010). A policymaker negotiating a climate treaty does not enter the room as a blank slate; they carry the weight of national interest, political alliances and ideological commitments. An activist demanding justice may carry the momentum of personal history, group loyalty and moral indignation. A business leader promoting sustainability may also be protecting reputation, market share or investor confidence. In each case, the content of consciousness with its images, memories and identifications, with their fears and desires shapes action. This influence scales across levels: Individually: In personal relationships, memory of past interactions can colour present perception, leading to mistrust, selective listening, conflict, jealousy or division. Institutionally: Bureaucracies inherit and act from past policies, norms and cultures that limit adaptability. Structures become more important than timely action, and attachment to procedure can obscure compassion and right action. Globally: International negotiations replay historical alignments, colonial legacies and competitive postures. Foreign policy shaped by insecurity and desire for control often results in fragmented action, driven less by necessity than by the self-interest of nations.

Consciousness and Global Patterns

These patterns operate both individually and collectively, meaning the global order often mirrors the same psychological dynamics we find in daily life: competition for status, fear of loss, the drive for control and the tendency to define oneself against an 'other.' This mirroring is visible in the SDGs, where progress on SDG 13 (climate action) or SDG 16 (peace, justice and strong institutions) is hampered by the same fear, division and competition that shape individual decision-making. The persistence of geopolitical blocs in climate negotiations mirrors interpersonal defensiveness in conflict resolution. Economic competition between nations, despite commitments to SDG 10 (reduced inequalities), reflects the self's tendency to measure worth comparatively. Short-term electoral or market cycles mirror the self's preference for immediate gratification over long-term sustainability.

Why Understanding Content Matters

Understanding the content of consciousness is therefore not an exercise in self-analysis for its own sake. It is essential to explaining why global crises persist despite technical knowledge, political agreements and moral frameworks. The self, as a movement of thought rooted in the past, is inherently limited in its capacity to respond to the present. It reacts according to memory and conditioning, which means it often perpetuates the very patterns it seeks to end. Without awareness of how the mind's structures constrain perception, attempts at reform, even those grounded in evidence, can paradoxically deepen the problems they seek to resolve. In this sense, seeing the content of consciousness directly is not a philosophical luxury; it is a foundational requirement for coherent governance, meaningful diplomacy and the achievement of the SDGs.

When we read from past consciousness, there is division, comparison and repetition. But when there is reading that is free from condemnation, acceptance or rejection, then there can be communion. This applies not only to the way we read texts, but to how we live and how we engage with global frameworks like the SDGs. To move forward requires insight at every step, not the accumulation of knowledge, but an understanding that is alive in the present. What we mean by 'knowing' will be explored more fully in Chapter 5, but the question can already be posed: can there be reading, or perception, that does not seek to achieve or to conclude, but simply to observe and find out?

This book itself is not outside that movement. It is not a static account of ideas but a reflection of the very inquiry it invites. To read it mechanically would be to repeat the past. To read it as an act of perception, without condemning, accepting or rejecting, opens the possibility of communion. In this way, the book becomes a mirror: not a body of knowledge to be accumulated, but a space of seeing together what is, because what you are observing here is not only the book, but yourself.

2.3 The Formation of the Self

The self, as we ordinarily experience it, is not an inherent essence but a construct, brought into being through the interaction of biology, language, culture and relationship. From infancy, there is sensation, perception and response, but not yet a fixed identity. The 'I' emerges gradually, shaped by naming, recognition and the mirroring of others.

Relational Foundations

Developmental psychology has long noted that a child's sense of self arises in relationship. In early life, caregivers reflect back signals of approval, disapproval, affection or neglect. These reflections begin to form an image of 'who I am' in the mind of the child (Erikson, 1968; Mead, 1934). Over time, this image becomes more elaborate, integrating not only personal experience but also the roles, expectations and narratives inherited from family, community and nation.

Language as Boundary-Maker

Language plays a central role in the formation of the self. By naming ourselves and others, we begin to draw conceptual boundaries between 'me' and 'not me.' Words crystallise experience into categories, allowing for memory, planning and comparison but also solidifying divisions that may not exist in the living movement of reality (Korzybski, 1933; Lakoff & Johnson, 1980).

Krishnamurti (1969/2004), in *Freedom from the Known*, warned that naming can become a barrier to perception: when we call a flower a 'rose,' the word comes between us and the act of direct seeing. Recognition through language often prevents us from encountering what is before us freshly, without the filter of memory and classification. He further illustrated that when a child is taught the name of a bird, the child may cease to truly look at the bird; the word replaces the wonder of direct observation.

Language thus does not merely describe reality, it shapes, limits and often distorts it. It structures the self into a narrative, carried forward as identity and frames how societies organise meaning and power.

Beyond individual naming, language also carries cultural metaphors that unconsciously shape perception. Military metaphors dominate political discourse, market metaphors pervade education and war metaphors appear in public health. Such framing not only structures individual self-concepts but also informs governance priorities, influencing decisions that ripple across domains.

Cultural Scripting and Institutional Reinforcement

Culture then supplies the scripts. Religion may define the self in terms of virtue, sin, salvation or liberation. Nationalism may define it in terms of citizenship, heritage and allegiance. Economics may define it in terms of productivity, ownership and consumption. Education reinforces these definitions by rewarding conformity to certain patterns of thought and behaviour, and by measuring worth through performance and competition (Bourdieu, 1984; Foucault, 1977).

This cultural scripting is not neutral; it encodes historical power relations, class structures and gender norms.

The Self as Social Product

The self is not simply a private psychological construct; it is a social product. It carries the imprints of collective history, ideology and power (Anderson, 1983; Said, 1978). A person born into a conflict zone may inherit narratives of victimhood or heroism that shape and condition their perception of the world. A person born into privilege may inherit narratives of entitlement or merit that equally shape perception. Both operate from conditioning, though the content differs. Both the private and the social are the same; there is no self separate from society. This very division is itself a product of identification, which creates 'us and them,' nations and identities. The content of consciousness is not personal but common to all human beings. We all suffer, fear and identify. The objects of fear, desire or skill may differ, but the nature of thought, memory and identification, how it divides and fragments is the same.

Psychological Time and Continuity

Crucially, the self is sustained through psychological time, the continuity of memory projecting itself into the future (Krishnamurti, 1983). We remember past experiences, interpret them as part of 'my' story and imagine futures in which 'I' will succeed, fail, suffer or be fulfilled. The rememberer here is not separate from memory or thought. This movement of the past and its projection generates both hope and fear, and it is within this movement that much of human action takes place. We see that we are violent, and instead of ending our violence, we create an idea of nonviolence while continuing to act violently. We speak of God and brotherhood, yet in the name of God and the flag we are ready to kill. Ideas, as thoughts act as psychological time and escape, and prevent actual ending of the self.

Identification as Continuity

This continuity is also what allows for identification, binding the self to groups, roles and beliefs. Through identification, personal survival becomes tied to the survival of the group or idea. Criticism of a belief can feel like a personal attack; perceived threats to a group identity can provoke intense defensiveness or aggression (Tajfel & Turner, 1986). At the scale of nations, this mechanism fuels nationalism, sectarianism and ideological polarisation, by making collaboration contingent on identity preservation.

When we enquire into why we identify, we begin to see that identification arises from insecurity, the need to belong, to feel larger than oneself, to avoid the fear of nonexistence, for self-gratification. Identification provides illusionary temporary security, but in doing so it breeds attachment and division.

Perception Through the Lens of the Self

Once formed, the self becomes the lens through which all perception is filtered. We interpret events in ways that confirm identity and avoid what contradicts it. We seek experiences that reinforce it. Neuroscience frames this as predictive processing: the brain generates hypotheses about reality based on prior experience, then updates or resists updating those hypotheses in response to sensory input (Friston, 2010). The result is a self-reinforcing loop in which perception serves identity, and identity shapes perception (Gupta, 2025).

Adaptive Advantages and Inherent Limits

While the structure of self has adaptive advantages, it provides continuity, allows for planning and supports social cooperation, it also carries inherent limits, and this adaptation becomes self-preservation and division, in movement and action. Because it is built from the past, it can only respond to the present through the filter of memory, and its action is based upon its own preservation. Because it is organised around separation, it easily gives rise to fear, comparison and conflict.

And because it is identified with particular roles and narratives, it resists change that threatens those identifications.

In governance, development and global cooperation, these limitations manifest as structural inertia, ideological rigidity and the prioritisation of self-interest over collective well-being. The same mechanism that allows a policymaker to represent their nation effectively also makes it difficult to perceive global challenges outside the frame of national interest. The same mechanism that motivates activism for a cause can also entrench division between 'us' and 'them.'

In short, the self is not merely a personal phenomenon; it is the psychological infrastructure of global disorder. To understand the persistence of the polycrisis, we must understand how this infrastructure operates not only within leaders and institutions, but within each of us. Without this insight, frameworks like the SDGs will continue to be shaped by the very conditioning that sustains the crises they aim to solve.

2.4 The Mechanisms of Fragmentation

Once the self is formed, its underlying structure naturally produces fragmentation, inwardly in intellect, emotion and action, and outwardly in relationships, institutions and global systems. These mechanisms operate so subtly that they are often mistaken for natural or inevitable features of human life. They underlie not only interpersonal misunderstandings but also the breakdown of governance, institutional trust and sustainability agendas, including the SDGs. Similar insights appear in Bohm's (1992) work on thought's divisive nature, McGilchrist's (2019) hemispheric theory and Capra's (1997) systems-based critique of modern institutions.

Identification and Division

The self sustains itself through identification: with beliefs, roles, possessions, achievements and affiliations. Identification turns abstractions into extensions of 'me' and 'mine.' This is why criticism of a political ideology, religion or nation can provoke the same defensive reaction as a personal insult (Said, 1978; Tajfel & Turner, 1986). At the collective level, identification fuels nationalism, sectarianism and ideological polarisation, making cooperation across boundaries conditional and fragile.

In governance, this manifests in negotiation deadlocks, selective application of international law and policies driven by loyalty to domestic constituencies rather than to humanity as a whole. The Israel–Gaza crisis, for example, is sustained not only by territorial and security concerns but by deeply embedded group identities on both sides, making compromise appear as betrayal (Amnesty International, 2024; UN OCHA, 2024). Such identity-based fragmentation undermines SDG 16 (peace, justice and strong institutions) by entrenching cycles of mistrust, while also limiting progress on SDG 17 (partnerships for the goals). At its root lies the illusion of separateness and the self's interests, which continue to be satisfied through conflict. Countries profit from selling weapons, companies secure contracts and actors cooperate only until their immediate needs are met. There is no true relationship in such identification, only action driven by convenience and self-interest. This self-interest becomes selfishness, corruption and division, because concern and care are restricted to the identified fragment rather than the whole.

Comparison and Competition

The self's sense of worth is often derived from comparison. Whether between individuals, organisations or nations, comparison creates a psychological hierarchy, someone must be ahead and someone behind. This mechanism underpins the global race for economic growth, technological dominance and military superiority (Meadows et al., 2004). Girard's (1977) theory of mimetic desire adds that humans often imitate the desires of others, producing rivalry and escalating competition.

Gupta (2025) observes that the self's search for psychological security through fear and desire paradoxically creates greater insecurity in the world. At the level of consciousness, he further notes (Gupta, 2024) that this is not a linear chain of causes and effects but a single movement of the self. In the dimension of space and time, this movement manifests as a positive feedback loop, a *self-replicating fragmentation*, in which fear produces division, division generates insecurity and insecurity reinforces fear. From a holistic perspective, however, beyond the confines of linear causality, this is simply the nature of the self: fundamentally insecure, identifying in order to feel secure and thereby creating the very divisions that sustain insecurity and conflict.

In international politics, this dynamic is visible in the 'security dilemma,' where one state's pursuit of security is perceived as a threat by others, fuelling an arms race and mutual suspicion (Herz, 1950). Competition easily blinds actors to what is actually required. Once caught in the race of becoming, nations prioritise relative advantage rather than solving emergent problems directly. While competition can spur innovation, it also entrenches inequality and perpetuates cycles of insecurity.

In the SDG context, many countries prioritise rankings and indicators over substantive transformation, producing policies aimed at statistical improvement rather than structural coherence (United Nations, 2023). This can distort SDG 8 (decent work and economic growth) into a competition for GDP figures and SDG 10 (reduced inequalities) into a relative performance metric rather than a shared human priority.

Psychological Time and Continuity of Conflict

The self is built from memory projecting into the future, sustaining both hope and grievance, fear and desire (Bohm, 1992; Krishnamurti, 1983). In personal life, unresolved hurts from the past shape current relationships; in geopolitics, historical injustices are carried forward as collective memory, justifying retaliation or resistance generations later (Bar-Tal, 2007; Lederach, 1997). Examples abound: colonial histories reproduced in postcolonial governance structures, cycles of ethnic violence in the Balkans and Rwanda, or ongoing disputes in Kashmir and Palestine.

Because the self experiences itself as continuous in time, it perpetuates patterns. Nations once colonised may replicate hierarchical governance structures introduced by colonisers. Communities that have experienced violence may adopt security measures that provoke further violence. Such feedback loops, self-reinforcing cycles of perception and response, obstruct SDG 16 and SDG 17 by keeping actors locked in adversarial postures.

The Division Between Observer and Observed

Perhaps the most fundamental mechanism of fragmentation is the way thought divides reality into 'the observer' and 'the observed.' Jiddu Krishnamurti (1969/2004) was uncompromising on this point:

> So long as there are the thinker and the thought, the observer and the observed, the experiencer and the experienced, there is no integration. ... We now think the thought is separate from the thinker; but is that so? ... If I realize that I am greedy, that there is not the observer who is greedy but I am myself greed, then our whole question is entirely different. (*The First and Last Freedom*, pp. 108–111)

Here, Krishnamurti directly challenges the assumption that there is a separate self who controls or corrects thought. The supposed distance between observer and observed is itself the source of conflict and deterioration.

In later dialogues, he insisted this indivisibility also dissolves the boundary between inner and outer:

> The outer is the inner, the inner is the outer. There is not the difference between the outward and the inner. They are totally related to each other. ... The world is you and you are the world. There is no other. (*A Wholly Different Way of Living*, 1974, pp. 7–9)

Failure to grasp this indivisibility perpetuates alienation. Nature becomes 'the environment' to be managed, society becomes 'the public' to be governed and people are reduced to 'stakeholders' or 'target groups.' A child in Gaza is seen as a 'Gazan,' a child in Ukraine as a 'Ukrainian', filtered through identity and ideas rather than encountered as a human being. Thought narrows sensitivity and conditions compassion through labels. Yet as Krishnamurti emphasised, true compassion has nothing to do with nationality, religion or identity; it arises only when the division between observer and observed ends.

Bohm (1992) argued that thought, by positing an external vantage point, fragments reality into categories and oppositions. Once this division is assumed, the illusion arises that the observer stands apart from life itself, a position that enables control and measurement but simultaneously produces alienation.

Fragmentation Through Specialisation

Modern knowledge systems, institutions and governance structures mirror the self's fragmentation by dividing expertise into silos. Education separates disciplines, the government divides ministries and global organisations partition issues into thematic clusters (Capra & Luisi, 2014). This leads to contradictory policies such as subsidising fossil fuels while investing in climate adaptation (UNEP, 2023) or expanding intensive agriculture under SDG 2 (zero hunger) while committing to SDG 15 (life on land).

While specialisation enables technical depth, it obscures the whole. Specialists become deeply skilled in one domain but may lack the systemic awareness to see how their work interacts with other fields. Decision-making thus becomes a patchwork of partial views, each advancing its own objectives without integrating into a coherent whole, undermining the integrative vision of SDG 17.

The Underlying Process

These mechanisms of identification, comparison, psychological time, observer–observed division and specialisation are not separate problems to be solved individually. They are expressions of the same underlying process: thought organising experience around a separate 'me' (Bohm, 1992). As long as this process remains unexamined, it will reproduce itself in every domain of life.

Efforts to address climate change, inequality or conflict will remain partial, because the very actor attempting to solve the problem is structured by the same fragmentation that sustains it. As Gupta (2025) puts it, *'The crisis in governance is not merely a failure of systems, but the absence of inward clarity in those who build and maintain them.'* You cannot solve all these problems separately because they are interrelated, and at their root lies conditioned thought, identified, separated, fragmented, that has divided the world into nations and religions and created both the atomic bomb and the climate crisis we now face.

Among these mechanisms, the division between the observer and the observed is perhaps the most fundamental. It does not merely fragment perception; it conditions the very stance from which we act in governance, conflict and development. Because of its centrality, this division requires closer examination in a separate section.

2.5 Psychological Time, Memory and the Perpetuation of Crisis

Psychological time is not the chronological sequence of days and years measured by a clock. It is the movement of thought that carries the past into the future, sustained by memory and projection. Physical time is necessary for practical action, to build, to grow, to coordinate. But psychological time sustains the continuity of the self and, with it, the continuity of conflict.

The Nature of Psychological Time

The self is essentially the story told about 'me,' woven from memory. These memories are never fixed; they are continuously reshaped by interpretation, emotion and narrative. From this storehouse of experience, the mind projects futures – the person I want to become, the goals I must achieve, the threats I must avoid. This movement between remembered past and imagined future is what keeps the present moment subordinate to an unfolding timeline, a means to some anticipated end.

In ordinary life, this appears as clinging to resentment after an argument or longing for a lost 'golden age.' In politics, it manifests as revanchism, the attempt to reclaim past glory or territory, or as utopian planning that sacrifices present well-being for an idealised future. Both patterns bind attention to a movement of becoming, preventing direct encounter with what is. Zimbardo and Boyd (2008) describe this as a 'future time perspective' that can distort judgement, pulling attention away from present responsibility into imagined horizons of fear or hope. Ricoeur (2004) similarly argued that memory and narrative weave past, present and future into a single temporal fabric, shaping identity itself. Yet when narrative becomes rigid, it confines action within inherited stories rather than opening to new possibilities.

Historical Memory and Geopolitical Stalemates

Psychological time plays a decisive role in sustaining geopolitical crises. Nations carry forward historical traumas of colonial exploitation, genocide, occupation, and use them as justification for present action. Acknowledging injustice is necessary, but turning it into the foundation of identity entrenches division.

Conflicts such as Kashmir, Israel–Palestine and the Korean Peninsula endure less because of present realities than because of collective memory. Education, media and ritualised commemorations keep grievances alive, passing them into new generations and obstructing reconciliation. Scholars like Lederach (1997) and Bar-Tal (2007) have shown how intractable conflicts persist because each side sustains a moral memory: one that positions itself as victim and the other as aggressor. For both, reconciliation appears as a threat to identity itself. Arendt (1963) warned that when history is carried forward without renewal, politics becomes trapped in cycles of vengeance, unable to generate new beginnings.

The SDG Timeline and the Illusion of Deferred Transformation

This same mechanism distorts global governance. The SDGs are framed around a 15-year horizon (2015–2030), with periodic measurement of progress. While such targets aid coordination, they also risk creating the illusion that transformation is something to be achieved later, rather than lived now.

Carbon neutrality pledges for 2050 or biodiversity goals for 2040 become symbolic promises that defer immediate responsibility. Meanwhile, ecosystems degrade and inequalities widen (IPCC, 2023; United Nations, 2023). This temporal gap weakens SDG 13 (climate action), SDG 15 (life on land) and SDG 10 (reduced inequalities), as urgent realities are subordinated to distant milestones. When a fragmented mind constructs linear frameworks of change, attachment to goals and methods becomes more important than responding to the living present. The COVID-19 pandemic revealed this vividly: despite the novelty of the moment, responses were largely filtered through predetermined structures and timelines.

Memory as the Continuity of Conflict

Thought operates through stored knowledge and memory. Even attempts at reform or reconciliation are shaped by the very past that created the problem. Peace negotiations often fail because each side arrives with fixed narratives of history and justice. Dialogue becomes an exchange of past-based positions rather than a meeting in the present.

At the personal level, old hurts rarely end. They are remembered, reinterpreted and carried forward into every new interaction. In both personal and political life, the inability to end the past psychologically ensures that the future is only a modified continuation of it.

2.6 The Observer–Observed Division

The mechanisms of fragmentation outlined above including identification, comparison, psychological time and specialisation are all expressions of one deeper dynamic. At their root lies perhaps the most fundamental division: that between the observer and the observed. As long as thought sustains the illusion that the 'I' stands apart from what is seen, a centre is created that seeks to control, reform and manage the 'other.' This split not only shapes inner life but also structures governance, education, economics and global systems (Bohm, 1992; Krishnamurti, 1983).

In ordinary perception, we experience ourselves as separate from what we see: *'I am the one thinking, feeling, and deciding, while the world and even my own thoughts and emotions are objects to be managed, changed, or controlled.'* This dualism gives rise to the very patterns of alienation, mistrust and conflict that persist despite reforms.

The Illusion of an Independent Actor

In governance, the observer–observed division appears when policymakers act as if they are independent of the consequences they produce. A government may legislate environmental protections while simultaneously maintaining fiscal policies that incentivise deforestation or fossil-fuel extraction, as though the policymaker's role exists outside the ecosystems they inhabit (Capra, 1997).

In development, interventions are often designed from the vantage point of external expertise. Communities are treated as 'beneficiaries' of aid, while the experts remain untouched by the structures of inequality and power they represent. This mirrors the psychological split in the mind, the self as a controller standing apart from life (Ostrom, 2010). As long as the actor perceives themselves as separate from the problem, they cannot be in touch with reality as it is; they continue to respond from what was or what should be, rather than from direct perception of what is.

In Conflict and Security

The observer–observed split is stark in war and security discourses. Nations consistently frame themselves as innocent responders to aggression, while casting others as the sole cause of conflict. Each side sees itself as the neutral observer, objectively defending justice, while failing to see how its own history, actions and fears are woven into the very hostility it condemns (Bar-Tal, 2007; Lederach, 1997).

At the psychological level, this division rests on the fact that the 'observer' is itself knowledge, the accumulated record of past experiences, resentments and fears. When someone is perceived as 'a Muslim,' 'an American' or 'white,' the perception is not of the living human being before us but of a projection from memory, bound to identity, history and attachment. What is seen is not what is but what the self brings to perception which is its own conditioning, motivations and desires. Thus, the field of vision in conflict is largely a projection of the observer's past, not a direct encounter with reality.

The Israel–Gaza crisis illustrates this distortion vividly. Israel justifies repeated offensives as acts of self-defence, yet international observers, including the UN Office for the Coordination of Humanitarian Affairs (2024) and Amnesty International (2024), document widespread civilian harm, infrastructure destruction and violations of humanitarian law. Even the International Criminal Court issued an arrest warrant against political leaders implicated in war crimes. Still, each actor clings to the narrative that it is the innocent observer acting upon a dangerous 'other,' while failing to see its own participation in the cycle of violence.

The Russia–Ukraine war, is structured by the same split. Both sides claim to be responding defensively to aggression, upholding a story in which 'we' are the justified observers and 'they' are the aggressors. This mutual stance sustains escalation while foreclosing the possibility of shared perception. In both Gaza and Ukraine, external powers whether through arms sales, strategic alliances or selective humanitarian concern reinforce these narratives, profiting from the continuation of conflict while presenting themselves as detached arbiters of peace.

This division undermines the very possibility of SDG 16 (peace, justice and strong institutions). So long as each party clings to the illusion of being a separate, righteous observer confronting an external problem, violence is perpetuated. True reconciliation requires seeing through this illusion, recognising that the observer and the observed are inseparable, and that conflict is not 'out there' but inseparable from the conditioning that sustains it in each of us.

In the Sustainable Development Framework

The same division permeates the SDGs. Governments and agencies often treat poverty (SDG 1), inequality (SDG 10) or climate change (SDG 13) as problems 'out there' in need of management, while ignoring how their own economic models, consumption habits and institutional cultures generate those very crises (Sachs et al., 2022).

This results in contradictory policy commitments. Many countries simultaneously promote endless GDP growth while pledging to cut emissions. For example, the *IPCC's Sixth Assessment* Report (2023) documents how G20 economies, responsible for 75% of global emissions continue to expand fossil fuel subsidies even while committing to net-zero targets. India and China have scaled renewable energy rapidly, yet both also approved record new coal projects in 2022–2023, reflecting the persistence of developmental models that equate growth with fossil energy (IEA, 2023).

Similarly, under SDG 9 (industry, innovation and infrastructure), governments often fund extractive or carbon-intensive infrastructure, such as highway expansion or oil pipelines, while simultaneously pledging under SDG 15 (life on land) to conserve biodiversity. Canada's 2022 approval of the Trans Mountain oil pipeline expansion, projected to increase annual emissions by up to 13–15 megatons, occurred alongside its biodiversity commitments at COP15, exposing how observer–observed separation permits contradictory action.

Agriculture and food systems provide another example. The *FAO's* State of Food Security and Nutrition Report (2023) shows that nearly 30% of global greenhouse (GHG) gas emissions come from food systems, yet many national strategies pursue agricultural intensification to meet SDG 2 (zero hunger) while undermining SDG 13 (climate action) and SDG 15 (life on land). The European Union's (EU's) Common Agricultural Policy has long been criticised for subsidising industrial farming practices that degrade soils and ecosystems, even as the EU presents itself as a leader in ecological transition.

Even in inequality reduction (SDG 10), contradictions persist. The *World Inequality Report* (Chancel et al., 2022) shows widening wealth gaps within and across countries, yet many governments advance tax cuts, deregulation or austerity policies that exacerbate inequality while simultaneously reporting progress through narrow statistical measures.

The underlying issue is that the observer as the policymaker, the institution, the development expert remains unexamined. Policies are designed as if the actor is separate from the system that produces the crisis. Poverty is treated as an external condition to be alleviated rather than the direct outcome of economic structures sustained by the same governments. Climate change is treated as a technical challenge while high-consumption lifestyles, militarisation and unsustainable trade models remain untouched.

As long as this separation persists, SDG interventions will carry the same fragmentation that produced the problem. What is measured and reported may improve, but the deeper contradictions ensure that systemic incoherence remains intact.

Why the Division Persists

Bohm (1992) explained that thought generates the division by projecting an image of 'me' as distinct from 'the world.' This makes thought appear objective and neutral, even as it filters perception through its own past conditioning. The observer feels independent, but both observer and observed are part of the same movement of life.

This mechanism is reinforced on multiple levels. Psychologically, it offers the comfort of control, the sense that 'I' am standing apart from the problem and can manage it. Culturally, education systems and political institutions teach individuals to see themselves as agents acting upon an external reality, rather than as participants within it. Institutionally, bureaucracies and states are designed to sustain the fiction of neutrality, positioning themselves as detached arbiters of justice or managers of resources. Because the division is reinforced in this way, it remains largely invisible. Reformers and activists can act with sincerity while still reinforcing the separation, fighting for justice without perceiving how their own identities, fears and ambitions are embedded in the very conflicts they oppose.

2.7 The Limits of Frameworks and Reform

The tendency to design ever-new frameworks arises directly from the observer–observed division explored earlier. When policymakers and reformers see themselves as standing outside the problems they aim to solve, reform becomes the obvious response: rearrange the parts of the system, negotiate stronger treaties or design more sophisticated strategies. What remains unquestioned is the stance of the actor, the fragmented perception from which reform itself arises.

Reform Within the Same Paradigm

Reform often operates within the very assumptions that created the crisis. The Paris Agreement, for instance, relies on nationally determined contributions

(NDCs) that respect sovereignty but reinforce competitive self-interest. States pledge reductions while balancing economic advantage, leading to a patchwork of symbolic commitments without binding enforcement (Falkner, 2016; UNFCCC, 2023).

The same pattern holds in economics. International institutions prescribe sustainable growth as the solution to inequality, yet the paradigm of growth itself is tied to extraction, consumption and rivalry. Progress is equated with GDP, even as GDP growth exacerbates ecological breakdown and inequality (Stiglitz, 2019). As Morin (2008) notes, systemic transformation is impossible if we retain the epistemology that produced the problem.

The SDGs: Ambition Without Holistic Perception

The SDGs illustrate both the promise and the limitation of reform. With 17 goals, 169 targets and over 230 indicators, they constitute one of the most ambitious frameworks ever designed for human and planetary well-being. Yet implementation reveals fragmentation:

- Ministries responsible for climate action (SDG 13) operate separately from those driving economic growth (SDG 8) or agricultural expansion (SDG 2).
- India expands coal to fuel growth while simultaneously pledging net-zero by 2070 (IEA, 2021).
- International Monetary Fund (IMF) loan conditionalities impose austerity measures in the Global South while SDG 1 (ending poverty) demands expanded social safety nets (Kentikelenis et al., 2016).
- The Organization for Economic Co-operation and Development's (OECD's) Programme for International Student Assessment (PISA) rankings push nations towards competitive educational performance, even when SDG 4 calls for inclusive, quality education (Sellar & Lingard, 2013).

The contradiction is not in ambition or data collection but in perception. Actors rarely embody interdependence; they operate from mandates, political incentives and disciplinary silos (Sachs et al., 2022). The framework calls for integration, yet the observer implementing it remains divided.

When Reform Becomes Control

Even well-intentioned reforms can harden into mechanisms of control. Education reforms often prioritise measurable outcomes like standardised testing, rankings, metrics over genuine learning, creating what Nussbaum (2010) calls 'a crisis of the humanities,' where students are trained for utility rather than understanding.

Illich (1971) argued that institutional reforms frequently colonise life itself, producing dependency rather than empowerment. Freire (1970) warned that reforms in education reproduce the 'banking model,' where students are passive recipients of knowledge, reinforcing domination under the guise of progress.

In governance, reforms emphasising transparency and accountability may introduce layers of surveillance without addressing the deeper erosion of trust. Foucault (1977) showed how such reforms often function as techniques of discipline, normalising control under the name of order.

Peace processes reveal the same flaw. The Oslo Accords promised resolution of the Israeli–Palestinian conflict, but without addressing the deeper structures of identity, memory and fear, the agreements collapsed. As Lederach (1997) observed, peace cannot endure if it rests on institutional design alone while the psychological roots of division remain intact.

Fragmented Seeing, Fragmented Solutions

The deeper problem is that the observer rarely sees the crisis as a whole. Policymakers, experts and reformers often attend only to the aspects of reality that are visible, measurable or aligned with their institutional mandates and personal motivations. What does not fit these lenses is overlooked. Because the observer's understanding is itself fragmented, the solutions they produce are equally fragmented. Climate policies are reduced to carbon metrics, ignoring the wider patterns of consumption, militarisation or inequality that sustain emissions. Education reform is framed around standardised performance, while the deeper issues of relationship, fear and creativity remain untouched. This narrowness is compounded by the fact that solutions are drawn from past knowledge, established frameworks and inherited assumptions. Reform then becomes less a response to what is, and more a projection of what the observer already knows or desires, guaranteeing continuity rather than transformation.

Why Frameworks Alone Cannot Transform

The limitation of frameworks is not sophistication but standpoint. Even the most adaptive systems maps or participatory strategies reproduce fragmentation if those applying them remain caught in identity, fear and competition (Meadows, 2008). Reform assumes that rearranging structures is sufficient, when in fact the real crisis lies in the quality of perception within those structures.

Frameworks drawn by divided observers become more elaborate versions of the old order, maps drawn by the same hand that created the terrain. Without transformation in perception, reform is continuity disguised as progress.

2.8 The Architecture of the Self

If the crisis lies not only in systems but in the actor who shapes them, then the next question is unavoidable: What is this actor? What is the structure of the self that sees, decides and acts?

The self, as it is commonly experienced, is not a singular, stable entity. It is a dynamic construct built from layers of memory, conditioning, belief and image (Gupta, 2024, 2025). It carries a personal history, our experiences, successes and failures, a collective history with our culture, religion, nationality and a projected future of ambitions, fears and ideals. These personal histories are themselves shaped by collective pasts and by futures once imagined and pursued by the collective. This psychological structure appears continuous, giving the impression of a permanent 'me,' yet it is constantly reinforced and reshaped by thought. As Meadows (2008) observed of systems more generally, persistent structures endure because their feedback loops are unexamined, and the self may be the most persistent structure of all.

Identity as Division

At the heart of the self lies identification. We identify with our name, profession, ideology, ethnicity, gender and nation. These identifications are not inherently problematic in a functional sense, and they do help us navigate social and institutional life, but when they become psychological anchors, they generate division (Gupta, 2024; Sen, 2006). Sen's *Identity and Violence* shows how reducing people to a single dominant identity, whether religious, national or cultural fuels violence and undermines pluralism.

In governance, identity-driven decision-making is visible when national interest overrides global well-being, or when political ideology takes precedence over scientific evidence. For example, climate negotiations under the United Nations Framework Convention on Climate Change (UNFCCC) remain shaped by the 'developed' versus 'developing' country divide, despite the reality that the atmosphere does not recognise such categories (Ostrom, 2010; UNFCCC, 2023). Similarly, in global human rights forums, voting patterns often align along geopolitical blocs rather than the universality of SDG 16 (peace, justice and strong institutions).

Identity can unify within a group, but it almost always divides between groups. This is as true in personal life as in geopolitics. The self is defined as much by what it is not as by what it is.

Memory as Continuity

The self is also the continuity of memory. Past experiences, whether personal or collective, are stored and recalled as reference points for present action. Memory enables learning, but it also sustains prejudice, resentment and fear (Bohm, 1992). Galtung's (1996) concept of 'cultural violence' shows how historical narratives become embedded in institutions, normalising structural harm and perpetuating cycles of conflict.

A policymaker who has faced political backlash for compromise may approach future negotiations defensively, even when conditions have changed. A nation that has suffered historical occupation may distrust international mediation regardless of the merits of a proposal. This continuity of memory is what makes conflict resolution so resistant to lasting success, from Kashmir to Cyprus to the Korean Peninsula.

In SDG terms, this continuity of unresolved grievances undermines the spirit of SDG 16, slowing progress in peacebuilding and reconciliation. The facts may change, but the self carries the imprint of the past into every new moment, interpreting the present through old wounds and victories (Krishnamurti, 1983).

Desire and Fear as Drivers

Beneath identity and memory lie two powerful motivators: desire and fear. Desire seeks security, recognition, achievement and permanence. Fear resists loss, uncertainty, humiliation and death. Together, they form the emotional core of the self's movement (Gupta, 2024; Nussbaum, 2010).

In policy and governance, desire may appear as the drive for economic growth (SDG 8), geopolitical influence or personal legacy. Fear may appear as resistance to reform, reluctance to share power or the prioritisation of control over cooperation. Both operate from the assumption that the self must be protected and enhanced, even at the expense of the whole.

This dynamic is visible in environmental governance, where fear of short-term economic loss delays climate action (SDG 13), and in education reform (SDG 4), where the desire for measurable performance metrics overrides the need for deeper learning and critical inquiry.

The Fear of Ending and the Quest for Permanence

Though the self as thought is impermanent, it clings to the illusion of permanence. Memory, belief and image are all transient, yet because thought becomes aware of its own fragility and the inevitability of death, it seeks continuity. This fear of ending drives the self to attach to roles, traditions, possessions and ideas, hoping to secure its own survival through identification. The search for permanence expresses itself in multiple ways: Personal: the desire to leave behind a legacy, to be remembered through family, reputation or achievement. Institutional: the construction of enduring structures such as constitutions, monuments or traditions that claim stability in the face of change. Collective: nations and religions that promise continuity through identity, territory or eternal truths. Yet these efforts never bring true security. The more the self identifies with what appears permanent, the more insecure it becomes, because it is trying to stabilise what is inherently unstable. The fear of death intensifies attachment, and attachment in turn generates conflict when permanence is threatened. Nations go to war to preserve borders, communities exclude to preserve traditions and individuals sacrifice freedom to preserve image. This quest for permanence is mirrored in development frameworks. SDG 8 (decent work and economic growth) often becomes an attachment to endless GDP expansion as a symbol of continuity, even when it undermines ecological balance. SDG 13 (climate action) is weakened when nations cling to fossil-fuel-based legacies to preserve short-term stability. SDG 16 (peace, justice and strong institutions) is compromised when states protect their institutional permanence at the expense of adaptability or justice. In each case, the actor prioritises the preservation of the 'self,' whether individual, institutional or national, over the needs of life as a whole. In this way, the architecture of the self is not only impermanent but also self-perpetuating. Its quest for permanence ensures its continuity, yet never resolves the fundamental insecurity that drives it.

The Illusion of Separation

The architecture of the self rests on an illusion: that the observer is separate from the observed. This is the root of fragmentation. When we believe we are outside the problem, we attempt to control it as an object, whether that object is the economy, the environment or another person.

But the reality, as Krishnamurti and Bohm both pointed out, is that the observer is part of what is being observed. Our thoughts, fears and desires are woven into the very fabric of the crisis. This illusion allows governments to speak of protecting SDG 15 (life on land) while enabling extractive industries, or to pledge SDG 5 (gender equality) while maintaining cultural and institutional structures that perpetuate exclusion. Thich Nhat Hanh (2017) called this *inter-being,* the recognition that separation is an error in perception.

The Self in the Mirror of Crisis

The polycrisis reflects the architecture of the self on a global scale:

- Climate breakdown mirrors the self's exploitation of nature for personal gain (SDG 13).
- Geopolitical conflict mirrors the self's identification with tribe, nation or ideology (SDG 16).
- Economic inequality mirrors the self's drive for comparison and accumulation (SDG 10).
- Institutional paralysis mirrors the self's fear of losing control (SDG 17).

Recognising this mirroring is not an abstract exercise; it is the beginning of responsibility. If the self is the root of fragmentation, then transformation cannot be achieved by rearranging structures alone; it must involve seeing the self in action.

Observation Without the Observer

This raises a crucial question: Can the self be observed without the interference of its own projections? If the observer is the observed, then the act of seeing the self is itself the dissolution of division.

This is not a matter of analysis, where thought examines itself, but of direct perception, where the movement of thought, memory and identity is seen as it happens. Such observation is rare because our education, culture and governance systems train us to strengthen the self, not to understand it. Leadership programs build confidence in identity; governance reforms optimise decision-making processes; social movements mobilise around shared grievances. In all these, the self is the unquestioned starting point.

Yet without understanding the architecture of the self, every reform remains bound by its limits. It is like redesigning the façade of a building while ignoring the cracks in its foundation (Meadows, 2008).

If the same patterns that structure the self also structure our collective crises, then the distinction between 'personal' and 'global' becomes artificial. The very divisions we see in the world are the ones we enact in our own thought and relationship. This recognition opens the inquiry of the next section: What if the self is not merely in the world, but is the world?

2.9 Seeing the Self as the World

If the architecture of the self mirrors the architecture of global crisis, then the separation between 'personal' and 'political,' 'inner' and 'outer,' begins to dissolve. This is not merely a philosophical claim; it is an observable fact. The same patterns that create conflict in a family create conflict between nations. The same impulses that drive exploitation in the marketplace drive the extraction of natural resources. The same fear that prevents an individual from facing loss prevents societies from confronting systemic collapse (Bohm, 1992; Krishnamurti, 1983).

The Continuity of Fragmentation

What we call the 'world' is not an independent reality detached from human consciousness. It is the outcome of countless actions rooted in identity, memory, desire and fear. When these patterns operate at scale, they create institutions, economies and governance systems that reflect the same logic. Bateson (2002) argued that mind and environment form a single systemic whole, meaning that any attempt to divide them is itself a distortion.

The polycrisis is not simply the interaction of climate change, inequality and geopolitical instability. It is the collective expression of the human mind's fragmentation:

- *Nationalism* is the collective form of personal identification.
- *Militarisation* is the scaled-up form of personal defence mechanisms.
- *Economic exploitation* is the institutionalised form of personal greed.
- *Bureaucratic paralysis* is the procedural form of personal indecision and fear of loss.

This continuity is visible in the SDGs. Goals like SDG 13 (climate action), SDG 16 (peace, justice and strong institutions) and SDG 17 (partnerships) cannot be achieved in isolation because they mirror one indivisible whole. Without addressing the psychological ground of fragmentation, progress in one goal often undermines another.

Breaking the Illusion of Two Worlds

The common approach to crisis assumes two separate realms: the inner psychological world and the outer political, economic and ecological world. This assumption underpins much of development and governance: change the outer systems and human behaviour will follow. Yet Meadows (1999) showed that the most effective leverage point for systemic change is at the level of paradigm and mindset. As long as the self remains structured by fear, comparison and division, it will inevitably recreate the same crises in whatever new system it inhabits.

This is why revolutions so often fail to deliver on their promises. Political regimes change, constitutions are rewritten and economies are restructured, but the underlying patterns of corruption, power accumulation and exclusion reappear. The actors have changed roles, yet the architecture of the self remains intact.

Responsibility Without Borders

Seeing that the self is the world transforms the meaning of responsibility. Responsibility is no longer confined to the scope of one's role, nation or profession; it becomes total. Galtung's (1969) concept of structural violence shows how systemic inequities are sustained by divisions of identity. Sen (1999) warns that reducing people to a single identity fuels exclusion and conflict. Thich Nhat Hanh (1998) describes this inseparability as *interbeing*: harming another is ultimately harming oneself.

The destruction of a forest is not merely the failure of environmental policy; it is the reflection of the same exploitative relationship we maintain with our time, attention and even with one another (Capra, 1997; Ostrom, 2010; Shiva, 2016). The outbreak of war is not only a failure of diplomacy; it mirrors the same inability to listen and the same attachment to identity that operates in daily life.

This recognition is not meant to induce guilt but to reveal possibility. If the self is the world, then transformation does not depend on waiting for others to act. It can begin wherever one is, in any moment, because the root of crisis is present in how we see and act right now.

From Awareness to Action

This shift does not bypass policy, governance or collective mobilisation; it deepens them. When action is grounded in the perception of wholeness, rather than in the defence or enhancement of the self, its quality changes (Krishnamurti, 1983; Meadows, 1999):

- Policy becomes not about competing interests but about what sustains life as a whole.
- Negotiation becomes not about compromise between fixed positions but about exploring what is true together.
- Leadership becomes not the projection of authority but the embodiment of coherence.

Historical figures who have acted from such perception in politics, science or social movements show that clarity of perception can have systemic impact without being rooted in ideology or identity. Their work was not reactive; it arose from seeing the fact of interdependence directly.

Why This Matters for Governance

Governance frameworks today largely focus on process, structure and measurement. While these are necessary, they are insufficient if the actor's perception remains fragmented. The SDGs aim for integration but are implemented through compartmentalised ministries and agencies. Peace processes are designed to resolve disputes without dissolving the identities that sustain division. Climate action frameworks speak of planetary limits while still treating nature as an external object to be managed (United Nations, 2023).

If the self is the world, then governance must evolve beyond systems thinking into perception-based *governance*: where the starting point is not 'How do we manage complexity?' but 'Can the actor see without fragmentation?'

A Single Movement

To see the self as the world is to see that change is one movement. The end of exploitation in the mind is the beginning of ending exploitation in the market. The end of division in relationship is the beginning of ending division between nations. There is no timeline for this because it begins now, in how we see and act in this very moment (Bohm, 1992; Thich Nhat Hanh, 1998).

2.10 The Self as a System

If the self mirrors the world, it can also be understood as a system in its own right: a complex, adaptive network of thoughts, emotions, memories and identities that interact in patterned ways. This perspective does not reduce the self to biology or psychology alone; rather, it recognises that the mechanisms through which we perceive, interpret and act share the same dynamics as the governance, economic and ecological systems we attempt to reform (Capra & Luisi, 2014; Gupta, 2024, 2025). From this lens, the patterns of fragmentation in the individual are not incidental; they are structurally homologous to the fragmentation in our institutions, policies and global frameworks.

Feedback Loops of Identity

In systems theory, a feedback loop occurs when a system's outputs are fed back as inputs, reinforcing or balancing the system's behaviour (Meadows, 2008). The self operates through similar loops:

Identity → Perception → Action → Reinforcement.

We perceive the world through the lens of who we think we are. This perception shapes our actions, which produce outcomes that confirm our original identity.

- If I see myself as a protector of my group, I interpret events as threats to that group, act defensively and then view the consequences as proof that my group is under attack.
- At the national level, a state that sees itself as a regional leader will interpret global events through that lens, enact policies to maintain dominance and treat the resulting geopolitical tensions as evidence of the need to assert its leadership role.

This dynamic entrenches the challenges of SDG 16 (peace, justice and strong institutions) by locking policy into identity-driven responses, and it undermines SDG 17 (partnerships for the goals) by making cooperation conditional on affirming one's status.

The System's Operating Code

Like any system, the self has an implicit operating code: organising principles that guide its functioning (Gupta, 2024):

- *Self-preservation:* The drive to maintain one's image, status or role, even at the cost of truth.
- *Comparison:* Measuring oneself against others to secure identity and value, mirrored in how nations are ranked on SDG performance indices.
- *Control:* Attempting to shape outcomes to ensure the self's continuity.
- *Certainty-seeking:* Preferring fixed beliefs and familiar patterns over uncertainty, mirrored in governance resistance to adaptive policy shifts (Ostrom, 2010).

These tendencies may have been adaptive for biological survival, but in governance, education and sustainability they distort perception. For example:

- Fear of losing competitiveness delays climate action, weakening SDG 13 (climate action) (IPCC, 2023).

- Certainty-seeking keeps education systems bound to outdated curricula, slowing transformation in SDG 4 (quality education).

Path Dependency and Psychological Time

In policy studies, path dependency describes how past decisions constrain present options, making certain outcomes more likely regardless of current conditions. The self operates in the same way through psychological time, memory projecting itself into the future. A past hurt is carried as an image, shaping how the present is met.

- A past success generates fear of loss, prompting protective behaviour even when conditions have shifted.

This continuity preserves both personal and collective conflict. Just as entrenched institutions resist radical change, the self resists disrupting its established pathways, leading to the repetition of old mistakes under new labels. This dynamic perpetuates cycles of inequality (SDG 10) and protracted geo-political disputes (SDG 16).

Boundaries and Separation

All systems define boundaries by distinguishing what is 'inside' from what is 'outside.' For the self, this is the psychological line between *me* and *not-me*. While useful in the physical world, these boundaries harden into divisions that distort relationship:

- My interest vs yours.
- My nation's security vs another's.
- Humanity vs the natural world.

When taken as absolute, such boundaries make cooperation conditional, empathy selective and sustainability secondary to self-preservation, undermining SDG 15 (life on land) and SDG 14 (life below water).

Self-Regulation and Breakdown

Healthy systems regulate themselves through feedback that maintains balance. In the self, regulation is distorted by cognitive bias (Kahneman, 2011), selective attention and emotional reactivity:

- *Negative feedback* – signals that might correct imbalance are ignored if they threaten identity.
- *Positive feedback* – reinforcing signals are amplified when they flatter belief.

This mirrors governance failures: inconvenient climate data is minimised, while economic narratives that protect legitimacy are amplified, delaying SDG 13 (climate action) and reinforcing contradictory SDG 8 (economic growth) policies.

Why Seeing the Self as a System Matters

Understanding the self as a system shifts the frame from moral blame to structural comprehension. The issue is not simply 'bad' individuals or 'corrupt' institutions; it is that the operating system of the self, with its feedback loops, boundaries and biases, is mirrored in the very systems it designs.

Implications include:

- *In governance:* Reforms must address the actor's perception, not just institutional mechanics (SDG 16, SDG 17).
- *In education:* Students must learn to observe thought, emotion and identity processes in real time (SDG 4).
- *In sustainability:* Ecological exploitation must be seen not only as a technical or economic challenge but as an extension of the self's exploitive relationship to life (SDG 15).

The Possibility of Systemic Transformation

In natural and human systems, transformation often requires a phase shift, a qualitative change in the system's state that cannot be achieved by incremental adjustment (Meadows, 2008). For the self, such a shift is not about refining identity or belief but about seeing the system in its entirety as it operates.

This seeing *is* the transformation. Once fully observed, the self-system can no longer function unconsciously. No external mechanism can 'fix' the self from the outside, because the self is both the observer and the observed (Bohm, 1992; Krishnamurti, 1954). Bohm emphasised that this is not thought applying itself to change thought, but the awakening of an intelligence not bound by the system of thought itself.

2.11 Conclusion: From System to Wholeness

Seeing the self as a system is not the endpoint of inquiry. It is a step, a way to recognise the mechanisms through which thought, identity and perception operate. But if we stop there, we risk reducing the self to another object of analysis, something to be mapped, categorised and managed. This would simply repeat the same error seen in governance and policy: treating life as an external system to be controlled.

The deeper movement is to see that the self, while functioning like a system, is not a fixed entity at all. It is a process: a flow of memory, image, sensation and reaction that appears solid only because it is constantly in motion. As Capra

(1997) notes in systems thinking, what we call 'structure' is often just a pattern of relationships sustained through continuous activity. When this is seen directly, the question of 'how to fix the system' falls away. What remains is the possibility of perceiving without the filter of the system.

The Limits of Systemic Self-Improvement

Many personal and leadership development approaches operate within the metaphor of the self-as-system. They offer methods to 'upgrade' the system, reframing beliefs, managing emotional triggers, improving decision-making. While these may yield incremental benefits, they rarely dissolve the underlying structure of fragmentation. Instead, they make the self more efficient at pursuing conditioned goals: more persuasive in argument, more strategic in competition, more resilient in defending identity. In effect, they polish the lens without questioning the act of looking through a lens.

Perception Without a Centre

Wholeness is not the result of optimising the self-system. It emerges when perception is no longer organised around a centre, the 'me' who experiences, interprets and acts for its own continuity (Gupta, 2025). Without the centre, perception is not divided into observer and observed, nor distorted by the need to protect or project an image. Action, then, no longer carries the residue of self-interest disguised as service. Krishnamurti (1983) described this as choiceless awareness; Bohm (1992) saw it as the movement of intelligence itself, unbound by the structures of thought.

From Boundaries to Relationship

The shift from system to wholeness dissolves the psychological boundaries that separate 'me' from 'not-me.' This is not the erasure of individuality, but the recognition that life is fundamentally relational. Just as ecosystems are webs of interdependence, so too is human life woven from relationship (Meadows, 2008). In wholeness:

- Relationship with people is no longer mediated by identity but lived directly (SDG 16, SDG 17).
- Relationship with nature is no longer 'management of the environment' but participation in its flow (SDG 13, SDG 14, SDG 15).
- Relationship with thought itself becomes transparent, as one sees how memory and fear generate division.

Action Beyond Psychological Time

Systems operate within the framework of time, with past experience shaping future planning. Wholeness, by contrast, acts from the immediacy of the present. While chronological time remains necessary for practical tasks, psychological time, the self's movement of becoming, is absent. Decisions guided by wholeness respond to what is, not to projections of what was or what might be.

This distinction is vital in governance. Leaders acting from wholeness are not bound by the need to protect legacies or secure future images. Their responsibility is immediate, extending to the long-term well-being of life as a whole rather than short-term gains for a part (SDG 13, SDG 15).

Wholeness as the Ground of Transformation

The movement from system to wholeness is the essence of transformation. Complexity science demonstrates interdependence, but wholeness allows us to live it. Policy frameworks call for integration, but wholeness removes the very divisions that make integration necessary. Education teaches sustainability, but wholeness is the perception from which sustainable living is the only possible way (SDG 4).

If wholeness is the ground, then governance must evolve into perception-based governance, where the starting point is not 'How do we manage complexity?' but 'Can the actor see without fragmentation?' From such seeing, action is no longer strategic but natural: the environmentalist is the environment, the reformer is society, the peacebuilder is the conflict.

References

Amnesty International. (2024). *Israel/OPT: 2024 report on civilian impact of conflict*. Amnesty International. https://www.amnesty.org

Anderson, B. (1983). *Imagined communities: Reflections on the origin and spread of nationalism*. Verso.

Arendt, H. (1963). *On revolution*. Viking Press.

Bar-Tal, D. (2007). Sociopsychological barriers to peace making: The case of the Israeli–Palestinian conflict. *Social Issues and Policy Review*, *1*(1), 63–109. https://doi.org/10.1111/j.1751-2409.2007.00004.x

Bateson, G. (2002). *Mind and nature: A necessary unity*. Hampton Press.

Bohm, D. (1992). *Thought as a system*. Routledge.

Bourdieu, P. (1984). *Distinction: A social critique of the judgment of taste*. Harvard University Press.

Capra, F. (1997). *The web of life: A new scientific understanding of living systems*. Anchor Books.

Capra, F., & Luisi, P. L. (2014). *The systems view of life: A unifying vision*. Cambridge University Press.

Chancel, L., Piketty, T., Saez, E., & Zucman, G. (2022). *World inequality report 2022*. World Inequality Lab. https://wid.world

Damasio, A. (2010). *Self comes to mind: Constructing the conscious brain*. Vintage.

Erikson, E. H. (1968). *Identity: Youth and crisis*. W. W. Norton.

Falkner, R. (2016). The Paris agreement and the new logic of international climate politics. *International Affairs*, *92*(5), 1107–1125. https://doi.org/10.1111/1468-2346.12708

FAO. (2023). *The State of Food Security and Nutrition in the World 2023: Urbanization, agrifood systems transformation, and healthy diets across the rural–urban continuum*. Food and Agriculture Organization of the United Nations. https://doi.org/10.4060/cc3017en

Foucault, M. (1977). *Discipline and punish: The birth of the prison*. Pantheon.

Freire, P. (1970). *Pedagogy of the oppressed*. Continuum.

Friston, K. (2010). The free-energy principle: A unified brain theory? *Nature Reviews Neuroscience*, *11*(2), 127–138. https://doi.org/10.1038/nrn2787

Galtung, J. (1969). Violence, peace, and peace research. *Journal of Peace Research*, *6* (3), 167–191.

Galtung, J. (1996). *Peace by peaceful means: Peace and conflict, development and civilization*. Sage.

Girard, R. (1977). *Violence and the sacred*. Johns Hopkins University Press.

Gupta, D. (2024). *Eternal movement*. Emerge Publications. https://doi.org/10.65169/q7d2z6

Gupta, D. (2025). *Governance without the self: Consciousness and the end of fragmentation, systems, and complexity*. Emerge Publications. https://doi.org/10.65169/a9m3p4

Herz, J. H. (1950). Idealist internationalism and the security dilemma. *World Politics*, *2*(2), 157–180.

IEA. (2021). *India energy outlook 2021*. International Energy Agency. https://www.iea.org

IEA. (2023). *World energy outlook 2023*. International Energy Agency. https://www.iea.org/reports/world-energy-outlook-2023

Illich, I. (1971). *Deschooling society*. Harper & Row.

Intergovernmental Panel on Climate Change. (2023). Sections. In H. Lee & J. Romero (Eds.), *Climate change 2023: Synthesis report. Contribution of working groups I, II and III to the sixth assessment report of the intergovernmental panel on climate change* (pp. 35–115). IPCC. https://doi.org/10.59327/IPCC/AR6-9789291691647

Kahneman, D. (2011). *Thinking, fast and slow*. Farrar, Straus and Giroux.

Kentikelenis, A. E., Stubbs, T. H., & King, L. P. (2016). IMF conditionality and development policy space, 1985–2014. *Review of International Political Economy*, *23*(4), 543–582. https://doi.org/10.1080/09692290.2016.1174953

Korzybski, A. (1933). *Science and sanity: An introduction to non-Aristotelian systems and general semantics*. Institute of General Semantics.

Krishnamurti, J. (1954). *The first and last freedom*. Victor Gollancz.

Krishnamurti, J. (1969/2004). *Freedom from the known*. HarperCollins.

Krishnamurti, J. (1983). *Education and the significance of life*. Krishnamurti Foundation India.

Krishnamurti, J. (1991). *The awakening of intelligence*. HarperCollins.

Lakoff, G., & Johnson, M. (1980). *Metaphors we live by*. University of Chicago Press.

Lederach, J. P. (1997). *Building peace: Sustainable reconciliation in divided societies*. United States Institute of Peace Press.

McGilchrist, I. (2019). *The master and his emissary: The divided brain and the making of the Western world* (2nd ed.). Yale University Press.

Mead, G. H. (1934). *Mind, self, and society*. University of Chicago Press.

Meadows, D. H. (1999). *Leverage points: Places to intervene in a system*. Sustainability Institute. https://donellameadows.org/wp-content/userfiles/Leverage_Points.pdf

Meadows, D. H. (2008). *Thinking in systems: A primer*. Chelsea Green Publishing.

Meadows, D. H., Meadows, D. L., & Randers, J. (2004). *Limits to growth: The 30-year update*. Chelsea Green Publishing.

Morin, E. (2008). *On complexity*. Hampton Press.

Nussbaum, M. C. (2010). *Not for profit: Why democracy needs the humanities*. Princeton University Press.

Ostrom, E. (2010). Polycentric systems for coping with collective action and global environmental change. *Global Environmental Change*, *20*(4), 550–557. https://doi.org/10.1016/j.gloenvcha.2010.07.004

Ricoeur, P. (2004). *Memory, history, forgetting*. University of Chicago Press.

Sachs, J. D., Kroll, C., Lafortune, G., Fuller, G., & Woelm, F. (2022). *Sustainable development report 2022*. Sustainable Development Solutions Network. https://www.sdgindex.org

Said, E. W. (1978). *Orientalism*. Pantheon.

Sellar, S., & Lingard, B. (2013). The OECD and global governance in education. *Journal of Education Policy*, *28*(5), 710–725. https://doi.org/10.1080/02680939.2013.779791

Sen, A. (1999). *Development as freedom*. Alfred A. Knopf.

Sen, A. (2006). *Identity and violence: The illusion of destiny*. W. W. Norton & Company. https://wwnorton.com/books/9780393329296

Shiva, V. (2016). *Earth democracy: Justice, sustainability, and peace* (2nd ed.). North Atlantic Books.

Stiglitz, J. E. (2019). *People, power, and profits: Progressive capitalism for an age of discontent*. W. W. Norton.

Tajfel, H., & Turner, J. C. (1986). The social identity theory of intergroup behavior. In S. Worchel & W. G. Austin (Eds.), *Psychology of intergroup relations* (pp. 7–24). Nelson-Hall.

Thich Nhat Hanh. (1998). *The heart of the Buddha's teaching*. Broadway Books.

Thich Nhat Hanh. (2017). *The art of living: Peace and freedom in the here and now*. HarperOne.

UN OCHA. (2024). *Occupied Palestinian territory: Humanitarian update 2024*. United Nations Office for the Coordination of Humanitarian Affairs. https://www.ochaopt.org

UNEP. (2023). *Emissions gap report 2023: Broken record—Temperatures hit new highs, yet world fails to cut emissions (again)*. United Nations Environment Programme. https://www.unep.org

UNFCCC. (2023). *Nationally determined contributions synthesis report 2023*. United Nations Framework Convention on Climate Change. https://unfccc.int

United Nations. (2023). *The sustainable development goals report 2023*. United Nations. https://unstats.un.org/sdgs/report/2023/

Varela, F. J., Thompson, E., & Rosch, E. (1991). *The embodied mind: Cognitive science and human experience*. MIT Press.

Zimbardo, P. G., & Boyd, J. N. (2008). *The time paradox: The new psychology of time that will change your life*. Free Press.

Chapter 3

Beyond Goals and Frameworks: Seeing the Root of Global Crisis

Abstract

Chapter 3 examines the Sustainable Development Goals (SDGs) and other global governance frameworks as case studies of how noble intent becomes entangled in structural fragmentation. It shows how the architecture of goals, metrics and timelines transforms the living movement of care into targets of measurement. Drawing on four decades of sustainable development thinking, it demonstrates that the global agenda's reliance on quantification and compliance reveals an underlying continuity of the same consciousness that created the problem. Each reform effort becomes a mechanism of delay, institutionalising postponement under the language of progress. This chapter does not reject global frameworks but repositions them: they can coordinate not transform. It invites a shift from integration to perception, seeing that coherence cannot be engineered through policy but must emerge from clarity of mind. This chapter concludes that when perception precedes prescription, reform gives way to transformation grounded in awareness.

Keywords: SDG critique; global metrics; policy fragmentation; integrative frameworks; governance delay; measurement paradigm; seeing beyond reform

3.1 Introduction: The Crisis Beneath the Crisis

The 21st century has produced a remarkable proliferation of global frameworks aimed at addressing humanity's most urgent challenges. From the United Nations (UN) Sustainable Development Goals (SDGs) to regenerative development agendas, circular economy models and well-being indexes, these initiatives promise a future of integration, equity and ecological balance. They bring together governments, businesses, civil society and academia under shared visions, measurable targets and timelines for action.

From Fragmentation to Wholeness, 73–94

Published under exclusive licence by Emerald Publishing Limited
doi:10.1108/978-1-80686-677-920261015

Yet, despite their sophistication and global endorsement, the crises they aim to solve are deepening. Climate change accelerates even as renewable energy capacity expands (IPCC, 2023). Inequality widens despite inclusive growth strategies (Piketty, 2014). Armed conflict proliferates despite decades of investment in peacebuilding institutions (UN, 2023). The persistence of these patterns suggests that the failure is not only in the execution of frameworks but in the perception and psychology of the actors who design and implement them and in forgetting that the actor is inseparable from the very world they act upon.

This becomes evident in practice. Governments pledging climate neutrality continue to subsidise fossil fuels, treating emissions as an external problem while their own policies sustain the crisis (IEA, 2023). Peacebuilding accords are signed while arms are sold to the same regions, as if violence were 'out there' rather than embedded in the structures of economy and identity that the actors themselves uphold. In both cases, the actor assumes the role of detached observer, while in reality, their decisions are part of the very dynamics they seek to change.

The core thesis of this chapter is simple: frameworks mirror the state of the actor. If the actor operates from a consciousness divided by identity, fear and competition, even the most integrated global framework will express that division in its design, priorities and implementation. This is not a rejection of the value of frameworks; rather, it is an insistence that their effectiveness depends on the quality of perception from which they arise and are enacted. To forget that the actor is part of the acted upon is to perpetuate the very fragmentation that sustains disorder. As Krishnamurti (1991) observed, 'The problem is not out there; the problem is the way we see.'

This chapter moves in two arcs. The first is an inward inquiry into life, action, change, development and the actor, to clarify what it is we are trying to transform. The second is a structural and psychological critique of the SDGs and related models, showing how their fragmentation reflects the inner fragmentation of the policymaker, leader or institution. By the end, we will return to the central proposition: that the only basis for coherent, sustainable global action is perception-based governance, action grounded in seeing the whole without distortion.

3.2 Foundational Inquiry

Before examining the SDGs and other global frameworks, we must clarify the terms that underwrite them. What do we mean by life, action, change, *development* and actor? Without this clarity, critique risks becoming another rearrangement of language without a shift in understanding.

What Is the Essence of Life?

Life, in its essence, is an indivisible movement: ecological, social, physiological and psychological dimensions in constant interaction (Capra & Luisi, 2014). Modern governance and development frameworks often fragment life into sectors: environment, economy, education, health, security. This segmentation may serve administrative convenience, but it distorts reality when treated as an ontological truth. A forest, for example, is not merely an environmental asset; it is simultaneously a climate regulator, a cultural landscape, an economic livelihood, a home for species and a spiritual presence for many communities.

According to Gupta (2024, 2025), the essence of life is non-division. Whatever separates from the Eternal Movement of life begins to perish, whether a fruit detached from the tree, water isolated from its source, a society divided by identity or a mind fragmented by thought. In this impermanence, whatever seeks permanence through control, thought or knowledge inevitably begins to wither. We can temporarily preserve aspects of life, but not life itself forever. To see this deeply is a kind of *psychological death of the past*, an ending of continuity that opens the freshness of perception, freedom from the known.

A tree is a tree, and a child is a child. To see nature as true, and identity as a construct of thought, is the beginning of living beyond fragmentation. How can a tree, which is part of nature, belong to an identity or an ideology? A flower does not discriminate; it gives fragrance without asking whether you are from Myanmar, Gaza or a powerful nation. The rain does not fall differently for Hindu or Muslim. Nature is eternally true to its meaning, while psychological discrimination arises only when perception is filtered by thought.

Biology reflects this indivisibility. Human beings are not apart from nature but expressions of its living systems. Our circadian rhythms align with the turning of the Earth; our cells respond to temperature, light and microbial life; our optic nerves extend into the world, receiving patterns of light; our bodies exchange oxygen with trees and oceans in a ceaseless cycle. We are participants in one whole metabolism of life, not observers standing outside it. Yet thought, in seeking security through identity, desire and fear, generates the illusion of separation. This illusion drives mechanical living, dulls bodily sensitivity and disturbs the natural systems of which we are part. Mountains collapse under extractive pressure, floods intensify with ecological imbalance and human bodies mirror the same fragmentation through stress, disease and alienation.

Recent explorations across disciplines point to the same indivisible ground. Neuroscience shows that consciousness is not a fixed entity but a dynamic process emerging from interdependent neural and bodily systems (Damasio, 2010). Research with psychedelics such as psilocybin and Dimethyltryptamine, and with ancient plant medicines like ayahuasca, suggests that states of consciousness in which ego-boundaries dissolve often reveal a direct sense of union with life (Carhart-Harris et al., 2018). Ancient Eastern wisdom traditions have long affirmed that this state of undivided awareness is the natural ground of being. Professor Richard Alpert, also known as Baba Ram Dass, an academic and psychologist who, with Timothy Leary, explored consciousness through both psychology and religion, noted in *The Only Dance* There Is (1974) that the higher states of consciousness are union.

In this light, the essence of life is not permanence, accumulation or division, but movement and relationship. To divide it conceptually or politically is already to misperceive it. To see it as indivisible is to live with a sensitivity that neither clings nor controls, but participates in the wholeness that is always already here.

What Is the Source of Action?

Action can arise from two fundamentally different sources: wholeness or fragmentation. Action from wholeness emerges from direct perception of the interrelated totality of life; it does not privilege one sector, identity or interest over the whole. Action from fragmentation, by contrast, arises from partial perception, often organised around self-interest, institutional mandates or ideological commitments. Bohm (1992) describes this as 'fragmentary action,' where efforts to solve a problem inadvertently reinforce it by ignoring its connections to other issues. For example, economic stimulus packages aimed at job creation may accelerate ecological collapse if designed without considering planetary boundaries (Rockström et al., 2009).

A critical question arises: are actors truly working for the same objective, with the same understanding and intensity? Until the world is seen and acted upon as a whole, there can be no communion even in defining the nature of the problem. When governments and institutions operate from fragmented ideas, total seeing is absent. The consequences are evident in proliferating crises, children dying in Yemen, Sudan, Palestine and South Sudan; wars in Ukraine, human rights violations in Afghanistan, Iran and Syria; entire islands facing submergence; ecological tipping points approaching; and institutions faltering in their ability to act coherently.

These realities compel us to ask: what is the source of right action, and can such action ever be collectively realised? And if not, will you, as the actor, remain unconditioned, not loyal to the known, to identity, or to policy, but loyal to the truth of clear perception, which begins with the ending of desire, fear and bias, and opens into pure observation?

What Is Change, What Change Is Not, What Is Development?

In most policy discourse, change is defined as movement, an improvement or deterioration within an existing structure, such as higher GDP, lower poverty rates or cleaner air, measured against baseline indicators. This is what Gupta (2025) calls 'better continuity': the same underlying patterns of division, control and exploitation producing temporarily improved conditions. Real change, by contrast, is discontinuity, the ending of the patterns that generate crisis in the first place. If a river is cleaned today but polluted tomorrow, there has been no net change in the underlying movement; only the measurement has shifted. Krishnamurti (1954) framed change as the ending of the psychological patterns of fear, greed and comparison that sustain external disorder.

As long as awareness is absent, reform merely rearranges disorder. Since the actor is part of that disorder, a mind conditioned by fear, desire and will inevitably reproduces the same patterns while believing it is solving them. Development, in such cases, becomes cyclical: growth and collapse, progress and regression. We build structures that appear stable, but when crisis comes, they collapse because the root insecurity of the self has not been addressed. During COVID-19, for example, national responses revealed this fragility. Governments prioritised their own citizens over global solidarity; within nations, people clung to protecting their families above all else. Religious, national and ideological identities gave way to the most primal identifications of survival, 'my child, my son,' revealing how thin the veneer of global development was when fear dictated action.

A striking contradiction is found in the domain of energy policy. Nuclear power is promoted under SDG 7 (affordable and clean energy) and SDG 13 (climate action) as a low-carbon solution to the climate crisis. France, for instance, derives over 70% of its electricity from nuclear energy and positions this as leadership in sustainability (IEA, 2022). Yet the same technology sustains nuclear weapons programmes and heightens global insecurity, directly undermining SDG 16 (peace, justice and strong institutions). What is celebrated as 'clean development' in one column of the SDG framework manifests as existential threat in another. This contradiction reveals the central problem: development is defined technically, not perceptually. The same atom can light a city or incinerate it; the difference lies not in the technology but in the consciousness of the actor who wields it.

The same paradox emerges in conflict. When war erupts, humanity brings all it has accumulated into the battlefield. The latest technology, the most advanced science and the brightest minds are mobilised for destruction. Satellites designed for communication and exploration are repurposed for surveillance and targeting. Chemistry and biology become chemical and biological weapons. Education (SDG 4), which is supposed to nurture learning and opportunity, is used to train new generations of soldiers. Institutions that claim to uphold peace and justice (SDG 16) legitimise or facilitate violence. What we call development is revealed as the accumulation of tools that can be deployed in direction, healing or harming or sustaining or annihilating. Can accumulation and access alone be considered development when their use is determined by conflict between fragmented identities?

True development cannot be separated from true change. Development is not the endless expansion of capacity or infrastructure; it is the unfolding of human and ecological potential in coherence with the whole. Change is not rearrangement but ending, the ending of fragmentation, fear and division that distort perception. Without such change, development remains superficial: one crisis is enough to collapse the façade. With such change, development is not an outcome imposed by policy but a natural expression of life lived without division.

Who Is the Actor?

Every framework is enacted by an actor: a policymaker, business leader, activist or educator. This actor is not an abstract rational agent but a human being conditioned by personal history, cultural narratives, institutional incentives and collective memory (Damasio, 2010; McGilchrist, 2009). The actor's perception is shaped by identity and filtered through the 'content of consciousness' (Krishnamurti, 1983), made up of the accumulated beliefs, experiences and assumptions that define the self. Unless the actor becomes aware of these inner determinants, their action will inevitably mirror their fragmentation, even in the pursuit of integration and sustainability.

The actor is not outside the problem; the actor *is* the problem. The observer is the observed. The discriminator is the discriminated. Thought itself creates the thinker. These are not metaphors but descriptions of cause and effect within consciousness. When one discriminates, the very act of discrimination arises from the divisions already carried within the self. In this sense, the discriminator is not separate from the discriminated, for the act of separation exposes the fragmentary structure of the one who sees. Similarly, when a person is hurt, there is no 'me' separate from the hurt; the hurt is the person, and from that hurt, further action arises.

This cycle plays out not only in individuals but also in nations. The protracted conflicts of Israel–Palestine and India–Pakistan reveal how collective identity becomes inseparable from hurt. Each side experiences itself as the victim of violence, displacement or betrayal, and acts from that hurt to secure itself, thereby producing new violence and new hurt in the other. The hurt is not confined to one people; it *is* the people, and as long as action emerges from that wound, conflict is endlessly reproduced. The actor and the acted upon are not two. It's the word which has divided people, the thought, the identity, the idea, the identification and its movement has been causing destruction.

This has direct implications for the SDGs. SDG 16 (peace, justice and strong institutions) cannot succeed if actors approach peace from the standpoint of accumulated hurt, past identities, because the very pursuit of justice then becomes another form of retaliation or identity-defence. SDG 13 (climate action) is weakened when nations act from competitive identities, each protecting its own energy security while treating planetary health as secondary. SDG 5 (gender equality) falters when institutions are run by actors whose perceptions are conditioned by patriarchal bias, reproducing discrimination in the very attempt to end it. Even SDG 10 (reduced inequalities) struggles when economic policies are designed from the standpoint of national prestige, market competition or electoral survival, rather than from compassion for human vulnerability.

From this perspective, the 'right actor' is not one who perfects knowledge, expands expertise or refines ideology, but one who has seen through the illusion of separation from their own conditioning. When the self is not the centre of action, then compassion, love and intelligence can operate. Only such an actor can give rise to action that does not reproduce division in new forms.

3.3 The Promise and Architecture of the SDGs

When the UN General Assembly adopted the SDGs in September 2015, it was heralded as an unprecedented moment of global consensus. One hundred and ninety-three member states committed themselves to a shared framework of 17 goals, 169 targets and more than 230 indicators, covering the full spectrum of human and planetary well-being: ending poverty and hunger, ensuring health, education and gender equality, protecting ecosystems, fostering peace and revitalising global partnerships (United Nations, 2015).

The SDGs were framed as the successor to the Millennium Development Goals (MDGs), but with important distinctions. They were universal (applying to all nations, not just the Global South), integrated (recognising the interconnections between issues) and transformative (explicitly committing to 'leave no one behind') (Sachs, 2015). Unlike the MDGs, the SDGs extended beyond social development to include economic, environmental and governance dimensions, acknowledging sustainability as a multi-dimensional challenge (Capra & Luisi, 2014).

A Framework for Integration

The 17 goals were deliberately designed to be interlinked. Climate action (SDG 13) was tied to food security (SDG 2), water access (SDG 6) and life on land (SDG 15). Gender equality (SDG 5) was recognised as central to economic growth (SDG 8) and institutional strength (SDG 16). Global partnerships (SDG 17) were positioned as the connective tissue enabling resource sharing, knowledge exchange and coordinated action.

This integrative architecture represented a significant step towards acknowledging systems thinking in global governance, aligning with scientific consensus that ecological, economic and social systems are interdependent (Rockström et al., 2009; Ostrom, 2010). In principle, it reflected the holistic view of life explored in Section 2.1. Yet, the translation from paper to practice often reveals the gap between integrated design and fragmented perception and the fallacy of attempting to integrate fragments without understanding the whole.

The Promise of a Shared Language

One of the SDGs' most celebrated achievements was the creation of a shared vocabulary for governments, businesses, civil society and academia. By codifying global priorities into a universally recognised set of goals, they established a common reference point for cooperation and accountability. Corporations could align sustainability strategies with the SDGs, NGOs could frame projects in SDG terms and governments could benchmark national development plans against global targets.

This standardisation of language had the potential to reduce duplication and fragmentation in development efforts, enabling resource pooling and the cross-border scaling of successful initiatives. In theory, it offered what Gupta (2025) calls a *'reference field,'* a symbolic north star around which diverse actors could orient their action.

Anchoring Transformation in Measurable Targets

Measurement was placed at the heart of the SDG design. Progress would be tracked through specific indicators, including the proportion of renewable energy in consumption, the percentage of the population living in poverty and biodiversity loss rates. This emphasis on data aimed to drive transparency, evidence-based policymaking and public accountability, while responding to criticisms that the MDGs lacked robust monitoring frameworks (United Nations, 2023).

By combining universality, integration and measurability, the SDGs embodied a vision of transformation that was both ambitious and structured. They sought to unite moral aspiration with institutional coordination, attempting to bring diverse actors into alignment around shared objectives.

Yet the promise of the SDGs rests on a critical assumption: that integration in design will naturally lead to integration in action. As later sections will show, this assumption falters when the actors implementing the framework operate from fragmented perception and when the one acted upon is treated as separate from the actor (Bateson, 2016).

3.4 The Structural Limits of the SDG Framework

The SDGs were conceived as a unifying architecture for global transformation, yet their implementation reveals deep structural limits. These limits are not merely operational such as funding gaps, data shortages or political resistance but are embedded in the very way the goals are institutionalised and pursued. They arise from contradictions between the integrative vision on paper and the compartmentalised realities of governance.

Siloed Implementation Despite Integrated Design

While the SDGs were designed as interlinked, governance systems remain institutionally fragmented. Ministries, agencies and international bodies typically operate within sectoral mandates:

- Health ministries focus on SDG 3 (good health and well-being).
- Environment ministries on SDG 13 (climate action) and SDG 15 (life on land).
- Finance ministries on SDG 8 (decent work and economic growth).

Coordination mechanisms may exist in theory, but decision-making authority and budget control are rarely shared in practice (United Nations, 2023). This produces policy incoherence: a government may promote renewable energy to advance SDG 13 while subsidising fossil fuels to protect GDP growth under SDG 8 (UNEP, 2023). Similarly, agricultural expansion for SDG 2 (zero hunger) may accelerate deforestation, undermining SDG 15 (life on land).

As Capra (1996) notes, managing interdependent goals through compartmentalised structures is a built-in contradiction.

The Growth–Sustainability Contradiction

A deeper structural tension lies in the assumption that economic growth (SDG 8) can be reconciled with ecological sustainability (SDG 13, SDG 14, SDG 15) without fundamentally rethinking the growth paradigm. The framework largely sidesteps post-growth and degrowth debates, embedding the language of 'sustained, inclusive, and sustainable economic growth' directly into its targets (Raworth, 2017).

This compromise reflects political pragmatism but perpetuates the economic drivers of ecological degradation. Despite gains in renewable energy and efficiency, global material consumption and emissions continue to rise (IPCC, 2023). The SDGs thus inherit a core paradox: they seek to resolve crises while preserving the very logic that produces them.

Quantification and the Illusion of Progress

Indicators enable benchmarking and accountability, but they can also reduce transformation to technical compliance. Governments may pursue incremental improvements in measurable areas, such as school enrolment (SDG 4) or forest cover (SDG 15), without transforming the deeper cultural, political or psychological structures that sustain inequity and exploitation (Bohm, 1992).

Examples include:

- Increasing renewable capacity while overall energy demand and emissions rise.
- Reducing poverty rates while wealth concentration accelerates, undermining SDG 10 (reduced inequalities).
- Expanding school access while reinforcing rote learning and social division.

Gupta (2025) warns that such '*image-based progress*' turns development into a scorecard exercise, masking and perpetuating underlying fragmentation.

Crucial questions emerge: how can an indicator or target ensure peace, compassion or non-discrimination? Who is the measurer of these reports, and through what lens are they defined? What does 'access to education' mean when the same knowledge is used to build weapons, borders and drones? Does killing those labelled as terrorists, be they Hamas' or 'ISIS,' count as peace under SDG 16? Are we simply playing the game of achieving and celebrating targets while the world burns? What does 'leave no one behind' mean when, in so many places, there is nothing left, no food, no nature, no safety and no love?

Dependence on Political Cycles and Voluntary Commitments

The SDG framework relies heavily on voluntary national action aligned with domestic political priorities. Commitments under the Paris Agreement (linked to SDG 13) are non-binding, vary widely in ambition and are easily reversed with changes in government (UNFCCC, 2023).

This reliance on voluntary action means that short-term political incentives frequently override long-term commitments. Sustainability goals are often deprioritised when they conflict with electoral calculations or economic pressures, a risk amplified by the absence of credible global enforcement mechanisms. The challenge is compounded by the broader fragility of international law: even binding conventions, such as the Geneva Conventions, are regularly violated, with accountability often determined not by principle but by political interpretation, will and voting power.

This raises a deeper question: how can there be genuine development of the whole when actors remain primarily concerned with advancing their own narrow interests? Even the notion of 'self-interest' is misunderstood. If actors were truly concerned with their own well-being, their actions would align with actual interest of humanity and not self-interest, which is a perceptual clarity that recognises the indivisibility of human and ecological systems. In contrast, conventional self-interest arises from attachment to past experience and identity, which corrupts and divides.

Gupta (2024) argues that intelligence's action is not the outcome of past desire or fear. It is action arising from direct perception in the present moment, which creates and sustains order. This order is not constructed by institutions or imposed through control but is the natural order of life itself, to live, to exist, to be. Importantly, this must not be confused with the ego's fear of death, which divides in order to preserve itself. Intelligence perceives actual danger and responds immediately, such as stepping aside when a bus approaches. By contrast, the ego projects imagined threats rooted in identity, memory and belief. The first is direct perception leading to coherent action; the second is conditioned reaction that sustains fragmentation.

Institutional Self-Preservation

International organisations tasked with advancing the SDGs are themselves bound by bureaucratic imperatives. Institutions often prioritise self-preservation, protecting reputation, maintaining funding and avoiding politically sensitive issues. This frequently results in:

- Rebranding existing programmes in SDG language without substantive change.
- Competing for visibility and resources instead of collaborating.
- Avoiding structural critiques that might threaten donor relationships.

This institutional tendency mirrors the psychology of the self: an identity-driven need to preserve the existing order, even when that order perpetuates crisis. Nations and actors who benefit from conflict often sustain insecurity because it guarantees their relevance and survival. If peace were to prevail, entire structures of power, economy and employment tied to conflict would lose their footing.

The global arms trade illustrates this clearly. Trillions of dollars are spent annually on weapons, justified under the logic of 'national security.' Yet these very industries depend on the continuation of war, tension and mistrust. Similarly, national armies, intelligence services and even political positions of authority are sustained by the perception of threat. If conflict were resolved, their centrality would diminish. In this way, insecurity itself becomes a resource, protected and reproduced by those whose identity and livelihood depend on it.

Despite decades of international peacekeeping, armed groups persist in the Democratic Republic of Congo, partly sustained by global demand for cobalt and coltan used in electronics and renewable technologies (Autesserre, 2010; Nest, 2011). Mission de l'Organisation des Nations Unies pour la stabilisation en République démocratique du Congo remains one of the UN's largest missions, yet it has been criticised for fostering dependency rather than resolving insecurity (Tull, 2009). At the same time, permanent members of the UN Security Council charged with maintaining peace, rank among the world's largest arms exporters, supplying weapons even to regions where peacekeepers operate (Stockholm International Peace Research Institute [SIPRI], 2023). This contradiction illustrates how institutional self-preservation undermines SDG 16 on peace, justice and strong institutions.

Such action is not truly intentional but mechanical. There is no independent 'thinker' behind it only thought itself, moving as desire, fear, ambition and insecurity. Institutions act as extensions of this movement: repeating patterns, defending identity and producing activity that appears purposeful but merely reproduces the conditions of disorder.

The irony is stark: the very institutions tasked with advancing SDG 16 are often undermined by the psychology of self-preservation. Instead of fostering trust, justice and stability, they may entrench mistrust, dependency and fragmentation, ensuring that peace remains an aspiration rather than a lived reality.

3.5 The Psychological Ground of Policy Failure

The structural contradictions of the SDG framework, including siloed governance, growth–sustainability tensions, measurement bias, political volatility and institutional self-preservation, are not random policy failures. They are the outward expression of the same psychological patterns that shape individual perception. The global governance crisis mirrors the movement of the self: identity-bound, competitive, time-bound and divided.

The Observer–Observed Division

Most SDG implementation strategies treat challenges like poverty (SDG 1), climate change (SDG 13), gender inequality (SDG 5) and biodiversity loss (SDG 15) as external problems to be managed. Policymakers, experts and activists position themselves as separate observers acting upon 'the environment,' 'the poor' or 'the marginalised.'

This division, described by Krishnamurti (1954) and Bohm (1992), produces a stance of control rather than relationship. It enables contradictory policies to coexist: subsidising fossil fuels while promoting renewable energy, expanding industrial agriculture while funding biodiversity conservation or promoting gender equality in some sectors while preserving exclusionary political structures.

This is not merely a governance flaw; it is the psychological default of a self organised around identity and control. The self imagines it stands outside the problem, allowing it to act without recognising its own participation in creating it.

Psychological Time and Deferred Transformation

The SDG timeline (2015–2030) was intended to create a coordinated horizon for global action. Yet it also reinforces what Krishnamurti (1954) called *psychological time*, the projection of transformation into the future. When change is imagined as something to be achieved later, present disorder is tolerated under the promise of eventual resolution.

This logic is visible across domains:

- *Climate commitments* pledge carbon neutrality by 2050, even as fossil fuel subsidies expand in India, China and the United States.
- Biodiversity strategies set restoration targets for 2040, while deforestation in the Amazon and Congo Basin accelerates today.
- Peacebuilding programmes promise reconciliation by 2030, while military expenditures and arms transfers hit record highs in the present.

By stretching responsibility into the future, actors preserve the comfort of current structures. Krishnamurti (1954) emphasised that continuity in psychological time is continuity of disorder, the patterns of fear, desire and competition that sustain fragmentation.

Gupta (2025) extends this to governance: by projecting action forward, institutions perpetuate harmful systems under the illusion of progress. The SDGs, in this sense, risk becoming an elaborate mechanism for deferral, allowing actors to 'promise away' transformation while sustaining extractive, competitive orders in the present.

Identification and Policy Selectivity

In policy practice, this is visible when:

- The United States proclaims global climate leadership while approving new oil and gas leases to preserve domestic energy security (Reuters, 2024; The Guardian, 2024).
- India expands its renewable energy portfolio while simultaneously increasing coal production in the name of national competitiveness (International Energy Agency [IEA], 2024; Ministry of Coal Government of India, 2024).
- *Governments* celebrate women's leadership in global forums while patriarchal laws and practices continue to define domestic life (UN Women, 2024; World Economic Forum [WEF], 2024).
- *The European Union (EU)* promotes human rights abroad while tightening borders against migrants, worsening humanitarian crises (Amnesty International, 2024; Human Rights Watch [HRW], 2024).

These contradictions are not accidental; they arise from the very structure of identification. The self maintains coherence by defending its chosen ideals and overlooking realities that expose its division.

This is not mere hypocrisy but the structural operation of identification. Krishnamurti (1954) observed that identification binds the self to continuity, blinding it to contradiction. Bohm (1992) framed this as thought defending its own image by suppressing inconsistencies.

Gupta (2025) describes this as *policy selectivity*: reform framed around identity-preserving choices rather than whole perception. Just as an individual protects self-image while ignoring behaviours that conflict with it, institutions protect their constructed identities, national pride, competitiveness, sovereignty, even at the cost of undermining collective survival.

Competition and the Metrics Race

The logic of comparison, central to the self's sense of worth, plays out in SDG rankings and indexes. Countries, like individuals, measure their value against others, turning progress into a reputational asset (United Nations, 2023). This creates a race for visibility rather than transformation.

In practice, this manifests in several ways:

- Superficial gains over substantive change: Nations prioritise highly visible infrastructure projects, digital dashboards or renewable capacity announcements, even when underlying consumption and emissions rise.
- Selective benchmarking: Governments emphasise indicators where they perform well: literacy rates, vaccination coverage, renewable capacity, while downplaying others like ecological loss, democratic decline or rising inequality.
- *Policy by optics:* Development plans are drafted to attract foreign investment, donor recognition or higher rankings, rather than to address the lived realities of citizens.

Gupta (2025) calls this *image-based action*, where appearance substitutes for substance. The competition for position mirrors the psychological mechanism of comparison: just as an individual derives worth by outdoing a neighbour, a country secures legitimacy by outperforming others in indices.

The result is that cooperation, ostensibly the heart of the SDGs, is eroded by rivalry. Instead of asking 'What is true for all of us here and now?' nations ask 'How do we perform better than others?' In this sense, the metrics race is not a neutral accountability tool but a structural reproduction of the self's fragmentary psychology.

Specialisation and the Self's Fragmentation

Modern governance mirrors the inner fragmentation of the self. Just as the psyche is divided into competing desires, fears and identities, institutions are fragmented into ministries, agencies and disciplines. Each part pursues its own mandate, often at cross-purposes with others.

Concrete examples abound:

- Infrastructure vs. ecology: Expansions under SDG 9 (industry, innovation, infrastructure) frequently undermine SDG 13 (climate action) and SDG 15 (life on land), when roads and dams cut through forests and wetlands.
- Agriculture vs. water: Food security programmes under SDG 2 (zero hunger) often expand irrigation and fertiliser use, degrading water quality and availability (SDG 6).
- Economic growth vs. social equity: Policies advancing SDG 8 (decent work and growth) often privilege industrial expansion over SDG 10 (reduced inequalities), deepening rural–urban divides.

Even efforts at integration like inter-ministerial task forces, multi-stakeholder platforms, 'whole-of-government' approaches often degenerate into territorial negotiations: who gets credit, who controls funding, who sets priorities. The very structures meant to integrate become new arenas of competition.

Capra and Luisi (2014) emphasise that systems cannot be understood by analysing parts in isolation; yet governance is built on specialisation. Without a shift in perception from fragmented mandates to the indivisibility of life, 'integration' remains technical, producing temporary alignments rather than coherence.

This is the institutional mirror of the fragmented self: thought dividing reality into compartments, then struggling to stitch them back together. Just as a divided mind cannot resolve inner conflict through more thought, a fragmented governance system cannot produce coherence through more frameworks.

3.6 Case Studies of Perceptual Failure in SDG Implementation

The structural and psychological patterns described above including the observer–observed division, psychological time, identification, competition and specialisation are not abstract. They operate in measurable ways across diverse arenas of SDG implementation. The following cases show how these perceptual failures undermine coherence.

Climate Policy and the Economic Growth Paradigm

Relevant SDGs: SDG 13 (Climate Action), SDG 8 (Decent Work and Economic Growth), SDG 12 (Responsible Consumption and Production).

The Situation

The EU's Green Deal is among the world's most ambitious climate strategies, aiming for carbon neutrality by 2050 with binding targets for renewable energy, emissions reduction and biodiversity protection (European Commission, 2023). Yet several member states continue to subsidise fossil fuels, exempt aviation fuel from taxation and shield energy-intensive industries in the name of competitiveness.

The Perceptual Problem

This reflects identity fragmentation: the EU positions itself as both a climate leader and an industrial bloc. The dual identity sustains contradictory policies like investments in renewables coexist with protections for structures incompatible with deep decarbonisation. This is not simply hypocrisy but the psychological split of identity, enacted at the level of policy.

Outcome

Climate action remains parallel to economic policy rather than integrated with it. As Capra and Luisi (2014) argue, unless the economy is perceived as embedded within ecology, strategies for sustainability will continue to be undermined by the growth paradigm.

Education Reform and the Metrics Trap

Relevant SDGs: SDG 4 (Quality Education), SDG 5 (Gender Equality), SDG 10 (Reduced Inequalities).

The Situation

India's National Education Policy (NEP) 2020 sets out ambitious reforms, promising universal access, multilingual education and digital literacy (Government of India, 2020). While enrolment rates and gender parity indicators have improved in several states (ASER, 2023), critical thinking skills, relational capacity and ethical awareness remain stagnant.

The Perceptual Problem

Gupta (2025) warns that image-based action, improving visible metrics while leaving structural and psychological foundations untouched is a hallmark of institutional reform. Here, knowledge acquisition is prioritised over awareness, empathy and the ability to perceive without bias.

This dynamic mirrors Foucault's (1975) analysis: modern institutions discipline through exams, rankings and compliance, producing efficient but conformist subjects rather than transformative agents. In the NEP, education reform remains locked within the same exam-oriented paradigm that perpetuates social division.

Outcome

Access expands, but the consciousness of education remains unchanged. Inequality persists in subtle forms: marginalised students may be 'included' statistically but remain excluded in lived experience, as the structures of hierarchy, conformity and competition are reproduced.

Peace Processes and Identity Entrenchment

Relevant SDGs: SDG 16 (Peace, Justice and Strong Institutions), SDG 17 (Partnerships for the Goals).

The Situation

The 2016 Colombian Peace Agreement between the government and the Fuerzas Armadas Revolucionarias de Colombia promised land reform, political reintegration and rural development (UN, 2016). While the agreement formally ended decades of armed conflict, violence against social leaders persists and new armed groups have emerged.

The Perceptual Problem

From a perception-based governance lens, the agreement dealt with external arrangements of disarmament, participation, development projects without dissolving identity-based mistrust. The observer–observed division persisted:

- The government viewed ex-combatants as subjects to be ‘reintegrated’ into a system they had long distrusted.
- Communities continued to perceive the state through the lens of neglect and exploitation.

As Lederach (1997) argues, sustainable peace requires relational transformation, not only institutional change.

Outcome

Without altering the psychological architecture of mistrust, formal mechanisms struggled to change lived realities. Conflict mutated into new forms, showing how structural settlements collapse without perceptual transformation.

Synthesis of Patterns

Across these cases, the same perceptual dynamics recur:

- Dual identities sustain contradictions (climate leadership vs. competitiveness).
- Metrics-driven reforms obscure deeper issues (access without transformation in education).
- Structural agreements collapse without relational change (peace without trust).

These failures are not incidental. They mirror the same psychological patterns outlined in Section 5: observer–observed division, projection into future time, identification, comparison and fragmentation. Without perceptual clarity, the SDGs risk becoming a patchwork of projects rather than coherent action from wholeness.

3.7 The New Paradigms: Regeneration, Systems Thinking, Doughnut Economics and Why They Also Fail Without Perceptual Transformation

In response to the perceived limitations of the SDGs, a wave of alternative frameworks has emerged over the last decade. Regenerative development, systems thinking, doughnut economics, planetary boundaries and nature-based solutions promise to move beyond the linear, siloed and technocratic tendencies of conventional policy.

These models enrich sustainability discourse, but they remain vulnerable to the same underlying problem that constrains the SDGs: they attempt to integrate a world still perceived through the lens of division.

Regenerative Development

Regenerative frameworks (Mang & Reed, 2012; Wahl, 2016) emphasise restoring ecosystems, revitalising communities and aligning with natural systems rather than working against them. They call for place-based, participatory approaches that move beyond sustainability's 'do less harm' ethos towards net-positive contributions.

Yet in practice, regenerative initiatives are often co-opted by the same growth and branding logics that undermine the SDGs:

- Corporations launch 'regenerative agriculture' programmes while continuing deforestation elsewhere.
- Cities develop 'regenerative districts' as real estate strategies to attract investment, rather than to transform systemic relationships with land and community.

Without a perceptual shift in how humans relate to the whole of life, regeneration risks becoming sustainability 2.0: a more inspiring narrative layered onto the same extractive paradigm.

Systems Thinking and Complexity Frameworks

Systems thinking (Capra & Luisi, 2014; Meadows, 2008) and complexity-informed governance (Boulton et al., 2015) encourage seeing interconnections, identifying feedback loops and embracing adaptive learning.

However, systems thinking is often applied methodologically, not perceptually:

- It maps linkages without interrogating how the mapper's own identity, fears or institutional incentives fragment the field.
- It treats integration as a coordination problem, rather than a transformation in the quality of perception of those acting.

Bateson (2016) calls this 'symptomatic coordination': creating the appearance of systemic awareness without altering the mental models that sustain fragmentation. In such cases, systems maps become instruments of control, not mirrors of wholeness.

Doughnut Economics

Kate Raworth's *Doughnut* Economics (2017) reframes development around two boundaries: the ecological ceiling and the social foundation. Cities such as Amsterdam and Copenhagen have adopted the doughnut as a planning tool.

Yet its application often reproduces old tensions:

- Municipalities celebrate doughnut adoption while maintaining high-carbon consumption patterns through global supply chains.
- The framework becomes a technical optimisation tool, not a reorientation of the desires and identities that drive overshoot and deprivation.

Without perceptual transformation, the doughnut risks becoming another dashboard for managing symptoms rather than a space for questioning the structures of desire, consumption and separation that cause them.

Planetary Boundaries and Science-Based Targets

The Planetary Boundaries framework (Rockström et al., 2009) and the Science-Based Targets initiative (SBTi, 2023) provide quantifiable thresholds for ecological safety. They clarify biophysical limits and offer benchmarks for policy and corporate action.

Yet the same observer–observed division persists. 'Nature' is treated as an external system to be managed, while economic expansion continues. Policy-makers often treat thresholds as negotiable margins rather than inviolable boundaries. Governments commit to science-based targets even as they expand fossil infrastructure; corporations pledge alignment while lobbying against regulation.

Why These Paradigms Still Fail Without Perceptual Clarity

Across these models, common problems emerge:

- Identity Capture: New frameworks are used to reinforce institutional image or competitive advantage.
- Method Without Self-Knowledge: Integration tools are applied without examining how thought itself fragments the whole.
- Optimising Within the Old Paradigm: The system is improved without questioning whether it is itself the embodiment of crisis.
- Deferred Transformation: Visions remain tied to future targets, preserving contradictions in the present.

As Gupta (2025) notes, 'Frameworks can extend the vocabulary of change, but only perception can alter its quality.' Without the dissolution of identity-bound action, even the most advanced paradigms are absorbed into repetition.

Implications for the Post-2030 Agenda

If next-generation frameworks are to avoid the fate of the SDGs, they must be rooted in perceptual transformation:

- Seeing the economy as embedded within the biosphere.
- Relating to others without comparison or control.
- Acting from clarity in the present, not from projections into the future.

Otherwise, SDG dashboards will simply be replaced with regenerative scorecards, changing the vocabulary while preserving fragmentation.

Conclusion: The SDGs and New Paradigms as Mirrors of the Self

The SDGs were the most comprehensive articulation of global ambition in human history, drawing on decades of development planning and integrating social and ecological priorities into a shared language for action. Newer frameworks (regeneration, systems thinking, doughnut economics, planetary boundaries) emerged to address their shortcomings.

Yet the diagnosis remains: these models are mirrors of the perception from which they arise.

- The SDGs fragment life into 17 sectors, each managed in silos.
- Regeneration becomes branding while extraction continues.
- Systems thinking maps interconnections but rarely examines the mapper's role.
- Doughnut economics reframes prosperity but is used as a technical template.
- Planetary boundaries quantify limits but still frame nature as an external object.

The problem is not the sophistication of these frameworks, but the state of mind that applies them. A mind organised around identity, fear and control will instrumentalise even the most holistic model for advantage, image management or deferred responsibility. It will treat integration as negotiation among fragments, not recognition of the whole.

This is why dashboards can glow green while ecosystems collapse, why peace agreements can coexist with militarisation, why education access can expand while conformity deepens. These are not errors of design but inevitable outcomes of action rooted in fragmentation.

Seen in this light, frameworks are neither failures to discard nor blueprints to perfect. They are diagnostic instruments. They reveal the structures of thought, identity and relationship that shape collective action. They show how noble intentions are absorbed into repetition.

With the escalating polycrisis, breached planetary boundaries and tipping points approaching, it becomes essential to look not at goals but at the processes of mind that create crisis. Only perception freed from attachment to goals, identity and control can release the intelligence to step out of the fire.

The next phase of this inquiry, therefore, is not to design the 'perfect' framework but to examine the actor who designs frameworks at all. If the outer world reflects the inner, then governance and development must begin not from targets but from perception. 'Without such a shift in perception, the post-2030 agenda will inherit the same contradictions – recycled in new vocabulary, new metrics, new delays.'

Part II now turns inward – not to ask what we must do, but to ask: from where does action arise? so that action may arise from a perception no longer bound to the very divisions it seeks to heal.

References

Amnesty International. (2024). *Europe: Containment, not compassion – EU external migration policies and human rights*. https://www.amnesty.org

ASER. (2023). *Annual Status of Education Report (ASER) 2023*. ASER Centre. https://www.asercentre.org

Autesserre, S. (2010). *The trouble with the Congo: Local violence and the failure of international peacebuilding*. Cambridge University Press.

Bateson, N. (2016). *Small arcs of larger circles: Framing through other patterns*. Triarchy Press.

Bohm, D. (1992). *Thought as a system*. Routledge.

Boulton, J. G., Allen, P. M., & Bowman, C. (2015). *Embracing complexity: Strategic perspectives for an age of turbulence*. Oxford University Press.

Capra, F. (1996). *The web of life: A new scientific understanding of living systems*. Anchor Books.

Capra, F., & Luisi, P. L. (2014). *The systems view of life: A unifying vision*. Cambridge University Press.

Carhart-Harris, R. L., Erritzoe, D., Haijen, E., Kaelen, M., & Watts, R. (2018). Psychedelics and connectedness. *Journal of Psychopharmacology*, *32*(1), 1–9. https://doi.org/10.1177/0269881117748903

European Commission. (2023). *The European Green Deal: 2023 progress and priorities*. Publications Office of the European Union. https://commission.europa.eu

Government of India. (2020). *National Education Policy 2020*. Ministry of Education. https://www.education.gov.in

Gupta, D. (2024). *Eternal movement: An exploration into self help, spirituality, psychology, society, relationships, education, self transformation, consciousness change, love and action*. Emerge Publications. https://doi.org/10.65169/q7d2z6

Gupta, D. (2025). *Governance without the self: Consciousness and the end of fragmentation, systems, and complexity*. Emerge Publications. https://doi.org/10.65169/a9m3p4

Human Rights Watch. (2024). *EU: Migration policies contributing to abuse and death*. https://www.hrw.org

International Energy Agency. (2022). *France 2022: Energy policy review*. https://www.iea.org

International Energy Agency. (2023). *World energy outlook 2023*. https://www.iea.org/reports/world-energy-outlook-2023

International Energy Agency. (2024). *Coal 2024: Analysis and forecast to 2026*. https://www.iea.org

Intergovernmental Panel on Climate Change. (2023). *Climate change 2023: Synthesis report of the sixth assessment report of the intergovernmental panel on climate change*. IPCC. https://doi.org/10.59327/IPCC/AR6-9789291691647

Mang, P., & Reed, B. (2012). Designing from place: A regenerative framework and methodology. *Building Research & Information*, *40*(1), 23–38. https://doi.org/10.1080/09613218.2012.621341

McGilchrist, I. (2021). *The matter with things: Our brains, our delusions, and the unmaking of the world*. Perspectiva Press.

Meadows, D. H. (2008). *Thinking in systems: A primer*. Chelsea Green Publishing.

Ministry of Coal, Government of India. (2024). *Annual report 2023–24*. https://coal.nic.in

Nest, M. (2011). *Coltan*. Polity.

Piketty, T. (2014). *Capital in the twenty-first century* A. Goldhammer, Trans.. Harvard University Press.

Raworth, K. (2017). *Doughnut economics: Seven ways to think like a 21st-century economist*. Chelsea Green Publishing.

Reuters. (2024, March 14). *Biden administration approves new offshore oil and gas leases despite climate goals*. https://www.reuters.com

Rockström, J., Steffen, W., Noone, K., Persson, Å., Chapin, F. S., III, Lambin, E., Lenton, T. M., Scheffer, M., Folke, C., Schellnhuber, H. J., Nykvist, B., de Wit, C. A., Hughes, T., van der Leeuw, S., Rodhe, H., Sörlin, S., Snyder, P. K., Costanza, R., Svedin, U., … Foley, J. A. (2009). A safe operating space for humanity. *Nature*, *461*(7263), 472–475. https://doi.org/10.1038/461472a

SIPRI. (2023). *SIPRI yearbook 2023: Armaments, disarmament and international security*. Stockholm International Peace Research Institute. https://www.sipri.org

The Guardian. (2024, April 7). *US climate credibility questioned after new drilling approvals*. https://www.theguardian.com

Tull, D. M. (2009). Peacekeeping in the democratic Republic of Congo: Waging peace and fighting war. *International Peacekeeping*, *16*(2), 215–230. https://doi.org/10.1080/13533310802685729

UN. (2016). *Final agreement to end the armed conflict and build a stable and lasting peace (Colombia)*. United Nations. https://peacemaker.un.org

UN. (2023). *The Sustainable Development Goals report 2023*. United Nations. https://unstats.un.org/sdgs/report/2023/

UN Women. (2024). *Progress on the sustainable development goals: The gender snapshot 2024*. https://doi.org/10.18356/9789211067248

UNEP. (2023). *Emissions gap report 2023: Broken record—Temperatures hit new highs, yet world fails to cut emissions (again)*. United Nations Environment Programme. https://www.unep.org

UNFCCC. (2023). *Nationally determined contributions synthesis report 2023*. United Nations Framework Convention on Climate Change. https://unfccc.int

Wahl, D. C. (2016). *Designing regenerative cultures*. Triarchy Press.

Part II

Inward Inquiry: Understanding the Source of Fragmentation

The crises of our time: ecological collapse, inequality, violence and systemic fragility are not merely external failures. As Part I revealed, they mirror the fragmentation of human consciousness. The world we experience is shaped by the mind that perceives it. If the outer world is divided, the source of that division lies within the structure of thought itself. This part turns inward, not in escape but in necessity, for without understanding the self, no reform can bring coherence. Frameworks fail because they begin after perception, rearranging symptoms rather than ending their cause.

The chapters in this part follow the movement of the self. They begin with the nature of the 'me', built on comparison and conflict, move through the traps of knowledge and belief, the intrusion of memory into relationships and the continuity of psychological time. Together they reveal a single process: thought seeking security through continuity. When this process is observed directly, without the interference of analysis or will, it comes to an end.

Inward inquiry is not accumulation but observation. It demands stillness rather than effort. The mind that sees its own disorder completely becomes quiet, and in that quietness a different intelligence operates, one that is whole, compassionate and unconditioned by fear or desire. The end of the self is not loss but the beginning of love, not as emotion but as the ground of relationship and right action in the world.

Chapter 4

The Self as the Source and Effect of Fragmentation, Conflict and Violence

Abstract

Turning inward, this chapter explores how the self sustains conflict and violence across every domain of life. It shows how identity divides the perceiver from the perceived, creating opposites that feed comparison, competition and aggression. The inquiry traces this movement from personal insecurity to collective ideology, from family and classroom to nation and state. Fear, desire and image preserve psychological continuity and become the basis for systems of domination and exclusion. This chapter argues that no structural reform can end conflict because structure itself arises from the same fragmented consciousness. To see the movement of the self, without naming, condemning or justifying, is to dissolve its continuity. When the mind observes division without resistance, fear and comparison lose energy, and intelligence acts without a centre. This chapter thus bridges the outer analysis of crisis with the inner recognition that conflict is not solved but ended in seeing.

Keywords: Identity and division; psychological conflict; collective violence; nationalism critique; conditioned ideologies; fear and desire; inner revolution; intelligence beyond self

Part I traced how global disorder mirrors inner disorder. Chapter 2 opened a space of perception and communion, helping the reader engage with insight rather than fall back into the patterns of knowledge and conditioning. With that foundation, this chapter turns inward to observe more directly.

This chapter explores how the self is both the source and the effect of fragmentation, conflict and violence. The self is not separate from the world it creates; its psychological divisions are mirrored in institutions, relationships and systems. What follows is not analysis for its own sake, but an invitation to observe the

From Fragmentation to Wholeness, 97–120

doi:10.1108/978-1-80686-677-920261004

movement of the self in daily life. Transformation does not lie at the end of inquiry but in the seeing itself.

4.1 Introduction

Chapter 2 established that institutions are not merely technical structures but projections of fragmented consciousness. Identities rooted in memory and conditioning shape policy, governance and international systems. Yet fragmentation alone does not explain the persistence of violence, whether in war, domestic abuse, ecological exploitation or systemic injustice. To understand this, we must examine the architecture of the self more deeply.

This chapter makes a decisive claim: violence is not imposed upon the self from the outside; it is inherent in the structure of the self itself. Not only in its outward expressions (war, domination, exploitation), but in its very construction: separation, comparison and continuity in psychological time. Without understanding this root, all outer reforms merely rearrange symptoms. The very idea that 'violence is not imposed upon the self from the outside' comes from a mind which sees violence as separate from the self.

Thinkers across traditions converge on this insight. Krishnamurti (1954) described the 'me' as the accumulated content of thought, which necessarily divides and thus breeds conflict. Bohm (1992) showed how abstraction and categorisation distort perception, reinforcing illusion and separation. Damasio (2010) demonstrated how the self is reconstructed through memory and bodily states – an adaptive function that can also harden defensiveness and rigidity.

These insights point to a crucial realisation: the self is not a neutral agent acting on a neutral world. It is itself a product of conditioning and experience, and therefore structurally incapable of perceiving wholeness. Where there is no wholeness, there is division; where there is division, conflict is inevitable; and where there is conflict, violence is always latent.

This matters profoundly for governance and sustainability. Sustainable Development Goal (SDG) 16, for example, seeks peace, justice and strong institutions. Yet peacebuilding investments remain marginal compared to military expenditures. Agreements often entrench divisions, as in Bosnia or Sudan, where settlements reinforced ethnic or religious identities as political categories. Likewise, SDG 5 aims for gender equality, but policies often operate within patriarchal logics of control and comparison. These contradictions show that policies derived from a divided self cannot bring about the end of division, they can only manage it.

This management mindset is the self's signature. It cooperates only for gain. As long as surface order is preserved, disorder at the root is tolerated. This order is conditional – based on rules, interests and appearances. When interests shift, it collapses. What is addressed is not the whole, but isolated symptoms, mistaken for the root. The conditioned mind identifies fragments, proposes disconnected solutions and thus reproduces its own disorder. Even well-intentioned action, arising from this structure, perpetuates the crisis.

Contemporary conflicts reflect this logic clearly. In Gaza and Israel, both military strikes and retaliatory hostage-taking operate from the same conditioning: reward, punishment, coercion. These are not solutions, but reinforcements of fragmentation. They act only on symptoms, leaving the structure intact.

Violence also manifests more subtly. Foucault (1975) described how institutions exert 'micro-violence' through surveillance, normalisation and discipline. Today, this is amplified through algorithms, managerialism and data regimes that standardise human behaviour and suppress deviation. Ecological destruction – deforestation, emissions, overfishing – is not merely an economic failure, but the extension of a self that seeks control over nature. Gupta (2024, 2025) argues that this violence emerges directly from the self's architecture of fear, control and continuity.

In this light, the continuation of global crises becomes unsurprising. A fragmented actor reproduces fragmentation: education becomes competition (SDG 4), economy becomes exploitation (SDG 8), politics becomes rivalry (SDG 16) and sustainability becomes branding. Trying to solve violence without transforming the self ensures that the same patterns persist.

This chapter examines the self's structure as the root of violence. Chapter 4 begins by examining the structure of conflict within the human mind. Section 4.2 explores how identity, by its very nature, separates 'I' from 'other,' making conflict inherent in relationship. Section 4.3 traces how comparison and competition sustain this violence, while Section 4.4 examines fear, desire and psychological time as the energies that perpetuate it. Section 4.5 turns outward to critique collective identities: nation, religion and ideology as amplifiers of the self's fragmentation. Section 4.6 then examines how governance and institutional frameworks mirror the same logic of control and fear that sustains violence, showing that transformation can begin only when this movement is seen and ended at its root. Section 4.7 extends this inquiry across education, economy, gender and ecology, revealing how awareness ends division and gives rise to intelligence that acts without the centre of the self. Yet even such seeing must end, for when it becomes memory, it turns into knowledge, giving continuity to the self and authority to the past. This movement prepares the ground for the next chapter's exploration of knowledge, belief and the subtle continuity of division that sustains the self and its world.

This is not a sociological analysis, nor a biological one. It is not about aggression or group dynamics. It is about the structure of the self as constructed by thought. To see this is not to despair, it is to begin, for the first time, to act without recreating the disorder we seek to end.

4.2 Separation and the Structure of the Self

At the root of human consciousness lies the experience of being a separate entity. From early childhood, one learns to distinguish 'me' from 'not-me': my body, my toys, my family, set against the backdrop of others who do not belong to me.

This sense of separation is reinforced through language, culture and institutions. Over time, the experience of being a separate centre crystallises into what is called the 'self.' The self appears as a stable entity, but on closer examination it is a bundle of memories, experiences, desires and fears, all organised by thought into the narrative of 'I.'

Krishnamurti (1974/1991), in dialogues with Professor Anderson (1991), observed that separation is the foundation of conflict: *'Where there is division, there must be conflict.'* The self is divided between what it is and what it seeks to be, between its own security and the perceived threat of the other. Separation is thus the very structure of violence. Bohm (1992) explained that thought perpetuates this division by generating distinctions, self/other, us/them, past/future, which return to perception as if real. Mistaking abstractions for reality, the mind creates alienation from others and from life itself.

Separation in Governance and Policy

The structure of separation is reflected in global governance. Nations define themselves through borders and sovereignty, sustaining a 'national interest' often at odds with global well-being. Climate negotiations illustrate this: the Paris Agreement marked unprecedented cooperation, yet its nationally determined contributions (NDCs) reinforced separation by framing action as compromise among competing actors rather than recognition of shared planetary life.

SDG 16, which calls for peace and strong institutions, is similarly pursued through law, policing and enforcement – frameworks that presuppose ruler and ruled, authority and subject. Foucault (1975) showed how such institutions rely on surveillance and discipline, perpetuating division. Peace within this paradigm becomes not the ending of violence but the managed coexistence of separate entities under restraint.

Separation in Society and Culture

This structural separation is evident across social and cultural life. In caste systems, hierarchies of purity and pollution fragment communities into rigid categories that persist in education, employment and marriage, contradicting SDG 10 on reducing inequalities. Gender divisions show the same pattern: while SDG 5 promotes equality, roles and valuations of male and female lives remain conditioned by separation. Consumer culture multiplies subtler divisions, where brands and lifestyles promise individuality through conformity, turning globalisation into a force that amplifies separation as status and distinction.

Ecological Separation

Perhaps the most dangerous expression of separation lies in humanity's relationship to nature. Modern development treats the environment as an external resource to be managed, extracted, or conserved. SDG 13 (climate action) and

SDG 15 (life on land) embody this dual stance: they seek to protect ecosystems while simultaneously advancing economic growth (SDG 8) that depends on exploitation. This contradiction arises from a perception of nature as separate from humanity, an object to be used rather than a living whole of which we are a part. Such mechanistic thinking, rooted in the illusion of separateness and thought's nature of constant division for self-gratification and preservation leads inevitably to ecological crisis. As long as the self perceives nature as 'other,' sustainability will remain an exercise in management rather than harmony.

Case Studies of Separation

India–Pakistan

Partition in 1947 left memories of violence and displacement that hardened into opposing national identities. Education, media and politics sustain this divide, so treaties remain fragile. SDG 16 is weakened not by the absence of agreements but by the persistence of 'us' versus 'them' (Khan, 2007).

Rwanda

Colonial rule codified Hutu and Tutsi identities into rigid categories that erupted in the 1994 genocide. Reconciliation efforts continue, yet trauma and mistrust endure. SDG 16 cannot be realised if psychological investments in division remain unexamined (Mamdani, 2001).

Brexit

The United Kingdom's withdrawal from the European Union (EU) emphasised sovereignty and identity over shared interdependence. Cooperation on trade, climate and migration was fractured. SDG 17's vision of partnership falters when separation prevails over recognition of common life (Oliver, 2018).

Separation as Psychological Time

Beyond external divisions, separation also exists inwardly as the division between what is and what should be. The self is perpetually caught between its actual state and its projected ideal, between the present and the imagined future. This movement in psychological time sustains conflict: the effort to become better, more successful, more secure creates a perpetual tension. Krishnamurti (1954) insisted that this inward separation is itself a form of violence, for it pits the self against itself. Development frameworks mirror this logic in their timelines and targets, progress by 2030, net zero by 2050. These promises perpetuate delay, embedding division into the very structure of global action.

Implications

To recognise separation as the root structure of the self is to see why outer efforts fail to bring about coherence. Policies may reduce inequalities, treaties may pause conflicts and metrics may show progress, yet the underlying division remains intact. As long as the self perceives life through separation, every action will carry the seeds of violence. SDG frameworks cannot transcend this limitation because they are conceived and implemented by the same consciousness of division.

This does not mean that goals like peace, equality or sustainability are meaningless. Rather, it means that their realisation depends on a transformation deeper than policy design. It requires a perception of wholeness in which the very logic of separation dissolves. Without this, development remains a series of temporary arrangements within a structure of violence. Or, as we have seen, it becomes management of symptoms rather than transformation of life.

4.3 Comparison and Competition as Sustainers of Violence

Human beings are conditioned to compare. From childhood, children are measured through grades, rankings and prizes. This pattern, rooted in thought's divisive movement, extends into adulthood through career, wealth, status and power. Comparison is not neutral; it establishes inequality and sustains a climate of competition in which fear, ambition and conflict thrive. Though often praised as a driver of progress, comparison is a fundamental structure that perpetuates violence in both individual and collective life.

Krishnamurti (1954) observed that *'the moment you compare yourself with another you destroy yourself.'* Comparison fractures perception by setting up conflict between what one is and what one should be. Gupta (2024) explains that this conflict is internal, since both the fact and the ideal are products of thought. The mind projects fulfilment into the future, measuring worth by the distance between itself and its image of success, producing dissatisfaction and rivalry. Bohm (1996) reinforced this by showing how thought creates systems of measurement as a form of control, embedding comparison into the very structure of consciousness.

Modern psychology echoes these insights. Festinger's (1954) social comparison theory shows that individuals derive self-worth by evaluating themselves against others. While adaptive in evolutionary contexts, this process generates chronic anxiety and resentment in societies saturated with metrics of success. Clark and Hogue (2020) further demonstrate that comparison erodes trust in relationships, increasing conflict and reducing cooperation.

Competition in Governance and Policy

Comparison and competition shape not only individual lives but also global governance. Nations are ranked by GDP, military expenditure, education scores and sustainability indexes. International institutions reinforce this race through comparative metrics such as the Human Development Index, the World Bank's Ease of Doing Business and SDG dashboards. Presented as neutral, these instruments frame cooperation in competitive terms: who is ahead, who is behind, who merits funding or sanction.

Even the SDGs, designed as a framework for collective transformation, are embedded in this logic. Countries compete for positions in rankings to signal legitimacy and attract investment, encouraging 'gaming' of indicators. Gains in enrolment under SDG 4 may be celebrated, while quality of learning and structural inequalities remain unaddressed.

Competition also weakens collective action. In climate negotiations, states calculate commitments relative to others, fearing loss of advantage. Conference of Parties debates on climate finance revealed this tension: wealthy nations pledged adaptation support for the Global South but consistently under-delivered, citing domestic priorities. The result is a lowest-common-denominator approach – agreements diluted while planetary boundaries continue to be breached (Rockström et al., 2009).

Comparison in Education

Education, which ought to nurture learning and relationship, has become a primary site of comparison. Standardised testing, competitive entrance exams and performance rankings reduce students to numerical hierarchies. This culture cultivates fear of failure, envy and division. Even SDG 4, which aspires to 'quality education for all,' is often interpreted through enrolment rates and standardised outcomes rather than the flowering of intelligence.

In India, competition manifests in extreme forms, with millions of students vying for limited university seats or government jobs. The result is despair, with rising rates of anxiety, depression and suicide among youth (Deb et al., 2015). Education equates worth with rank, entrenching division and perpetuating psychological violence. Education pursued in fear and comparison cannot flower into intelligence, for it is bound to the self's movement of becoming and hence, education framed by performance and rank, mirrors the same fragmented psyche that structures governance.

Competition in Economy and Society

Modern economies are founded on competition. Capitalist markets reward efficiency and innovation, but they also institutionalise rivalry. Corporations compete for market share, workers compete for jobs and nations compete for foreign investment. SDG 8 (decent work and economic growth) exemplifies this paradox. While it promotes inclusivity, it remains rooted in competitive growth that drives inequality and ecological destruction. Piketty (2014) demonstrates how wealth accumulation in competitive economies generates structural disparities that destabilise societies.

Consumer culture intensifies this rivalry by linking identity to possessions. Advertising thrives on comparison, promising distinction through products while ensuring conformity to market trends. Social media amplifies this dynamic, creating an endless loop of curated images against which individuals measure themselves. The result is a crisis of self-esteem, particularly among youth, whose sense of worth becomes tethered to external validation (Twenge, 2017).

Competition in International Relations

At the geopolitical level, competition manifests as arms races and struggles for dominance in trade and technology. SDG 16 (peace, justice and strong institutions) is consistently undermined by military rivalry. States justify massive defence budgets as deterrence, but the effect is escalation: each perceives the other's build-up as a threat, necessitating its own expansion. This cycle, described by Mearsheimer (2001), sustains perpetual insecurity.

Competition for resources further destabilises peace. Water scarcity in regions such as the Nile Basin or Indus Valley fuels interstate rivalry, threatening SDG 6 (clean water and sanitation). Instead of recognising water as a shared lifeline, nations hoard and weaponise it as leverage, deepening division.

The Violence of Comparison

Comparison and competition sustain violence not only externally but inwardly. A mind trained to measure itself against others cannot rest in the present; it lives in the tension of 'more' and 'better.' This psychological violence manifests outwardly as rivalry, exploitation and war. Gupta (2025) observes that 'the same mind that compares is the mind that divides nations, races and faiths.' The personal and the political are not separate; both are expressions of the same conditioned consciousness.

Beyond Comparison: The Possibility of Wholeness

To end violence, one must question whether comparison is necessary at all. Krishnamurti (1991) proposed that learning is possible without comparison, just as love is possible without rivalry. Bohm (1996) suggested that dialogue, free of competition, can reveal a common ground beyond division. Neuroscientific research supports this: empathy and relational awareness activate cooperative neural circuits, reducing the grip of comparison (Singer & Klimecki, 2014).

For governance, this implies moving from competitive benchmarking to cooperative perception, seeing crises not as arenas of rivalry but as expressions of a shared condition. For education, it means shifting from grades and rankings to spaces of inquiry where intelligence flowers without fear. For economies, it means redefining prosperity not as competitive accumulation but as collective well-being within ecological limits.

Implications for the SDGs

If the SDGs are to become more than a scoreboard of national performance, they must be liberated from the logic of comparison. This requires:

- Recognising that progress cannot be measured by ranking one nation against another.
- Ending the competitive framing of development finance and climate commitments.
- Reimagining education indicators beyond enrolment and test scores.
- Questioning economic growth as the foundation of well-being.

Without such shifts, the SDGs risk perpetuating the very violence they seek to end. As long as comparison structures action, conflict will remain inevitable.

4.4 Fear, Desire and the Roots of Conflict

If comparison and competition sustain conflict outwardly, fear and desire sustain it inwardly. These are not secondary emotions; they are the very ground from which much of human action arises. To understand violence, whether in the psyche, in education, in governance or in international relations, we must first understand the movement of fear and desire within consciousness.

Fear: The Shadow of Thought

Fear is one of the oldest and most pervasive psychological forces. Evolutionarily, it protected humans from predators and danger, but in the modern world it extends far beyond immediate threats. We fear failure, rejection, loss of reputation, poverty, war and death itself. Though often triggered by external conditions, such fears are largely sustained by thought.

Krishnamurti (1983) observed that fear is always a projection: thought recalls painful experiences of the past or imagines threats in the future, creating a state that governs the present. Fear is therefore inseparable from time, what happened yesterday and what might happen tomorrow. Neuroscience reinforces this insight. Damasio (2010) shows how the brain integrates memory, bodily states and imagination to generate anticipatory anxiety even without actual danger.

Fear manifests in personal life as avoidance, withdrawal or aggression. In governance, it appears as policies of control, walls, weapons, surveillance. Fear narrows perception and blocks creative or cooperative response.

Desire: The Pursuit of Fulfilment

Desire is the counterpart of fear. Where fear pushes away, desire pulls towards. Desire promises completion: if I attain success, wealth, love or recognition, I will be fulfilled. But like fear, desire is rooted in thought. The mind imagines a state of satisfaction, projects it into the future and then strives to reach it.

Nietzsche (1887/1994) described this insatiable drive as the *will to power* – the impulse to affirm oneself, overcome lack and extend one's force into the world. In contemporary societies, this drive is institutionalised through advertising and media, which manufacture cravings by presenting idealised images of beauty, wealth and status. Girard (1977), in *Violence and the Sacred,* deepened this analysis by showing how desire is mimetic: we want what others want, not because of the object itself, but because of the meaning attached to it by others. This mimetic structure generates rivalry and, when rivalry escalates, societies often resolve tension through scapegoating, projecting inner violence onto an external 'enemy.'

Desire is not only personal but also collective. Nations desire prestige, economic dominance and military superiority. Institutions desire legitimacy and growth. Even social movements, when driven by the desire for recognition or ideological victory, risk reproducing the very cycle of conflict they aim to resolve.

Fear and Desire as Interlinked Movements

Though often treated separately, fear and desire are two sides of the same coin. Desire often arises from fear: the desire for wealth may stem from fear of poverty; the desire for power from fear of vulnerability; the desire for belonging from fear of exclusion. Conversely, fear is often shaped by desire: fear of losing what one has, fear of not attaining what one longs for. Together, they form a self-reinforcing cycle in which the psyche is never at rest.

This cycle operates not only in individuals but also across global systems. Fear of losing competitive advantage fuels economic desire. Desire for dominance fuels military build-ups, which in turn generate fear in rivals. This interplay sustains an endless spiral of insecurity and violence.

Fear and Desire in Education

Education, ideally a space for discovery, is pervaded by fear and desire. Students fear failure, punishment and exclusion. They desire success, recognition and upward mobility. Parents and institutions reinforce this by equating self-worth with achievement. The result is a system where learning is secondary to performance, and inquiry is suffocated by anxiety. Studies in India and East Asia show that exam-related fear is one of the leading contributors to student depression and suicide (Deb et al., 2015).

Even reforms framed under SDG 4 (quality education) often reinforce these dynamics. Efforts to expand access frequently emphasise competitive testing and global rankings, which perpetuate the same cycle of fear and desire under a different name.

Fear and Desire in Governance

Fear and desire are embedded in the structures of governance. Leaders manipulate fear to consolidate power: fear of terrorism, migration, disease or economic collapse. At the same time, they promise fulfilment of desires such as prosperity, security or national greatness. This dual strategy is as old as politics itself, but in the modern era it is amplified by media and technology.

Internationally, fear of decline drives nations to pursue endless growth (SDG 8), while desire for influence drives intervention in conflicts abroad. Policies around migration (SDG 10) are shaped less by humanitarian commitment than by fear of demographic change and the desire to protect national identity. Environmental negotiations (SDG 13) are similarly driven by fear of ecological collapse coupled with desire to maintain economic advantage. The result is paralysis: agreements are diluted and commitments postponed.

Fear and Desire in Economy and Society

Economic systems are perhaps the most explicit embodiments of fear and desire. Consumer markets thrive on fear of inadequacy and desire for status. Credit systems exploit fear of exclusion and desire for consumption, binding individuals and nations into cycles of debt. Piketty (2014) has shown how desire for accumulation among elites creates structures of inequality that fuel resentment and instability.

At the societal level, fear of the 'other' fuels xenophobia, racism and sectarianism, while desire for homogeneity or dominance reinforces authoritarianism. Girard (1977) demonstrated how fear and desire converge in scapegoating: communities project their internal violence onto an external 'enemy,' uniting through exclusion and aggression.

The Root of Conflict

If we look closely, conflict is not caused merely by external scarcity or competing interests. It is rooted in the psyche's dependence on fear and desire as motivators. As Gupta (2025) argues, 'action born of fear perpetuates the insecurity it seeks to escape, and action born of desire perpetuates the dissatisfaction it seeks to end.' Both movements bind human beings to a cycle of continuity.

Conflict resolution frameworks often fail because they address symptoms without touching this root. Peace treaties redistribute resources but leave fear and desire intact. Development programmes expand opportunity but reinforce the desire for endless growth. Even spiritual or psychological practices, when motivated by the desire for enlightenment or fear of suffering, entrench the very structures they aim to dissolve.

Beyond Fear and Desire

Is action possible without fear or desire? This is perhaps the most radical question. Krishnamurti (1983) proposed that true action emerges only when the mind sees fear and desire in operation and ends their hold, not through suppression or control, but through direct awareness. Bohm (1992) suggested that when thought ceases to project fulfilment or threat, perception becomes whole and relationship is no longer mediated by images.

In education, this would mean creating environments where students are not driven by fear of failure or desire for success, but are free to enquire without pressure. In governance, it would mean decision-making not rooted in electoral fear or the desire for legacy, but in perception of the whole. In economics, it would mean redefining prosperity not as accumulation but as sufficiency, relationship and care.

Neuroscience suggests that this is not utopian. Studies of compassion-based practices show reductions in amygdala reactivity (fear response) and increases in prosocial motivation not tied to personal desire (Singer & Klimecki, 2014). Yet the challenge is not to adopt practices mechanically, but to question the very ground of fear and desire in daily life.

Implications for the SDGs

If the SDGs are to contribute to real transformation, they must recognise how fear and desire shape their implementation. Otherwise, they risk becoming instruments of the very conflict they aim to resolve. For example:

- *SDG 8's* growth paradigm must confront how desire for accumulation generates ecological breakdown.
- *SDG 16's* peace agenda must acknowledge how fear-based security policies perpetuate violence.
- *SDG 4's* education targets must move beyond fear-driven competition and desire for rank.
- *SDG 13's* climate commitments must resist being negotiated through fear of loss and desire for advantage.

Without such recognition, the SDGs remain trapped in the same cycle of psychological violence. Metrics may show incremental progress, but the underlying consciousness of fear and desire ensures that conflict, inequality and ecological destruction will reappear in new forms.

True transformation requires questioning whether action can emerge outside this dual movement altogether. This means shifting the ground of policy and governance from reaction and pursuit to perception and responsibility. Unless this inner shift takes place, development frameworks will remain bound to the very forces they aim to overcome, producing temporary order while perpetuating deeper disorder.

4.5 Violence in Relationship: Family, Community and the Nation

Violence is often thought of as physical harm, but its roots extend much deeper. It manifests in the subtle structures of relationship: between individuals, within families, across communities and between nations. When conditioned by fear, desire and comparison, relationships become battlegrounds where conflict is sustained and reproduced. To understand war, terrorism or systemic oppression, we must first examine violence in the most intimate spaces of human life.

Violence in the Family

The family is often idealised as a space of love and care. Yet for many, it is also where psychological and physical violence first takes root. Patriarchal norms, rigid hierarchies and expectations of obedience create environments where children grow up internalising fear and comparison. Gender-based violence, marital abuse and coercive discipline are not isolated deviations but normalised practices across societies (UN Women, 2023).

In India, surveys reveal that girls are disproportionately denied nutrition and healthcare within households (Jha et al., 2006). In Rajasthan and other northern states, excess female under-five mortality persists due to neglect and gender discrimination, with an estimated 239,000 additional female child deaths occurring annually across India (Guilmoto & Saikia, 2018). Here, violence is not only physical but structural: it is encoded in the distribution of care, opportunity and recognition.

Krishnamurti (1954) emphasised that such violence begins when a relationship is mediated by an image – when parents project expectations onto children, or when spouses relate through roles rather than direct perception. Gupta (2024) observes that there is no true relationship in that state, only idea projecting itself: the observer is the observed, yet the image lives in the illusion of separateness. The child is then seen not as a living being but as 'my son' or 'my daughter,' an extension of the parent's ambition. This subtle violence of image precedes physical violence and lays the foundation for broader social divisions.

Community and the Politics of Belonging

What begins in the family extends into the community. Communities offer belonging and protection, but they also enforce conformity and exclusion. Identity is affirmed by defining insiders and outsiders, often leading to discrimination against minorities, migrants or lower castes.

Girard's (1977) theory of mimetic rivalry is instructive here. Communities stabilise themselves by projecting internal tensions onto scapegoats, those blamed for social unrest or decline. Pogroms, lynchings and riots are extreme expressions of this mechanism, but its subtler forms include bullying, marginalisation and stigma. Similarly, governments and policies often function in the same way: unrest or insecurity is displaced onto an 'other,' whether immigrants, minorities or dissenters, so that the deeper causes of disorder remain unexamined. In such cases, violence is not accidental but structural. By scapegoating, societies protect the illusion of coherence while perpetuating cycles of division.

In contemporary politics, this dynamic is visible in the targeting of refugees, religious minorities, or LGBTQ+ groups. The rise of nationalist movements across Europe, South Asia and the United States demonstrates how fear of difference and desire for homogeneity turn community into a tool of exclusion.

Such violence is not incidental. It is woven into cultural narratives, textbooks, rituals and media. Bourdieu (1991) described this as symbolic violence: the imposition of meanings that legitimise inequality without overt coercion. A caste system or racial hierarchy may persist for centuries not because of open brutality, but because its assumptions are naturalised in everyday life.

Nationalism and Institutionalised Violence

At the national level, violence becomes institutionalised through governance, law and policy. Nations define themselves against external others, producing borders, armies and defence budgets. The rhetoric of national security normalises surveillance, militarisation and sacrifice. In this way, fear and desire, for example, fear of invasion and desire for dominance become the organising principles of political life.

Foucault (1977) showed how modern states extend this violence through disciplinary institutions like schools, prisons and hospitals that regulate bodies and behaviour. Even welfare policies may carry coercive dimensions, imposing norms of productivity, citizenship and conformity.

SDG 16, which aims at peace, justice and strong institutions, often fails to confront these roots. Instead, it measures the absence of open conflict while ignoring the everyday violence of marginalisation, structural inequality and state repression. Peace agreements may reduce fatalities, yet surveillance states, systemic racism and gender hierarchies continue to reproduce violence in less visible forms.

International Relations: Violence Scaled up

Between nations, violence becomes overt and spectacular. Wars, sanctions, arms races and economic coercion are justified as necessary for security or development. Yet these conflicts mirror the same psychological structures visible in family and community: fear of loss, desire for dominance and the projection of threat onto an enemy.

Bohm (1992) argued that international relations are trapped in 'endless fragmentation,' with each nation acting as if its security were separate from the whole. Climate negotiations reveal this vividly. Even in the face of collective catastrophe, nations cling to self-interest (Raworth, 2017). Singh and Bhatnagar (2023) has described this impasse as a manifestation of wicked problems, where competing interests and fragmented values ensure that agreements remain elusive. Gupta (2025) extends this diagnosis, showing how governance frameworks conceived without intelligence and awareness become self-limiting, reflecting the very divisions they aim to resolve.

The contradiction is stark. SDG 13 demands urgent climate action, yet global military expenditures reached over $2.4 trillion in 2023 (SIPRI, 2024). These vast resources not only divert funding away from sustainability but also directly intensify ecological destruction through emissions, land use and technological waste. Violence at the national and international level thus sabotages the very goals proclaimed under the SDG framework.

The Cycle of Violence

Violence in relationships is cyclical. A child raised in fear often becomes an adult who governs through fear. A community forged through exclusion builds a nation that practices exclusion at scale. Nations organised around dominance reproduce global systems of inequality and coercion. At each level, the same psychological architecture – fear, desire and image propagates itself.

Freud (1930/2010) recognised this when he described civilisation itself as a compromise between individual aggression and collective restraint. Yet compromise does not end violence; it displaces it. Suppressed aggression re-emerges in scapegoating, nationalism or war. Without addressing the root in consciousness, violence is destined to recur in new forms.

Violence and the Failure to See Movement

Why does such violence persist, often unrecognised in its subtler forms? One reason lies in how human beings perceive. Most approach life not as a whole movement but from points: a fact, a role, an outcome, an isolated event. In this fragmented perception, the origin, the unfolding and the ending are not seen together.

Knowledge arises from such points. It is always of memory, conclusion or image. Knowing belongs to intellect, which divides. Understanding, by contrast, belongs to observation. It perceives movement as a whole – origin, process and ending seen in one flow. Knowledge may explain, but understanding sees. It is not of the intellect at all, but of being; it is born when observation is whole, not when thought offers an answer.

Philosophers and scientists across traditions have pointed to this distinction. Heidegger (1962) described truth as *aletheia*, not the possession of isolated facts, but an unconcealment that reveals itself as a movement of disclosure. Merleau-Ponty (1962) argued that perception is embodied and temporal, never reducible to static points. Whitehead (1929/1978) carried this further into process philosophy, insisting that reality itself is constituted by process, not by separate things. McGilchrist (2019) has shown neuroscientifically how the left hemisphere abstracts and fragments, while the right hemisphere perceives flow and wholeness. To live from the dominance of the fragmentary mode is to mistake the part for the whole, knowledge for understanding.

Because we fail to perceive life as movement, competition is praised as achievement rather than recognised as violence. Ideas are celebrated as solutions rather than seen as divisions. Thought is treated as neutral rather than seen as the movement of fragmentation itself. Krishnamurti (1983) insisted that to mistake thought for reality is already to create conflict. Bohm (1992) explained that thought breaks the flow of life into fragments and then mistakes those fragments for separate realities. Korzybski (1933/1994) warned against this error in his dictum that 'the map is not the territory.'

This failure of perception sustains violence across domains. In education, grades and rankings are treated as ends, while the movement of fear and conformity remains unseen. In governance, policy reforms are pursued as technical points, while the deeper movement of insecurity and identity is ignored. Even in spirituality, practices are adopted as methods without observing the whole movement of becoming that sustains the self. Foucault (1980) showed how institutions reinforce this by transforming life into categories of knowledge and power, concealing the deeper movement of human being beneath abstractions.

To see as a whole is to enquire. True inquiry is not the search for answers but direct observation in which beginning and ending are present together. When competition is seen in its entirety, it is revealed as division. When an idea is seen as image, it is revealed as violence. When thought is seen as fragmentary, its divisive nature is revealed. Awareness of this movement is already the ending of its hold. What falls away is false; what remains is true.

4.6 The Image and the Continuity of Division

At the root of human violence and disorder lies an act so subtle that it often escapes notice: the creation of images. By 'image,' we mean the mental pictures, representations and narratives through which we relate to ourselves, others and the world. These images are not neutral reflections of reality; they are conditioned by memory, culture, ideology and desire. Once created, they mediate perception, so that we rarely encounter life directly. Instead, we relate through images, and in this mediation lies the seed of division.

The Nature of Image

The 'image' is not simply a conscious thought but the accumulated residue of past experience. When we think of another person, what comes to mind is rarely the living presence before us but a bundle of memories, judgements and expectations. Similarly, when we think of ourselves, we recall achievements, failures, identities and aspirations, an image of 'who I am' or 'what I wish to be,' both rooted in the past.

McGilchrist (2019) suggests that the brain's hemispheric functions reinforce this dynamic: the left hemisphere tends to abstract, categorise and fix representations, while the right remains open to lived experience. Overreliance on the left-hemispheric mode produces rigid images that substitute for reality. Bohm (1992) called this the 'system of thought,' where abstraction is mistaken for fact.

Image in Personal Relationship

In personal relationships, images become the basis of expectation and conflict. A spouse relates to the image of their partner: the roles of 'husband,' 'wife,' 'provider' or 'caretaker,' rather than to the living person. Parents carry images of what their children 'should become,' producing pressure, disappointment or coercion. These images accumulate, distorting perception until relationship is no longer between two human beings, but between two sets of projections.

This process generates subtle violence. When the other fails to conform to the image, frustration and anger arise. When the image of self is threatened by insult, neglect or comparison, defensiveness emerges. The very act of relating through image ensures conflict, because no image can ever fully contain reality.

Image in Communities and Nations

The same mechanism operates collectively. Communities sustain cohesion by constructing images of 'us' and 'them' based on religion, caste, ethnicity or class. Girard's (1977) theory of mimetic rivalry shows how such images become repositories of blame, stabilising the group by projecting disorder onto outsiders. Later scholars, such as Redekop (2008), extended this towards a 'mimetic of blessing,' suggesting that imitation can also foster reconciliation. Yet even this constructive turn has limits. Compassion and love cannot be imitated or practiced; imitation, memory and practice all belong to thought, while compassion arises only in direct perception when desire and image end.

The problem is not merely that we imitate others' desires, but that desire itself is impersonal, arising from the common field of fragmented consciousness. Objects of desire may differ in wealth, status or belonging, but the structure is the same. Disorder lies in the mind's discontent: when desire is unfulfilled there is frustration and when it is fulfilled the fragment seeks repetition, generating new movements of identification and control. Within the individual, this produces contradiction as one desire pursues gratification while another attempts restraint. Desire thus fragments both outwardly and inwardly, because it is born of partial experience and sustains fragmentation in its very movement. To see this directly is education, not the cultivation of practices but the ending of imitation and projection.

Nations magnify this process. National identity depends on shared images: myths of origin, heroic histories, cultural superiority or moral destiny. Flags, anthems, textbooks and rituals reinforce solidarity while generating division. To be 'Indian,' 'American' or 'Chinese' is often defined against a threatening image of others, producing cycles of suspicion, rivalry and conflict. Foucault (1977) showed how institutions reinforce such images by disciplining bodies and minds:

schools cultivate the 'good citizen,' militaries create enemies and heroes and development programmes construct the 'deserving poor.' Each image channels behaviour into conformity with a pre-defined norm.

Image and the SDGs

The SDGs are themselves saturated with images. They promise a world 'free of poverty,' 'at peace' and 'sustainable.' These are powerful aspirations, yet they also function as collective images, symbols that inspire action but risk detaching from lived reality.

For example:

- *SDG 5 (Gender Equality):* Governments may celebrate rising numbers of women in parliaments while ignoring the persistence of patriarchal images that devalue women in households and workplaces.
- *SDG 13 (Climate Action):* Nations pledge carbon neutrality by 2050 while continuing to subsidise fossil fuels. The image of future sustainability masks present inaction.
- *SDG 16 (Peace and Justice):* Institutions report declines in battle deaths while ignoring everyday structural violence including racism, surveillance and exclusion.

The gap between image and reality becomes a source of disillusionment. When people perceive that promised change exists only as rhetoric, cynicism deepens. Yet institutions persist in managing images because they serve political legitimacy. Development reports become mirrors of aspiration rather than mirrors of reality.

The Continuity of Division

Images ensure the continuity of division in two ways. First, they fix identity. Once a person, group or nation is labelled, interaction is filtered through that label. A refugee becomes an 'illegal migrant,' a protestor a 'threat,' a nation an 'enemy state.' Such labels stick, shaping perception long after circumstances change. Second, images reproduce themselves. When we interact through images, we generate new experiences that reinforce them. If I meet someone with suspicion, my behaviour elicits defensiveness, which confirms my suspicion. Nations that define each other as rivals engage in arms races, which confirm rivalry. The cycle is self-perpetuating.

Freud (1921/2010) recognised this dynamic in his analysis of group psychology: individuals lose critical distance when submerged in collective images, leading to conformity and aggression. Contemporary social media amplifies this effect, producing echo chambers where images of self and other are intensified and weaponised (Zuboff, 2019).

Breaking the Cycle of Image

Can relationship exist without image? Krishnamurti (1954) argued that true relationship begins only when image ends. To see another without the mediation of memory or expectation is to encounter them directly, in the present. Bohm (1996) suggested that dialogue a conversation in which assumptions are suspended, can dissolve the hold of images and allow meaning to flow freely.

This does not mean erasing memory or denying culture. It means perceiving when an image arises and not confusing it with reality. Such perception is not analytical but immediate. It requires awareness: the capacity to see thought as thought, image as image, without identification. Awareness is not a method or practice, but a living seeing in which the observer and the observed are not divided.

History offers glimpses of this possibility. Truth and Reconciliation processes in South Africa and Rwanda sought, however imperfectly, to move beyond entrenched images of victim and perpetrator by creating spaces for shared humanity. Indigenous traditions of council dialogue often suspend hierarchical roles, allowing participants to speak not as representatives of power or status but as human beings in the present. These gestures do not erase the past, but they point towards a mode of relationship where image is not sovereign.

When the observer is changed, the physical record of the past may still exist, but it is no longer approached through the psychological residue of fear and desire. Freed from that content, the act of observation itself is transformed. What is seen is altogether different.

Implications for Transformation

If images sustain division, global transformation cannot be achieved by producing better images alone. Branding the SDGs with colourful logos or inventing new indexes may mobilise attention, but they cannot end fragmentation at its root. Transformation requires three orientations: perceptual honesty, relational awareness and institutional humility. These are not new frameworks but shifts in seeing and meeting, from which different kinds of action can arise.

Perceptual honesty means recognising when images dominate action. For global leaders, this does not mean discarding goals or data, but seeing their partiality. Indicators, targets and rankings can highlight progress, yet they also mask realities when treated as substitutes for life itself. Perceptual honesty refuses to let images become reality. It holds data as signals, not as truths, and lets lived experience and direct observation correct the distortions of abstraction.

Relational awareness means meeting others without the filters of role, identity or category. For global leaders, this does not mean ignoring positions or interests, but refusing to let them exhaust the field of relationship. It points to spaces where dialogue suspends images and adversaries meet not as enemies or allies, but as human beings confronting a shared crisis (Bohm, 1996). In education, this means classrooms where teachers and students meet without the mediation of rank, stereotype or exam score. In peacebuilding, it means moving beyond treaties that manage conflict towards conversations that dissolve enemy images. Relational awareness restores the human face to governance, where cooperation arises not from conditional interests but from the perception of shared being.

Institutional humility means recognising that no framework can substitute for direct human responsibility. For global leaders, this does not mean abandoning frameworks like the SDGs, but seeing their limits. Indicators, treaties and targets can mobilise cooperation, but they cannot substitute for perception of the whole. Guidance here is not another framework but an orientation: to approach negotiations as encounters, not transactions; to hold frameworks lightly, as tools for coordination rather than substitutes for reality; and to create spaces for direct dialogue where assumptions and images can be suspended, allowing common ground to emerge (Bohm, 1996).

This is not another roadmap or method. It is an orientation, a way of seeing and meeting that can open the door to action beyond repetition, and to a transformation that begins not with images but with awareness. And even this orientation cannot be adopted as a new framework or borrowed light. To look to another for guidance is still the same movement of the mind that seeks certainty in structures rather than learning directly. Once this dependence is seen, and the movement of seeking ends, one becomes a light to oneself, responsible, attentive and capable of action rooted in perception, not projection.

From such awareness, action naturally follows: to see the problem at its root, to create a plan, to share and act. But here the actor is not a framework, a role or a rule; it is the living intelligence of the moment. Consider a classroom. A teacher may enter with a lesson plan for mathematics, but if the child is distracted, the plan becomes ineffective the moment the teacher clings to it. Even a framework that prescribes 'be compassionate' becomes an idea, and when one tries to follow it, there is only imitation of compassion, not compassion itself. Humanity has carried values of nonviolence and compassion for centuries, yet when they remain ideas, they act mechanically and cannot transform relationship.

When there is direct meeting, awareness gives knowledge and tools their right place. Lesson plans, policies and frameworks may still be used as tools, but they do not dominate. Where there is awareness, one sees which tool matters in the moment, and if any tool is needed at all. Without awareness, tools become baggage: a hammer carried everywhere, even when someone simply needs a gentle touch. To be attached to a tool is to impose it regardless of the situation, turning it into a limitation rather than a support. Awareness frees action from this rigidity. Life is always new, and so action must be alive to the new. In such freedom, action is not dictated by rules or bound by images, but moves with the living reality of each moment.

The persistence of global conflict, inequality and ecological destruction cannot be understood without recognising the role of images. They organise identity, sustain rivalry and mask reality with aspiration. Images promise unity but perpetuate division.

To move beyond this continuity of division requires more than reforming institutions or reframing goals. It requires a revolution in perception, to see images as images, to end their hold over relationship and to meet life directly.

Only then can family, community, nation and world be woven together not by projection, but by genuine encounter.

4.7 The Mirror of Governance and Life: Awareness Beyond the Self

The self, in seeking security and order, constructs outer systems that mirror its own inner movement. The same thought that divides the world within divides it without. Governance, education, economy, gender relations and humanity's relationship with nature all reveal this pattern: the desire for control born of fear, the pursuit of order through domination and the creation of frameworks that reflect the mind that built them.

In governance, this appears as bureaucratic and legal control that promises order but breeds conformity. Reform follows failure, yet each reform begins from the same ground of division. Education reproduces the same pattern: the teacher projects an ideal student, the student an image of success or failure and learning becomes imitation. In the economy, growth and competition mirror the same comparison that drives the individual mind. Gender inequality and ecological destruction arise from the same inner fragmentation, the idea of separation between man and woman, humanity and nature. Even the SDGs, though noble, often treat interrelated crises as separate silos. Dashboards of progress mirror the very fragmentation they seek to end. The outer world is the inner made visible. Reforming the mirror does not change the reflection.

If knowledge perpetuates division, awareness ends it. Awareness is not an extension of knowledge or a practice to be cultivated. It arises when thought is seen in operation without condemnation or justification. Krishnamurti (1969) called this choiceless observation; Heidegger (1962) described it as truth unconcealed, where reality reveals itself without mediation. Awareness is not achievement but perception, an intelligence beyond effort.

When awareness is present, relationship ceases to be governed by image. Communication becomes communion. In education, awareness replaces comparison with inquiry. In governance, it prevents action from being rooted in identity or fear. Awareness reveals thought as thought, image as image and fear as fear. In that seeing, the false falls away. What remains is intelligence, not of the thinker, but of life itself.

Violence persists because we try to end it through the same divided mind that creates it. The self that seeks peace sustains conflict, for becoming itself is struggle. Awareness is the ending of becoming, the ending of the self as centre. From that ending, love acts.

The SDGs can still serve as mirrors, not blueprints. They reveal the consciousness that produced them. The essential question is not what new framework we should design, but from where we act. Knowledge, conditioned by the past, cannot end the disorder it describes. Awareness ends division at its root by revealing the mind's movement in real time. In that stillness, action is whole.

And yet, even this seeing, has to be ended, otherwise if remembered, becomes memory. Experience becomes memory, and memory becomes knowledge, giving continuity to the self and authority to the past. From this continuity arise belief, ideology and order. The next chapter turns to this movement of knowledge, showing how the very instrument we rely upon for progress becomes the mechanism through which the self sustains itself and division continues.

4.8 Conclusion

The Ending of the Self and the Emergence of Awareness The inquiry into the self reveals that violence is not separate from the perceiver of violence. Every attempt to control, reform or transcend the self becomes another expression of its continuity. The self divides experience into observer and observed, and within that division, conflict perpetuates itself. No ideology, method or reform can end this cycle because they all operate from the same root – the movement of becoming. When the structure of the 'me' is seen completely, without choice or escape, its continuity ends naturally. In that ending, awareness is not cultivated; it simply is. Such awareness has no centre and therefore no conflict. It is from this silence that true relationship and order emerge. The next chapter turns to the movement of knowledge itself, to see how thought sustains continuity even when the self appears to have ended.

References

Anderson, B. (1991). *Imagined communities: Reflections on the origin and spread of nationalism* Rev. ed.. Verso.

Bohm, D. (1992). *Thought as a system*. Routledge.

Bohm, D. (1996). *On dialogue*. Routledge.

Bourdieu, P. (1991). *Language and symbolic power*. Harvard University Press.

Clark, M. S., & Hogue, M. (2020). Trust in close relationships. *Annual Review of Psychology*, *71*(1), 361–384. https://doi.org/10.1146/annurev-psych-010419-050754

Damasio, A. (2010). *Self comes to mind: Constructing the conscious brain*. Pantheon.

Deb, S., Strodl, E., & Sun, J. (2015). Academic stress, parental pressure, anxiety and mental health among Indian high school students. *International Journal of Psychology and Behavioral Sciences*, *5*(1), 26–34.

Festinger, L. (1954). A theory of social comparison processes. *Human Relations*, *7*(2), 117–140. https://doi.org/10.1177/001872675400700202

Foucault, M. (1975). *Discipline and punish: The birth of the prison*. Pantheon.

Foucault, M. (1977). *Discipline and punish: The birth of the prison* (A. Sheridan, Trans., p. 227). Vintage, 1979.

Foucault, M. (1980). *Power/knowledge: Selected interviews and other writings, 1972–1977*. In C. Gordon (Ed.), Pantheon.

Freud, S. (1921/2010). *Group psychology and the analysis of the ego*. Martino Publishing. (Original work published 1921).

Freud, S. (1930/2010). *Civilization and its discontents* J. Reddick, Trans.. W. W. Norton. (Original work published 1930).

Girard, R. (1977). *Violence and the sacred* P. Gregory, Trans.. Johns Hopkins University Press.

Guilmoto, C. Z., & Saikia, N. (2018). Excess under-5 female mortality across India: A spatial analysis using 2011 census data. *Lancet Global Health*, *6*(6), e650–e658. https://doi.org/10.1016/S2214-109X(18)30184-0

Gupta, D. (2024). *Eternal movement*. Emerge Publications. https://doi.org/10.65169/q7d2z6

Gupta, D. (2025). *Governance without the self: Consciousness and the end of fragmentation, systems, and complexity*. Emerge Publications. https://doi.org/10.65169/a9m3p4

Heidegger, M. (1962). *Being and time* J. Macquarrie & E. Robinson, Trans.. Harper & Row. (Original work published 1927).

Jha, P., Kumar, R., Vasa, P., Dhingra, N., Thiruchelvam, D., & Moineddin, R. (2006). Low male-to-female sex ratio of children born in India: National survey of 1·1 million households. *The Lancet*, *367*(9506), 211–218.

Khan, Y. (2007). *The great partition: The making of India and Pakistan*. Yale University Press.

Korzybski, A. (1933/1994). *Science and sanity: An introduction to non-Aristotelian systems and general semantics* (5th ed.). Institute of General Semantics.

Krishnamurti, J. (1954). *The first and last freedom*. Harper.

Krishnamurti, J. (1969). *Freedom from the known*. Harper.

Krishnamurti, J. (1974/1991). *The awakening of intelligence*. Harper.

Krishnamurti, J. (1983). *Krishnamurti's notebook*. Harper.

Krishnamurti, J. (1991). *The awakening of intelligence*. HarperCollins.

Mamdani, M. (2001). *When victims become killers: Colonialism, nativism, and the genocide in Rwanda*. Princeton University Press.

McGilchrist, I. (2019). *The master and his emissary: The divided brain and the making of the Western world* (2nd ed.). Yale University Press.

Mearsheimer, J. J. (2001). *The tragedy of great power politics*. W. W. Norton.

Merleau-Ponty, M. (1962). *Phenomenology of perception* C. Smith, Trans.. Routledge & Kegan Paul. (Original work published 1945).

Nietzsche, F. (1887/1994). *On the genealogy of morals* W. Kaufmann, Trans.. Vintage. (Original work published 1887).

Oliver, T. (2018). *Brexit and British politics*. Oxford University Press.

Piketty, T. (2014). *Capital in the twenty-first century* A. Goldhammer, Trans.. Harvard University Press.

Raworth, K. (2017). *Doughnut economics: Seven ways to think like a 21st-century economist*. Chelsea Green Publishing.

Redekop, V. N. (2008). *From violence to blessing: How an understanding of deep-rooted conflict can open paths to reconciliation*. Novalis.

Rockström, J., Steffen, W., Noone, K., Persson, Å., Chapin, F. S., III, Lambin, E., Lenton, T. M., Scheffer, M., Folke, C., Schellnhuber, H. J., Nykvist, B., de Wit, C. A., Hughes, T., van der Leeuw, S., Rodhe, H., Sörlin, S., Snyder, P. K., Costanza, R., Svedin, U., … Foley, J. (2009). Planetary boundaries: Exploring the safe operating space for humanity. *Ecology and Society*, *14*(2), 32. http://www.ecologyandsociety.org/vol14/iss2/art32/

Singer, T., & Klimecki, O. M. (2014). Empathy and compassion. *Current Biology*, *24*(18), R875–R878. https://doi.org/10.1016/j.cub.2014.06.054

Singh, N., & Bhatnagar, D. (2023). *Applied spirituality and sustainable development policy*. Emerald Publishing.

SIPRI. (2024). *Trends in world military expenditure, 2023 (SIPRI fact sheet, April 2024)*. Stockholm International Peace Research Institute.

Twenge, J. M. (2017). The costs of overprotecting the young-iGen: Why today's super-connected kids are growing up less rebellious, more tolerant, less happy-and completely unprepared for adulthood-and what that means for the rest of us. *Journal of Psychology*, 1–13.

UN Women. (2023). *Progress on the sustainable development goals: The gender snapshot 2023*. United Nations.

Whitehead, A. N. (1929/1978). *Process and reality* Corrected ed.; D. R. Griffin & D. W. Sherburne, (Eds.) Free Press.

Zuboff, S. (2019). *The age of surveillance capitalism: The fight for a human future at the new frontier of power*. PublicAffairs.

Chapter 5

The Trap of Knowledge and Belief

Abstract

Chapter 5 explores how knowledge, belief and authority sustain fragmentation by replacing perception with conclusion. It distinguishes between understanding as direct seeing and knowledge as accumulated memory. This chapter examines how institutions and belief systems transform living awareness into doctrine, converting curiosity into conformity. Knowledge becomes power, and belief becomes identity, giving continuity to the self. Even the pursuit of spirituality or science can become an extension of conditioning when guided by motive or method. The analysis reveals that knowledge cannot end confusion because confusion lies in the knower. When the mind sees the limits of its own knowing, it gets silent and that silence is understanding. This chapter closes by suggesting that transformation is not the accumulation of insight but the ending of dependence on the past. Awareness, not belief, becomes the foundation for right action.

Keywords: Knowledge trap; belief systems; authority critique; ideological conditioning; perception vs memory; institutional dogma; direct seeing; intelligence as awareness

5.1 Introduction

Chapter 4 showed that violence is woven into the structure of the self, sustained by separation, comparison, fear, desire and image. One element completes this picture: the role of knowledge and belief. Fear and desire generate conflict; images sustain division; but it is knowledge and belief that stabilise these fragments, granting them authority, continuity and institutional force.

Knowledge is indispensable in its right place. Without it, language, craft and science would be impossible. Yet when knowledge crosses from the practical to the psychological, it distorts. Knowledge is the past, the residue of memory filtering the present through what has been. When thought looks at the living

From Fragmentation to Wholeness, 121–132

doi:10.1108/978-1-80686-677-920261005

through memory, it does not meet it freshly; it meets an image. This is why Krishnamurti insisted that 'knowledge is always old' (1954): it cannot meet the present now. Gupta (2024) extends the point: even noble ideals like non-violence, compassion, when held as knowledge to become, belong to the same movement of violence, for the self that strives to become non-violent is still centred in becoming.

Belief compounds the trap. If knowledge is memory's residue, belief is memory hardened into certainty. Where knowledge may yet be revised, belief demands allegiance. It turns provisional descriptions into absolutes, defended as identity. Belief promises security, but it divides. It generates competing 'truths' as doctrines, ideologies, dogmas each partial, each defended. Nietzsche called such truths 'illusions we have forgotten are illusions' (1873/1994). Freud showed how groups bind themselves to shared beliefs to secure cohesion, suppressing dissent (1921/2010). Belief systems often stabilise communities by scapegoating another. To defend belief is to defend the self; and in that defence, conflict is inevitable.

Institutions built on knowledge and belief reproduce this pattern. Education becomes transmission rather than awakening. Governance becomes application of frameworks rather than perception of the whole. Sustainability is administered through dashboards and indicators even as ecological breakdown deepens.

This chapter examines the trap of knowledge and belief in seven steps.

- 5.2: Knowledge as memory and fragment – how it conditions perception and divides life and knowledge not merely as a lens, but as the very structure of the observer.
- 5.3: Belief – how fragments harden into identity.
- 5.4: Knowledge institutionalised as power – how it sustains exclusion and inequality.
- 5.5: The limits of knowledge in education, governance and sustainability – why frameworks manage but cannot transform.
- 5.6: The possibility of understanding – how it differs from knowledge and belief.
- 5.7: Conclusion and transition to relationship, memory and image in the next chapter.

The guiding question is simple but radical: Can knowledge and belief, indispensable in their proper domain, ever touch truth? Or does truth disclose itself only when knowledge and belief are seen as they are: limited, fragmentary and incapable of ending disorder?

5.2 Knowledge as Memory and Fragment

Knowledge is indispensable for human life. Without it, language could not be spoken, skills could not be learned and science could not advance. In its right place, knowledge makes survival and communication possible. But this indispensability conceals its danger. For knowledge is always of the past. It is the residue of experience stored in memory. When carried into the psychological realm, it ceases to be a practical tool and becomes a distorting lens. Knowledge divides the living present through the filter of what has been.

Hence, knowledge can never perceive freshly, because it is built out of memory. Looking at another through knowledge, through past hurt or pleasure, means meeting not the living person but an image constructed by thought. Eternal Movement (Gupta, 2024) makes the point more radical: even ideals like non-violence, when held in thought, are still part of violence. The 'me' that says, '*I will be non-violent*', merely projects itself into a new form, sustaining division rather than ending it.

David Bohm (1992) explained how thought carves life into categories as friend and enemy, past and future, success and failure. These fragments are then mistaken for separate realities and feed back into perception. Alfred Korzybski (1933/1994) expressed the same insight with his dictum '*the map is not the territory*'. Knowledge is always a representation, yet human beings mistake the abstraction for reality itself.

Neuroscience reinforces this. Damasio (2010) shows that the brain constructs an 'autobiographical self' by integrating memory, bodily states and narrative. This provides continuity but also filters every perception through the past. McGilchrist (2019) distinguishes between the hemispheres: the left isolates fragments for manipulation, while the right perceives context and wholeness. Both are necessary, but when the fragmentary mode dominates, knowledge usurps the place of perception. The result is a civilisation that privileges abstraction over relationship, measurement over meaning and control over coherence.

The limitation becomes stark in governance and conflict resolution. Trade negotiations and peace treaties can redistribute power or suspend hostilities, but if the root fragmentation in human beings remains untouched, these mechanisms become tools for managing division rather than ending it. Agreements are provisional, shaped by self-interest and ultimately fragile. The Palestine–Israel peace process illustrates this paradox: knowledge creates frameworks of order, but that order is temporary and born of the same division it claims to resolve.

Knowledge has its right place in the technical realm like flying a plane, conducting surgery, remembering an address or drafting a treaty, but in the psychological realm it is always divisive. When the past dictates relationships or action, the mind is never free to see. The same applies at the systemic scale: frameworks designed from accumulated knowledge may reform symptoms but inevitably reproduce the same fragmentation that created the need for reform.

Thus, knowledge can describe disorder, but it cannot end it. It can explain violence, but it cannot bring peace. Description belongs to memory, and memory belongs to the past. Disorder ends only in the present, in the direct seeing of what is. Any skill, framework, best practice or value, when applied by a fragmented mind will perpetuate fragmentation. Because the root of fragmentation lies in knowledge itself, no tool of knowledge can finally resolve it.

Knowledge as the Observer

To see the trap of knowledge more deeply, it is not enough to say that knowledge distorts perception. Knowledge does not merely stand outside as a lens; it is the very structure of the observer.

Krishnamurti (1983) insisted that '*the observer is the observed*'. What we call the 'me' that looks, judges and interprets is nothing apart from the content of consciousness. It is memory, images, fears and desires, organised by thought into a sense of identity. When the observer claims to stand apart, it is only thought dividing itself and projecting that division outward. Gupta (2025) warns that even the pursuit of ideals like non-violence sustains violence, because the self that seeks to change is itself violent. Instead of ending what one is, the mind projects ideals of non-violence, love, God and continues the same movement of violence.

The insight here is stark: the self is not separate from its content but is that very content. Fear, desire, habit and conditioning are not attributes of the self; they constitute it. Thus, when one says, '*I know I am violent, and I will become non-violent*', the 'I' who knows is itself knowledge sustaining violence. Fear is not something the self has; the self is fear when it arises.

Bohm (1992) explained this mechanism: thought projects fragments and then mistakes them for external facts. Merleau-Ponty (1962) described perception as embodied and temporal, never reducible to an observer standing outside life. Heidegger (1962) likewise pointed out that being is inseparable from perception; the idea of an observer apart from the observed is itself an illusion. Even the so-called 'silent witness' is a construct of memory. Concepts of soul, Brahman or God may sound noble, but they remain ideas generated by thought. True silence is only when all ideas end.

This recognition has radical implications. Every attempt to reform oneself or society through knowledge ends in continuity. The observer that seeks to change is the very structure it seeks to escape. Transformation begins only when the observer ends, leaving perception unmediated by memory or image. Such perception is awareness: direct seeing in which division ends and knowledge returns to its rightful place, as a technical tool, not the ground of identity.

The same distinction applies collectively. Frameworks, treaties and strategies are useful tools, but when mistaken for reality they perpetuate division. Without awareness, even the most refined knowledge becomes baggage, binding action to the past. With awareness, tools are applied where they serve life and set aside where they limit. Life is always new, and in that newness knowledge takes its proper place.

5.3 Belief as Hardened Knowledge

If knowledge is the residue of memory, belief is knowledge hardened into certainty. Where knowledge may still admit revision, belief demands allegiance. It transforms fragments into absolutes, defended as identity. Belief promises security, but in practice it sustains division.

Belief is never discovery; it is always a conclusion rooted in memory, fear or desire. People believe to feel secure, to belong or to maintain continuity. Yet this comfort conceals its danger: belief resists observation and replaces inquiry with certainty. Nietzsche (1873/1994) captured this when he wrote that what we call truths are 'illusions which we have forgotten are illusions'. Freud (1921/2010) showed how groups bind themselves through shared beliefs, suppressing dissent in the name of loyalty. Girard (1977) argued that belief systems stabilise communities by channelling rivalry towards scapegoats. To defend belief is to defend identity; in that defence, conflict is inevitable.

Beliefs about Brahman, heaven or enlightenment may offer comfort, but they remain images in thought, second-hand constructs of memory that sustain division rather than dissolve it. Truth cannot be stored as belief; it must be discovered directly, moment to moment.

This hardening process appears everywhere. Religious enquiries become dogma and persecution. National sentiment congeals into ideology that legitimises conquest. Political theories: fascism, communism, neoliberalism become total explanations that silence complexity. Each begins as interpretation, then calcifies into identity.

In personal life, belief distorts relationship. A parent who believes 'my child must succeed' no longer meets the child as a living being but as an extension of ambition. A person who believes 'I am unworthy' or 'I am superior' lives inside an image reinforced by memory. Belief becomes a prison, fixing identity in rigid forms.

The same is true of institutions and frameworks. Curricula become dogmas; national myths justify exclusion; development frameworks, including the Sustainable Development Goals (SDGs), are treated not as provisional tools but as articles of faith. Such frameworks, when treated as unquestionable, become identities in themselves and turn reform into continuity.

This is not to deny the value of monitoring, reviews or indicators. They generate accountability, cooperation and data. But when signals are mistaken for reality, they harden into belief. A country may show gains on SDG 4 (education) via enrolment while classrooms remain governed by fear and rote. SDG 13 (climate) can report commitments while fossil-fuel subsidies persist. Without awareness of these contradictions, assessments risk becoming rituals of legitimacy.

Dependence on frameworks dulls communion with what is. Frameworks are assembled from steps and parts; life does not unfold in prearranged sequences. The actor who clings to method cannot discern what this moment requires, because they are attached to procedure rather than awake to the living problem. The result is performance, 'doing the framework', instead of perception.

The SDGs illustrate these limits. SDG 16 may record fewer battlefield deaths while militarisation and enemy images intensify; SDG 5 may expand women's representation while patriarchal attitudes persist at home and work; SDG 10 may redistribute statistically while caste and racial hierarchies continue to shape daily life. These are not monitoring glitches but signs of frameworks operating from belief, not awareness.

And where crises are overtly political or violent, the gap is sharper. Which SDG target resolves Gaza? Which framework compels Responsibility to Protect when a veto blocks action? Refugee law grants rights yet perpetuates the division of citizen and non-citizen. Humanitarian law protects civilians yet normalises 'rules of war', managing violence rather than ending it. Frameworks coordinate and restrain, but they cannot substitute for the inner ending of fear, desire and image.

When frameworks are pursued for validation, funding or the prestige of global forums, they become symbols rather than transformations, colourful circles on banners that comfort the self while leaving its fragmentation untouched. The critique is not of tools but of the mind that clings to them. When belief is seen as the movement of memory and identity, its hold ends. What remains is awareness: choiceless, undivided, free of allegiance. In such awareness, relationship is no longer bound by belief, and action arises from wholeness rather than ideology.

5.4 Knowledge and Power

Knowledge, once hardened into belief, rarely remains personal. It enters institutions, doctrines and policies, becoming a structure of power. To define what counts as truth is to decide who is included, who is excluded and how life is organised. Knowledge ceases to be description and becomes regulation. It not only reflects reality but enforces the categories through which reality is perceived.

Foucault (1980) argued that knowledge and power are inseparable. Knowledge functions as a regime of truth, legitimising authority through categories such as 'normal' and 'deviant', 'developed' and 'underdeveloped', or 'citizen' and 'migrant'. Once embedded, these categories organise institutions, policies and punishments. They appear natural, yet they are instruments of control.

Bourdieu (1991) called this symbolic violence: the imposition of meanings that sustain inequality while concealing their constructed nature. A caste label, racial category or educational credential does not merely describe but orders hierarchy. Its power lies in invisibility: people forget it is constructed and accept it as reality.

Education offers a clear illustration. Paulo Freire (1970) described the 'banking model' in which knowledge is deposited into students as unquestionable truth. What appears as education is indoctrination, reproducing the worldview of the dominant order. When knowledge is mistaken for truth, it creates relative realities of 'heaven', 'progress', 'peace' but doesn't actually transform.

In governance and sustainability, this dynamic is amplified. Frameworks built from accumulated knowledge mirror the self's fragmentation. They may reform symptoms but leave untouched the deeper roots of fear, desire and identity. As a result, frameworks become self-limiting: they describe disorder but cannot end it. Even innovations framed as 'awareness' or 'competence' are often reduced to measurable skills or checklists. When awareness itself becomes another form of knowledge to acquire, its transformative essence is lost and institutions reinforce continuity under the guise of change.

The digital age sharpens this entanglement. Zuboff (2019) shows how surveillance capitalism transforms personal data into systems of prediction and control. Knowledge is commodified, behaviour rendered calculable and autonomy reduced to patterns of consumption. Knowledge, instead of illuminating, becomes economic violence.

Case studies make this vivid. In Israel and Palestine, textbooks embed competing narratives, each legitimising one side while denying the other (Podeh, 2002). In Afghanistan, successive regimes from Soviet occupation to the Taliban to US-backed governments rewrote curricula to secure loyalty rather than inquiry. Textbooks shifted from Marxist-Leninist ideology to religious orthodoxy to liberal-democratic ideals (Burde, 2014). What appeared as education was in fact the shaping of identity to mirror the ruling order.

The danger lies not only in the misuse of knowledge but in mistaking knowledge for truth. When categories are treated as reality and frameworks as transformation, power and violence hide in plain sight. Knowledge institutionalised as power is not truth; it is belief embedded in systems, shaping how people see themselves and others, often without awareness.

The insight is clear: as long as knowledge is confused with truth, institutions will mirror the fragmented, defensive and violent self. Clarity cannot result from better frameworks; it is the precondition for right action. Without this recognition, knowledge as power may sustain order but at the cost of perpetuating division.

5.5 The Limits of Knowledge in Education, Governance and Sustainability

If knowledge and power are inseparable, then their limits are revealed most clearly in the very institutions that claim to advance human progress. Education, governance and sustainability all rely on knowledge to define goals, measure outcomes and legitimise authority. Yet in each domain, knowledge proves inadequate to transformation, for it manages symptoms while leaving the root disorder untouched.

Education

Education is celebrated as the foundation of equality and empowerment. Yet in most systems, what passes for education is the transmission of information, the accumulation of knowledge. Students are filled with facts, ranked against one another and judged by examinations. This appears as progress: higher enrolment numbers under SDG 4 and expanded access to schooling, but beneath the surface, it deepens conformity and fear.

Freire (1970) described this as the 'banking model' of education, where students are treated as passive receptacles. Knowledge is not truth, and the repetition of facts can never awaken intelligence. Even noble ideas such as 'peace' or 'non-violence', when carried as ideals, remain within the movement of violence.

The self that holds the ideal is the same self that sustains division. Education, when reduced to accumulation, therefore reproduces the psyche of comparison, fear and becoming.

Governance

Governance depends heavily on knowledge codified into frameworks, policies and treaties. Yet these frameworks, however sophisticated, are drawn from memory and identity. Peace agreements often codify divisions by fixing ethnic or religious categories. Development policies perpetuate dependency by defining communities through lack. Laws that define 'citizen' and 'non-citizen' institutionalise exclusion.

Frameworks built on accumulated knowledge reform structures but leave untouched the consciousness of fear and identity that shapes them. They describe crises but cannot end them, because they arise from the same divided psyche. As long as knowledge is mistaken for understanding, governance remains trapped in cycles of reform without transformation.

Sustainability and the SDGs

Sustainability frameworks such as the SDGs demonstrate this trap with clarity. They translate complex realities into dashboards and indicators. Poverty, health, education, gender equality and climate are treated as separate goals, mirroring the self's fragmentation into parts. Progress is celebrated when numbers improve, even as ecosystems collapse, militarisation expands and inequality deepens.

Beck (1992) described modernity as a 'risk society', where attempts to manage crises create new risks. Capra (1996) warned that sustainability framed in mechanistic terms perpetuates the very fragmentation it seeks to heal. Singh and Bhatnagar (2023) argue that sustainability policy often functions as a 'wicked problem', where competing values and fragmented interests prevent coherence (Rittel & Webber, 1973). The SDGs embody these contradictions: they manage symptoms but cannot touch the root, for they are designed from the same fragmented consciousness that created the crisis.

The Instrumentalisation of Awareness

Even frameworks that aim to move beyond traditional policy, such as the Inner Development Goals, Theory U or regenerative governance, often fall into the same paradox. They reduce awareness to a competency, a skill to be learnt, measured and performed. Yet the moment awareness is instrumentalised in this way, it is no longer awareness. It becomes another form of knowledge and time: a method to practice, a capacity to cultivate, a benchmark to achieve. What is presented as transformation is in fact continuity, the self reasserting itself through subtler forms of becoming.

As has been noted elsewhere, 'awareness is not an attribute. It is the ending of the self, not the refinement of it'. Theory U illustrates this contradiction vividly. While it gestures towards silence and presence, it treats them as steps in a process – 'downloading, sensing, presencing, crystallizing'. In doing so, it assumes that awareness can be organised, sequenced and achieved. But to map stillness is already to move away from it; to treat perception as a path is to turn it into becoming. Regenerative governance frameworks reveal the same crisis in new language. They embrace storytelling, indigenous wisdom and ecological restoration, but often as tools to be professionalised, networked and certified. A 'regenerative elite' emerges, projecting healing into the future as a journey of development. In this way, regeneration itself becomes another performance of the old mind, carrying the continuity of identity, only in more soulful forms.

The deeper issue is not the sincerity of these frameworks, but the fact that they remain grounded in the structure of the self. When presence becomes performance, fragmentation is subtly reinforced. Awareness is not time, method or measurement. It is direct seeing, and in that is the ending of the self as observer. Without this, even the most well-meaning frameworks risk becoming institutional theatre: rearrangements of knowledge mistaken for transformation.

The Crisis of Knowledge

This reveals what some scholars have described as a crisis of knowledge (Gupta, 2025). Humanity faces not only ecological collapse, inequality and institutional failure, but the deeper crisis that the very tools we use to interpret and address these problems: our knowledge systems, frameworks and indicators are themselves part of the disorder. Knowledge, when mistaken for truth, sustains division even as it claims to resolve it. Frameworks become relative realities: dashboards, competencies and indexes that give the appearance of progress while deepening the underlying fragmentation. This is why, no matter how sophisticated our reforms, disorder persists. The crisis is not only in the world, but in the mind that seeks to reform the world through knowledge.

Implications

The lesson is not that knowledge should be discarded, but that it must be seen in its right place. Technical knowledge is essential, including medicine, engineering and ecological monitoring. But in the psychological and relational realm, knowledge is inadequate. When treated as truth, it hardens into belief; when embedded in institutions, it sustains power and exclusion.

The limits of knowledge, therefore, are not technical but existential. Knowledge can describe disorder, but it cannot end it. It can measure poverty, but not dissolve comparison. It can quantify emissions, but not transform desire. It can set targets, but not awaken care. Transformation begins only when knowledge is seen for what it is – useful in its domain, but fragmentary in the psyche. Without this recognition, education, governance and sustainability will continue to mirror the same fragmentation they attempt to heal.

5.6 The Possibility of Understanding

If knowledge fragments and belief hardens those fragments into identity, then is there another ground from which to act? The possibility lies in understanding. But understanding must be distinguished from knowledge. It is not an extension or refinement of knowledge. It does not accumulate, analyse or conclude. Understanding belongs to a different order.

Knowledge arises from memory and past experience; understanding arises in the immediacy of perception. One may know the idea that 'anger is destructive', yet such knowledge does not end anger. Only when anger is observed directly as it arises, without justification, suppression or escape, does understanding take place. In that seeing, the movement of anger ends.

Bohm (1996) made a similar point in his reflections on dialogue: when assumptions are suspended and participants listen without defending images, meaning flows not from conclusions but from shared perception. Understanding here is not static; it is a movement of insight that cannot be stored in memory.

Whitehead (1929/1978) described reality as process rather than substance. Understanding perceives this movement, while knowledge freezes it into categories.

The implications are profound. In education, understanding means learning as inquiry, not accumulation, with classrooms where teachers and students meet without fear or comparison. In governance, it means decisions shaped not by inherited categories or policy inertia but by direct perception of the whole situation. In sustainability, it means seeing poverty, inequality and ecological destruction not as separate problems but as expressions of one disorder in consciousness.

Understanding cannot be produced by method or sustained by will. It cannot be stored as memory or codified as belief. It arises only in direct observation, when the mind sees without separation. When the false falls away, what remains is true.

5.7 Conclusion

This chapter has explored the trap of knowledge and belief. Knowledge, though indispensable in the technical realm, becomes divisive in the psychological realm. As memory, it conditions perception through the past. As fragment, it divides life into categories mistaken for reality. As belief, it hardens those fragments into identity and certainty. As power, it enters institutions and frameworks that sustain inequality and exclusion. Across education, governance and sustainability, the same pattern is visible: knowledge manages, measures and reforms, but it cannot transform.

The central insight is radical in its simplicity: knowledge is not truth. Knowledge is always old, the residue of experience. Truth is alive, discovered only in direct perception. To mistake knowledge for truth is to live in relative realities, acting from abstractions mistaken for the whole. Korzybski's (1933/1994) dictum that 'the map is not the territory' captures this clearly. Neuroscience reinforces this: Damasio (2010) shows how the autobiographical self filters every perception through memory.

Krishnamurti (1991) illustrated this when he said that the 'I' who claims to be afraid is not separate from fear, the 'I' is fear. Similarly, the self that aspires to non-violence is itself violent. Ideals of God, peace or love remain projections of thought, escapes from what is. As he noted in dialogue with Anderson (1974/2000, Dialogue 2), the ideal is a flight from reality, and the observer who creates the ideal is the very movement of disorder. In this way, the pursuit of ideals sustains the very violence or fear it seeks to overcome.

Belief compounds this trap. What begins as provisional knowledge hardens into certainty, defended as identity and imposed through institutions. Belief promises security but perpetuates division. Violence that follows, from family to nation, is not accidental but inevitable.

Humanity's condition may therefore be described as a crisis of knowledge. We face ecological breakdown, inequality and institutional fragility, but beneath them lies a deeper crisis: the very tools we use to interpret and address problems: knowledge systems, frameworks and indicators are themselves part of the disorder. Knowledge can describe violence but not end it, measure risk but not dissolve fear, project sustainability targets but not transform the mind that sustains fragmentation. Unless this crisis of knowledge is seen directly, disorder will persist, regardless of reform.

Yet there is another ground. Understanding is not knowledge refined but perception unmediated by knowledge. It arises when thought is seen as thought, image as image, belief as belief, fear as fear. In such direct seeing, the false falls away and what remains is true. This is not a method or path, but a quality of attention in the present. From there, the possibility of understanding and awareness opens.

This sets the stage for Chapter 6, which turns to relationship, memory and image. For it is in relationship that knowledge and belief most directly shape human life, and it is in relationship that the possibility of understanding can be lived.

References

Beck, U. (1992). *Risk society: Towards a new modernity* M. Ritter, Trans.. Sage.

Bohm, D. (1992). *Thought as a system*. Routledge.

Bohm, D. (1996). *On dialogue*. Routledge.

Bourdieu, P. (1991). *Language and symbolic power* (J. B. Thompson, (Ed.), G. Raymond & M. Adamson, Trans.). Harvard University Press.

Burde, D. (2014). *Schools for conflict or for peace in Afghanistan*. Columbia University Press.

Capra, F. (1996). *The web of life: A new scientific understanding of living systems*. Anchor Books.

Damasio, A. (2010). *Self comes to mind: Constructing the conscious brain*. Pantheon.

Foucault, M. (1980). *Power/knowledge: Selected interviews and other writings, 1972–1977* (C. Gordon, Ed.). Pantheon.

Freire, P. (1970). *Pedagogy of the oppressed*. Continuum.

Freud, S. (2010). *Group psychology and the analysis of the ego* (J. Strachey, Trans.). Martino Publishing. (Original work published 1921).

Girard, R. (1977). *Violence and the sacred* (P. Gregory, Trans.). Johns Hopkins University Press.

Gupta, D. (2024). *Eternal movement*. Emerge Publications. https://doi.org/10.65169/q7d2z6

Gupta, D. (2025). *Governance without the self: Consciousness and the end of fragmentation, systems, and complexity*. Emerge Publications. https://doi.org/10.65169/a9m3p4

Heidegger, M. (1962). *Being and time* (J. Macquarrie & E. Robinson, Trans.). Harper & Row. (Original work published 1927).

Korzybski, A. (1994). *Science and sanity: An introduction to non-Aristotelian systems and general semantics* (5th ed.). Institute of General Semantics. (Original work published 1933).

Krishnamurti, J. (1954). *The first and last freedom*. Harper & Brothers.

Krishnamurti, J. (1983). *Education and the significance of life*. Krishnamurti Foundation India.

Krishnamurti, J. (1991). *The awakening of intelligence*. HarperCollins.

Krishnamurti, J. (2000). Dialogue 2. In J. Krishnamurti & A. W. Anderson (Eds.), *A wholly different way of living: Krishnamurti in dialogue with Professor Allan W. Anderson* (Vol. 1974, pp. 25–47). San Diego: HarperSanFrancisco.

McGilchrist, I. (2019). *The master and his emissary: The divided brain and the making of the Western world* (2nd ed.). Yale University Press.

Merleau-Ponty, M. (1962). *Phenomenology of perception* (C. Smith, Trans.). Routledge & Kegan Paul. (Original work published 1945).

Nietzsche, F. (1994). On truth and lie in an extra-moral sense. (D. Breazeale, Trans.). In D. Breazeale (Ed.), *Philosophy and truth: Selections from Nietzsche's notebooks of the early 1870s* (pp. 79–97). Humanities Press. (Original work published 1873).

Podeh, E. (2002). *The Arab–Israeli conflict in Israeli history textbooks, 1948–2000*. Bergin & Garvey.

Rittel, H. W. J., & Webber, M. M. (1973). Dilemmas in a general theory of planning. *Policy Sciences*, *4*(2), 155–169. https://doi.org/10.1007/BF01405730

Singh, N., & Bhatnagar, D. (2023). *Applied spirituality and sustainable development policy*. Emerald Publishing.

Whitehead, A. N. (1978). *Process and reality* (Corrected ed.; D. R. Griffin & D. W. Sherburne, Eds.). Free Press.

Zuboff, S. (2019). *The age of surveillance capitalism: The fight for a human future at the new frontier of power*. PublicAffairs.

Chapter 6

Relationship, Memory and Image

Abstract

This chapter explores how relationship, whether between individuals, communities or institutions is sustained by memory rather than perception. We meet not the other but our image of the other. The same distortion shapes families, classrooms and political life, where expectation and judgement replace awareness. The chapter reveals how dependence and attachment, born of insecurity, generate conflict and illusion. Through examples in education, governance and everyday interaction, it shows that relationship without image is not detachment but deep communion. When memory no longer dictates perception, relationship becomes a mirror in which the self is seen and dissolved. Such seeing gives rise to care that is not duty and love that is not possession. In that space of non-division, cooperation and compassion emerge naturally. The chapter deepens the movement of inquiry by showing that the ending of image is the birth of right relationship.

Keywords: Relationship dynamics; memory in perception; image-making; interpersonal distortion; educational awareness; governance and image; unconditional relationship; compassionate seeing

6.1 Introduction

The previous chapter showed how humanity is caught in a crisis of knowledge. Knowledge fragments, belief hardens fragments into identity and institutions built upon them reproduce division while appearing to resolve it. Yet nowhere are these dynamics more visible or more consequential, than in relationship. It is in the relationship between parent and child, student and teacher, citizen and state, one nation and another, that knowledge and image most directly shape human life.

From Fragmentation to Wholeness, 133–140

doi:10.1108/978-1-80686-677-920261006

Relationship, in its essence, is the encounter between human beings in the present. But in practice, relationship is almost always mediated by memory and image. We see one another not freshly, but through accumulated experiences, roles, expectations and projections. The child is seen as 'my son' or 'my daughter'; the neighbour as 'Hindu' or 'Muslim'; the other nation as 'ally' or 'enemy'. These images, sustained by memory, filter perception and perpetuate division.

Krishnamurti (1954) insisted that 'the observer is the observed', meaning that the image we hold of the other is inseparable from the self that holds it. Bohm (1996) added that thought creates these representations and then mistakes them for reality, ensuring fragmentation. McGilchrist (2019) has shown how the left hemisphere's dominance reinforces this: abstraction and categorisation override direct encounter. In this way, relationship is rarely with another human being but with an image, loaded with characteristics liked and disliked, and thus violence, subtle or overt, becomes inevitable.

This has profound implications. In the family, memory and image create patterns of expectation and control. In education, the student is seen through grades, ranks and categories rather than inquiry. In community, belonging is defined by inclusion and exclusion, often scapegoating those deemed outsiders. In nations, collective identities amplify these divisions, producing conflict at scale. Earlier insights in this inquiry note: 'when relationship is mediated by image, there is no meeting'. Governance, too, mirrors this: frameworks rooted in identity institutionalise separation into policy and law.

This chapter therefore explores how memory and image condition relationship, how they sustain division and violence, and whether it is possible to meet without them. Section 6.2 examines the role of memory in shaping identity and perception. Section 6.3 turns to image, showing how it distorts relationship by substituting projection for reality. Section 6.4 explores how these dynamics play out in education, community and governance, linking them to the Sustainable Development Goals (SDGs). Section 6.5 raises the radical question: is relationship possible without image? And Section 6.6 concludes by preparing the movement into Chapter 7, which examines psychological time and inertia as the deeper ground sustaining memory and image.

The central question of this chapter is not abstract but intimate: Can we meet one another as human beings, or only as roles, categories and projections? Is relationship ever direct or is it always mediated by the past?

6.2 Memory as the Ground of Relationship

Memory is the foundation upon which most human relationships are built. It shapes how we recognise one another, how we respond and how we sustain continuity over time. Without memory, there would be no language, no names, no history, no shared culture. Memory provides stability. Yet in the psychological realm, the same memory that enables recognition also conditions perception. It makes a relationship an encounter not with another human being in the present, but with the past.

Krishnamurti (1954) warned that when thought looks through memory, it does not see what is, but what has been. The friend is remembered through past affection, the enemy through past hurt, the partner through accumulated expectations. Thus, memory creates an image that becomes the ground of perception. Bohm (1996) explained this mechanism: thought generates representations and then mistakes them for reality, so that the actual person is never encountered directly.

Neuroscience confirms this dynamic. Damasio (2010) describes the autobiographical self as a continuous reconstruction of identity from past experiences, bodily states and narratives. Every interaction is filtered through this autobiographical memory. McGilchrist (2019) adds that the left hemisphere's dominance favours abstraction and categorisation, so that memory tends to freeze fluid encounters into fixed patterns. Memory, then, while indispensable for survival, distorts relationship by substituting record for reality.

This distortion becomes especially clear in intimate life. Parents may approach their child not as the living being before them, but as the continuity of past moments: the child who once pleased, disappointed or frightened them. Partners may meet each other through accumulated grievances or idealised recollections. In such cases, what is present is never truly seen. Krishnamurti observed that 'fear is you', pointing out that fear is not separate from the one who experiences it but is the very structure of the self (Dialogue 2, Krishnamurti & Anderson, 1991). Memory sustains fear by replaying old experiences, and in relationship that fear becomes the lens of meeting.

At collective scales, memory operates as history, tradition and identity. Nations remember their victories and wounds, sustaining narratives that shape policy and culture. Communities preserve stories of belonging and exclusion, which define insiders and outsiders. These memories provide continuity but also perpetuate division. Frameworks built on historical memory inevitably mirror the self's violence: they institutionalise separations into law, borders and categories.

Memory is therefore double-edged. It is necessary for practical continuity, yet destructive when it dominates relationship. To remember where one lives or what language one speaks is functional. But when memory dictates how we meet another human being, it becomes a barrier. The present is no longer seen directly. The question, then, is whether relationship is possible without the weight of memory. Can one meet freshly, without the past dictating perception?

6.3 Image and the Distortion of Relationship

If memory is the ground upon which relationships are built, then image is the structure through which they are sustained. An image is more than recollection; it is a projection, a psychological construct shaped by memory, desire, fear and expectation. Once formed, images mediate perception so that we rarely meet another human being directly. Krishnamurti (1954) emphasised that this mediation is the essence of conflict. The parent who says, 'my child must succeed', is relating not to the child but to an image of ambition projected onto the child. The partner who sees their spouse through the lens of disappointment or idealisation is relating to the image, not the living person. In each case, what is present is obscured by what thought has constructed.

Bohm (1996) described this as the 'system of thought', where abstractions take on a reality of their own and feed back into perception. The result is fragmentation: one sees not the whole, but a part mistaken for the whole. McGilchrist (2019) reinforces this, showing how the analytic mode of the brain reduces the living flow of relationship into fixed categories that quickly become rigid.

Images also perpetuate subtle violence. When an image is contradicted, when the child fails to live up to the parent's ambition or the partner deviates from expectation, conflict arises. The disappointment is not with the living person, but with the failure of an image. There is no true relationship when image mediates perception; there is only idea projecting itself.

At collective scales, images harden into stereotypes, roles and categories. Communities construct images of insiders and outsiders. Nations sustain images of 'enemy' and 'ally'. Media multiplies these images, reinforcing cycles of fear and desire. Bourdieu (1991) called this symbolic violence: the imposition of meanings that appear natural but legitimise inequality.

Even noble images can distort. The image of 'peace', when pursued as an ideal, can be a form of violence, as it often legitimises coercion or conformity in the name of harmony. The image of 'sustainability' can conceal ongoing exploitation under the appearance of progress. Krishnamurti emphasised that 'the description is not the described', underscoring that words, concepts and images are not reality itself but only representations (Dialogue 4, Krishnamurti & Anderson, 1991). To mistake the image of a person, or an ideal of peace, for the living fact is to perpetuate illusion.

The question, then, is whether relationship is possible without image. Can one meet another without the filter of past impressions, categories or ideals? To encounter without image would mean to meet directly, in the present, without mediation. Krishnamurti (1991) proposed that such relationship is possible only when the mind sees the operation of image as it arises, and in that seeing, the image loses its hold. Awareness of image is already the ending of its dominance.

6.4 Memory, Image and Relationship in Education, Community and Governance

Education

Education is one of the clearest sites where memory and image dominate relationship. Teachers approach students not as unique beings in the present but as categories: 'bright', 'weak', 'average'. Students meet teachers as figures of authority rather than partners in inquiry. Grades and rankings create images of success and failure that students internalise as self-image: 'I am capable', 'I am worthless'. These images, once formed, condition the whole trajectory of relationship to learning.

Krishnamurti (1983) argued that learning without comparison is possible only when the mind is free from image. Yet education systems worldwide normalise image as the currency of value. Even SDG 4 (quality education) interprets progress through enrolment numbers, test scores and literacy rates, indicators that perpetuate conformity and competition. Education pursued through fear, desire and comparison cannot flower into intelligence. When education is framed through performance and rank, it mirrors the same fragmented psyche that structures governance itself.

Community

In communities, memory and image create belonging through exclusion. Identity is affirmed by who 'we' are and who 'they' are not. Caste systems in India, for example, persist through centuries of memory embedded in ritual and image, despite legal prohibition. Religious communities often define themselves through historical grievances, ensuring that memory fuels present division. Migrants, minorities or dissenters become images of threat, sustaining the illusion of unity in the majority. This dynamic is visible in the rise of nationalist movements across Europe, South Asia and the United States (US), where fear of the outsider becomes a rallying cry.

Governance

At the national and international level, memory and image scale up into powerful instruments of policy and conflict. Nations cultivate images of their historical destiny, often defining themselves against perceived enemies. Memory of past trauma like partition, colonisation or war conditions present action, sustaining cycles of suspicion and hostility.

Bohm (1992) described international relations as 'endless fragmentation', with each nation acting as if its security were separate from the whole. Climate negotiations, though framed as global cooperation under SDG 13, are structured by the image of 'national interest'. This ensures that agreements remain compromises of rivalry rather than expressions of interdependence. Singh and Bhatnagar (2023) call such tensions wicked problems, where competing interests and fragmented identities prevent resolution.

The SDGs as Mirrors of Image

The SDGs themselves reveal how deeply image conditions collective action. The logos and dashboards present an image of unity, progress and shared purpose. Yet beneath this image, contradictions abound. SDG 16 promotes peace and justice even as military expenditures escalate. SDG 8 champions growth while contributing to ecological collapse that undermines SDG 13. SDG 5 promotes gender equality, yet patriarchal images continue to shape household and workplace dynamics.

When awareness is reduced to a competency or indicator, transformation is replaced by performance. The image of 'awareness' becomes a metric to achieve, rather than perception itself. The result is institutional theatre: colourful dashboards and ambitious frameworks that mask the persistence of fragmentation.

6.5 The Question of Relationship Without Image

If memory and image dominate human relationship, shaping how individuals, communities and nations perceive one another, then the central question arises: is relationship possible without them?

Krishnamurti (1954) argued that true relationship begins only when image ends. To see another without the mediation of memory, role or projection is to encounter them directly, in the present. Otherwise, what we call 'relationship' is merely the interaction of images: parent and child as roles, teacher and student as functions, nation and nation as abstractions. In such interactions, conflict is inevitable because no image can ever contain the fullness of life.

Bohm (1996) described dialogue as a way of meeting without image. When assumptions are suspended and participants listen without defending identity, meaning can flow freely. This suspension of image is not a method to be applied but an act of perception in the present. Relationship without image is not an ideal to achieve but a fact revealed when the mind observes the movement of image as it arises and ceases to identify with it.

Most attempts to 'improve' relationship rely on ideals: be more compassionate, more tolerant, more peaceful. But these ideals are themselves images, projections of what should be, which keep the self in movement. Frameworks that prescribe values or competencies inevitably turn relationship into performance. What results is not transformation but continuity, the same structure of self refined, not ended.

The question of relationship without image is therefore not about adopting new beliefs or methods but about observation. Can the mind see that it is relating through memory and image? Can it notice how desire, fear or past experience shape perception in the moment? In such observation, the image loses its authority. The false falls away, and what remains is direct encounter.

Krishnamurti noted that 'when there is no image, there is a different kind of silence in relationship', suggesting that true encounter emerges only when the projection of thought ends (Dialogue 6, Krishnamurti & Anderson, 1991).

6.6 Conclusion

This chapter has examined how memory and image shape the fabric of human relationship. Memory, while indispensable for continuity, conditions perception by filtering the present through the past. Image, born of memory, desire and fear, then mediates encounter so that we rarely meet another human being directly.

What we call 'relationship' is often the interaction of images, not the meeting of living beings.

At every scale, from family to community, from nation to governance, this mediation sustains division. Parents project ambitions onto children. Communities enforce belonging by scapegoating outsiders. Nations perpetuate cycles of hostility by clinging to historical wounds and enemy images. Even sustainability frameworks like the SDGs reproduce these distortions by presenting images of progress while underlying fragmentation deepens. Institutions built on identity mirror the very violence they are meant to resolve.

The insight is unsettling but necessary: relationship mediated by memory and image is never whole. Where there is image, there is division. Where there is division, there is conflict. Krishnamurti (1983) insisted that relationship without image is possible only when the mind sees the operation of image as it arises. In that seeing, the false falls away, and what remains is direct encounter.

This opens a radical question: can relationship exist without image? If so, it is not through ideals or methods, but through awareness. Awareness is not an added competency, not a skill to be cultivated over time, but the direct perception of thought in motion. To see the image as image is already to end its hold. In such seeing, relationship is no longer conditioned by the past but grounded in the present, becoming the space of care, responsibility and wholeness.

Krishnamurti emphasised that 'the description is never the described', a reminder that images and words are not the reality they point to (Dialogue 4, Krishnamurti & Anderson, 1991). To live this is to end illusion. He also noted that 'when there is no image, there is a different kind of silence in relationship', a silence that is not emptiness but the ground of true encounter (Dialogue 6, Krishnamurti & Anderson, 1991).

This inquiry prepares the movement into the next chapter. For what sustains memory and image at their root is psychological time: the self's endless movement from past to future, from what is to what should be. Chapter 7 will explore this deeper structure of time and inertia, showing how the continuity of the self sustains fragmentation and whether it is possible for action to arise outside the movement of becoming. If such an ending of time is possible, it may also reveal what relationship is when no image divides it, a flowering that is love itself, which Chapter 8 will unfold.

References

Bohm, D. (1992). *Thought as a system.* Routledge.

Bohm, D. (1996). *On dialogue.* Routledge.

Bourdieu, P. (1991). *Language and symbolic power.* (J. B. Thompson, Ed.; G. Raymond & M. Adamson, Trans.). Harvard University Press.

Damasio, A. (2010). *Self comes to mind: Constructing the conscious brain.* Pantheon.

Krishnamurti, J. (1954). *The first and last freedom.* Harper & Brothers.

Krishnamurti, J. (1983). *The transformation of man.* Harper & Row.

Krishnamurti, J., & Anderson, A. W. (1991). A wholly different way of living: Krishnamurti in dialogue with Professor Allan W. Anderson, San Diego 1974. HarperSanFrancisco.

McGilchrist, I. (2019). *The master and his emissary: The divided brain and the making of the Western world* (2nd ed.). Yale University Press.

Singh, N., & Bhatnagar, D. (2023). *Applied spirituality and sustainable development policy*. Emerald Publishing.

Chapter 7

Psychological Time and Inertia

Abstract

Chapter 7 examines how the projection of psychological time sustains both personal and institutional inertia. It distinguishes physical time, which is necessary for coordination, from psychological time, which projects becoming, success and reform into the future. This projection gives continuity to fear and desire, translating insecurity into ambition and progress into delay. This chapter demonstrates how development agendas and policy frameworks mirror this same postponement, promising transformation 'by 2030' while avoiding perception in the present. By seeing that psychological time is thought seeking permanence, one discovers that the future is a projection of the past. The ending of this movement frees the mind from continuity and opens the possibility of immediate transformation. Real sustainability, this chapter concludes, is not achieved in the future but lived now, when action arises from awareness rather than fear.

Keywords: Psychological time; future projection; reform inertia; temporal illusion; fear-based planning; present action; ending continuity; living now

7.1 Introduction

The previous chapter showed how memory and image shape human relationship, sustaining division by mediating perception through the past. Beneath these dynamics lies a deeper structure that gives them continuity: psychological time.

Time in its physical sense is essential. Clocks, calendars and measurement make possible coordination, agriculture, science and technology. Without physical time, there would be no order in collective life. But when time moves from the practical to the psychological, it becomes the ground of disorder. Psychological time is the self's projection of continuity: 'I was', 'I am', 'I will be'. It sustains identity through the past and drives becoming towards the future.

From Fragmentation to Wholeness, 141–162

Published under exclusive licence by Emerald Publishing Limited
doi:10.1108/978-1-80686-677-920261007

Krishnamurti and Anderson (1991, Dialogue 4) emphasised that this movement of time is the root of bondage. The past, stored in memory, shapes perception in the present, while desire and fear project imagined futures that govern action. Bohm (1996) explained that thought mistakes this continuity for permanence, reinforcing the illusion of an enduring self. Neuroscience supports this: Damasio (2010) shows how the autobiographical self is reconstructed moment by moment from past memory and projected into imagined futures.

At personal and collective scales, this structure is visible everywhere. Individuals postpone change by saying 'I will become better'. Communities carry forward historical grievances as living identity. Institutions, designed to manage risk, endlessly reform symptoms without touching disorder at its root.

The Sustainable Development Goals (SDGs) illustrate this paradox vividly. Launched in 2015, they embody unprecedented global mobilisation. Yet their design reveals how deeply time structures policy: targets are fixed for 2030, progress is reviewed periodically, and success is measured retrospectively. Action is constant, but results are deferred. The framework itself is temporal, organising transformation as something to be achieved 'by' a certain date. In this way, the SDGs mirror the psychology of postponement: acting within time while rarely questioning whether change is possible outside it.

The ground of such frameworks is the past itself: memory, identity and conditioned interest. Action shaped by the past is never fully present; it is selective, partial and defensive. What appears urgent is already filtered by convenience or political advantage. Negotiations and plans inherit this distortion, reflecting not direct perception of the whole but fragments bargaining for continuity. Coordination is not inherently difficult, cooperation flows naturally when there is perception of what is. The difficulty arises when coordination becomes bargaining among fragments.

Timelines designed from this field of fragmentation carry its instability. What is divided cannot generate wholeness, because division is inherently self-contradictory. Krishnamurti and Anderson (1991, Dialogue 4) observed, 'That which is broken up can never bring about the whole'. Frameworks built within fragmentation inevitably perpetuate disorder, no matter how ambitious they appear.

The central question of this chapter is therefore simple yet radical: can disorder end when action is always structured within time? Or is transformation possible only in the present, outside the movement of becoming?

7.2 Physical Time and Psychological Time

Time appears as a single continuum, but human experience reveals two distinct orders. One is physical, indispensable for survival. The other is psychological, sustaining conflict and division.

Physical Time

Physical time is measurement. It chronicles the rhythm of day and night, the seasons and the orbit of the earth around the sun. It makes agriculture, science and technology possible. Without it, there would be no calendars, no schedules, no coordination of daily life. It is neutral and necessary.

Philosophers across traditions have reflected on this. Augustine, in his *Confessions* (Book XI), noted the paradox that the past exists only as memory, the future only as expectation and the present vanishes when grasped. Bergson (1910/1991) distinguished 'clock time', measurable and divisible, from durée, the qualitative flow of lived experience. Whitehead (1929/1978) emphasised that reality itself unfolds as process, while Heidegger (1962) described temporality as fundamental to being-in-the-world. Physical time provides rhythm and order, but it is not the problem.

Psychological Time

Psychological time is the projection of thought. It is the continuity of the self carried through memory into the present and projected into the future. It manifests as becoming: 'I am this, I will be that'. Between what is and what should be lies effort, struggle and contradiction.

Krishnamurti and Anderson (1991) drew the distinction sharply, observing that while physical time is necessary, psychological time sustains conflict. Bohm (1996) explained that thought creates the illusion of permanence by stringing fragments of memory together. Damasio (2010) confirmed this neuroscientifically, showing how identity is continually reconstructed from memory and imagination. McGilchrist (2021) added that the brain's analytic mode reinforces linear stories of selfhood, even as life itself flows beyond categories.

The Trap of Confusion

Disorder arises when psychological time is mistaken for physical time. The mind treats becoming: 'I will change tomorrow', 'humanity will solve this by 2050' as if it were as real as the turning of the seasons. But change does not occur in time; it occurs only in the ending of time. The attempt to transform through postponement perpetuates disorder.

At the systemic level, this confusion is institutionalised. Frameworks such as the SDGs are structured around temporal targets: end poverty by 2030, achieve net zero by 2050. These timelines appear necessary for coordination but quickly become attachments as symbols defended for their legitimacy rather than questioned for their adequacy. They also become boundaries, confining action to what 'fits the target', even when reality demands responses beyond it.

The underlying assumption is that fragmentation will persist, yet within that persistence transformation can somehow be achieved. But what is fragmented can only replicate fragmentation. The opposite belief that once targets are achieved, fragmentation will dissolve into coherence is equally false. History shows that even in societies with advanced development, fear, division and violence persist. To build models on the outcomes of what an 'evolved' society should look like is to mistake surface for substance. Prosperity cannot coexist with fear and division. As long as borders and identities remain intact as ideas and boundaries, wholeness will not arise through targets.

Implications

The distinction between physical and psychological time is therefore not academic but existential. Physical time is indispensable; psychological time sustains fear, desire and division. To conflate the two is to entangle survival with disorder, ensuring that action is postponed and conflict perpetuated.

The question that follows is whether the mind can live in physical time without being trapped in psychological time. Can life be organised in calendars and clocks, yet free of the projection of becoming?

7.3 The Self as Continuity in Time

If psychological time is the projection of thought, then its primary manifestation is the self. The 'me' is experienced as continuity: a thread woven from memory, desire and anticipation. It feels stable and enduring, but on closer examination, it is nothing more than a narrative stretched across time.

The Narrative of Self

From early childhood, identity is constructed through memory. A child remembers praise or punishment, successes and failures, and begins to assemble these fragments into a sense of 'who I am'. This narrative is carried forward, reinforced by family, culture and society. Adults define themselves by past achievements, hurts and attachments, while projecting into imagined futures of career, relationship or legacy. In this way, the self is sustained as continuity in time.

Krishnamurti and Anderson (1991) argued that this continuity is the root of disorder: the self that says 'I was', 'I am', 'I will be' is never free to meet the present. It lives in becoming, which breeds conflict between what is and what should be. Bohm (1996) explained that thought mistakes this continuity for permanence. What is actually a series of reconstructed fragments is experienced as a stable entity.

Neuroscience confirms the illusion. Damasio (2010) describes the autobiographical self as a narrative continually reconstructed from past memories and projected futures. McGilchrist (2021) notes that the brain's analytic mode intensifies this illusion by privileging categorisation and stability over context and flow.

Identity as Historical Continuity

At collective scales, the same dynamic operates. Nations define themselves through victories, defeats, colonisations and revolutions. Communities preserve traumas and triumphs to sustain belonging. Religious traditions preserve stories that bind adherents to a past and project a promised future. All create the sense of continuity: 'we have always been', 'we will endure'.

Yet these continuities, like the personal self, are constructed. They are fragments selected and repeated until they form an image mistaken for permanence. Anderson (2006) described nations as 'imagined communities', sustained by shared narratives and rituals. Koselleck (2004) highlighted how modern societies stabilise identity through a 'space of experience' and a 'horizon of expectation', ensuring that belonging is always framed by past memory and projected future.

Continuity as Inertia

The problem is not continuity in practical life but continuity as psychological identity. Continuity gives the self stability, but also creates inertia. The past dictates the present; the future becomes repetition in new forms. Violence, competition and exclusion are carried forward under the banner of identity. Freud (1930/2010) described civilisation as a compromise between individual aggression and collective restraint yet compromise is itself continuity, not transformation. Psychological research confirms this inertia. Zimbardo and Boyd (2008) show how individuals locked in 'past-negative' or 'future-anxious' time perspectives tend to repeat cycles of fear and dissatisfaction, unable to act with freedom in the present. Trauma research echoes this: van der Kolk (2014) shows how trauma 'freezes' the body in time, forcing people to relive the past as if it were still present. What appears as stability is often paralysis. Arendt (1998) offered a counterpoint with her notion of natality, the human capacity to begin anew. Yet this capacity is obscured when continuity dominates, and action is reduced to repetition under the guise of progress.

Implications

To see the self as continuity in time is to question whether it has any reality beyond thought. If the self is merely memory projected forward, then it is not permanent but an illusion of permanence. This recognition opens a radical possibility: if the self is continuity in time, then the ending of psychological time may also be the ending of the self. Far from being a loss, this ending dissolves inertia, releasing relationship and action from repetition. When continuity no longer dictates perception, life is met directly in the present. Responsibility, clarity and care are no longer bound by the past, but arise as expressions of wholeness. The same insight applies collectively. Nations and communities are sustained by imagined continuities of belonging, history and destiny. The SDGs themselves, though mobilising unprecedented cooperation, are framed through these imagined continuities. They assume transformation can occur while nations remain locked in structures of identity and interest. Yet as long as continuity dominates, frameworks risk becoming vehicles of inertia, reinforcing separation under the guise of shared progress. Only when identity as continuity is seen for what it is can responsibility emerge in the present, grounded not in projection but in direct perception of the whole.

7.4 Time, Fear and Desire

Psychological time does not only sustain the continuity of the self; it also fuels the two central movements of the psyche: fear and desire. These are not separate forces but expressions of the same temporal structure. Fear projects past pain into the future; desire projects imagined fulfilment into the future. Both displace attention from the present into what has been or what might be.

Fear as Projection

Fear, in its original evolutionary role, was protective, helping human beings avoid predators and danger. Psychological fear, however, extends beyond immediate threats. We fear losing status, reputation, wealth, relationships or life itself. These fears are rarely about the present moment; they are projections of memory and anticipation.

Krishnamurti and Anderson (1991) emphasised that fear is inseparable from time: without 'yesterday' or 'tomorrow', fear has no continuity. Neuroscience supports this. Damasio (2010) shows how memory and imagination combine bodily states into anticipatory anxiety. van der Kolk (2014) describes trauma as a temporal loop, where individuals re-experience the past as though it were present. Zimbardo and Boyd (2008) add that people locked in 'future-anxious' or 'past-negative' orientations suffer heightened fear and stress, unable to inhabit the immediacy of now. Fear is therefore not simply an emotion but a temporal orientation, binding perception to time.

Desire as Projection

Desire mirrors fear. Where fear pushes away, desire pulls towards. Desire imagines completion in the future: 'If I achieve this, if I gain that, I will be fulfilled'. Yet fulfilment never endures. Once attained, desire shifts to a new object, perpetuating the cycle of becoming. Nietzsche (1994) described this as the 'will to power', the endless striving to overcome lack. Girard (1977) showed how desire is mimetic: we want what others want, because objects gain value through social meaning. Freud (2010) argued that the tension between pleasure-seeking and civilisational restraint ensures desire is never at rest. Consumer culture amplifies this dynamic, institutionalising desire through advertising and media, embedding dissatisfaction into everyday life. Psychology echoes this: Zimbardo and Boyd (2008) highlight 'future-positive' orientations where desire drives ambition, but at the cost of well-being in the present.

Fear and Desire as One Movement

Though often treated separately, fear and desire are inseparable. Desire for wealth may arise from fear of poverty; desire for love from fear of loneliness. Conversely, fear often stems from desire: fear of losing what one values, fear of not attaining what one longs for. Both are movements of psychological time. Bohm (1996) explained that thought sustains this duality by projecting continuity. The mind oscillates between avoiding pain and seeking pleasure, both stretched across past and future. In this oscillation, the psyche is never free. Only when fear and desire are seen as one movement of becoming does the possibility of ending arise.

Fear and Desire in Institutions

These dynamics shape institutions as much as individuals. Fear and desire are not merely private experiences but the very logic around which modern systems are designed.

Education: Fear and Desire Institutionalised

Education is often celebrated as the pathway to liberation, equality and progress. Yet on closer observation, it becomes evident that education mirrors the same movements of fear and desire that shape society as a whole. Instead of awakening intelligence, schools and universities often reproduce fragmentation: ambition, comparison, obedience and allegiance to insecure structures. From classrooms to debating circuits, from admissions to universities, from curricula to the military, education becomes less about truth and more about performance, control and survival.

Performance in Schools and Debating Circuits

Spaces like Model United Nations and debating platforms are often *theatres of projection* rather than fields of inquiry.

- *Fear* drives students to avoid vulnerability, cling to authority and silence dissent.
- *Desire* fuels competition, ambition and the need to impress adjudicators.

Children debate atrocities including rape, war, displacement while laughing, eating and posing for photographs. Chairs deny points of order to protect their egos or appease dominant schools. Students mimic global rhetoric without realising that the same movements of division and power operate within themselves. These simulations mirror the world not by teaching responsibility but by reproducing posturing, hierarchy and division.

Inequality in Admissions and Teacher Distribution

Elite schools, with resources and prestige, attract teachers who seek higher salaries and comfort. Meanwhile, government schools in rural areas remain underfunded and avoided by teachers who fear isolation or hardship. This produces *systemic inequality*: the children most in need of education receive the least.

- Desire for financial stability and prestige pulls teachers towards elite institutions.
- Fear of poverty and discomfort drives them away from rural schools.

Thus, education consolidates privilege and deepens exclusion.

Teachers as Students of Fear and Desire

Teachers themselves are caught in the same conditioning they are asked to overcome. Many cling to outdated methods out of fear of losing authority, or deliver rigid lessons because they are distracted by their own anxieties of financial strain, fractured relationships, societal expectations. In classrooms, they defend self-image rather than meeting students openly. Teachers too are students, students with fear and desire. Without transformation here, education simply transfers fragmentation from one generation to the next.

Universities, Knowledge and False Equality

Universities intensify the same cycle. PhDs and publications, instead of being paths to truth, become currencies for recognition and employment. Fear of irrelevance and desire for prestige push academics into compliance with journal metrics, rankings and grants.

Policy think tanks and research institutions weaponise knowledge to serve national or economic agendas. Even the idea of equality is distorted:

- As an idea within thought, equality is rooted in comparison, and comparison sustains division.
- True equality arises only in the ending of division, when consciousness itself is whole.

Yet universities reward fragmentation as achievement, codifying fear and desire into academic structures.

Curriculum as Commodity

Curriculum is rarely neutral. It is designed around demand and agenda.

- Parents and students desire career pathways and employability, so curricula emphasise technical and professional skills.

- Knowledge is fragmented into measurable 'units' to be acquired, ranked and certified.
- Coding, automation and efficiency are prioritised while inquiry into meaning, ethics and wholeness is ignored.

Curriculum reflects the same divided consciousness of society, driven by recognition, accumulation and control while silencing the questions that matter most.

The Nation and the Insecurity of Structures

At the deepest level, curriculum serves the nation. Nations are insecure constructs, built on fragile identities of history, territory and ideology. To preserve themselves, they require loyalty. Thus, education is designed to produce obedience, not intelligence.

History textbooks glorify national victories and vilify enemies. Civic education emphasises conformity over inquiry. Science and economics are aligned to productivity and growth. True education, which questions division and awakens compassion, is suppressed because like love it is dangerous to insecure structures.

Discipline, Fear and Desire in Military Training

The nation's insecurity finds its starkest form in military schools and academies. Here, inquiry is allowed only up to the boundary of nationalism. Discipline is enforced through fear, pain and punishment: drills, humiliation and regimentation to suppress individuality. Alongside this, desire for symbols and rewards is cultivated: medals, ranks, pensions, privileges and the imagined immortality of one's name inscribed on a memorial.

Even killing or being killed is framed as pride, service and permanence. The military mechanises human beings through fear and desire, justifying violence as sacrifice and obedience as patriotism.

Bias, Nationalism and Instrumentalisation

Fear and desire also seep through the biases of teachers and administrators. In the author's own interactions (Gupta, 2024), one teacher remarked: '*Look, this religious community has grown so much. They are now in our school also, now we have three of them*'. Such statements reveal how fear of the other enters pedagogy, dividing classrooms and shaping perception.

Nationalism instrumentalises education even further. Schools may request training programmes not to awaken intelligence but to make students 'efficient' to prepare them for roles in the military-industrial complex, designing missiles or managing weapons systems. Fear of insecurity and desire for dominance turn education into an apparatus of division and violence.

Education, Love and Wholeness

From schools to universities, from curricula to the military, education is shaped by fear and desire. It produces ambitious students, anxious teachers, fragmented academics, insecure nations and conditioned soldiers. It trains individuals to survive in a divided world but does not awaken wholeness.

True education, like love, is different altogether. It serves life, not the nation or the market. It refuses comparison, dissolves division and awakens compassion. This is why nations fear it, why institutions resist it and why societies often suppress it. For true education, like love, is dangerous to structures built on insecurity.

Without such education, no reform can heal fragmentation. Only when fear and desire end in individuals and institutions can education fulfil its true purpose: the awakening of intelligence and the flowering of union.

Governance: Fear and Desire as Instruments of Power

Governance is often presented as the rational organisation of collective life. Constitutions, laws and policies are framed as safeguards for justice and stability. Yet beneath this formal order, governance frequently operates by mobilising fear and desire as instruments of legitimacy. Political authority does not merely regulate resources; it shapes consciousness by institutionalising projections of insecurity and hope.

Fear in Political Authority

Fear of instability and insecurity is a primary tool of governance. In democracies and authoritarian regimes alike, fear of external enemies sustains nationalism and militarisation. Fear of migrants and minorities is often amplified to consolidate majoritarian identity. Anti-terror legislation, refugee restrictions and surveillance programmes illustrate how laws are drafted in response to projected threats rather than present perception (UNHCR, 2011). Fear is not simply a response to reality; it is institutionalised as a means of control.

Desire as Promise and Aspiration

Alongside fear, governance harnesses desire to sustain authority. Leaders promise prosperity, jobs, infrastructure and growth to inspire hope and secure votes. Citizens invest in these promises as projections of a better future, binding their loyalty to imagined gains. SDG 8 (Decent Work and Economic Growth) exemplifies this desire, presenting growth as the universal path to security. Yet this desire, when institutionalised, often legitimises exploitation of ecological and social systems (UNGA, 2015).

Law, Reform and Postponement

Fear and desire also shape how laws and reforms are enacted. Peace agreements are framed with future timelines: ceasefires to be implemented 'by 2025' or elections promised 'by 2030'. Development plans such as SDG 1 (No Poverty) and SDG 10 (Reduced Inequalities) are structured around 2030 targets, creating the illusion of gradual transformation while deferring present responsibility. Governance Without the Self (Gupta, 2025) critiques this reliance on frameworks, showing how temporal targets reform symptoms but stabilise disorder by leaving the root of division untouched. Koselleck's (2004) distinction between a 'space of experience' and a 'horizon of expectation' captures this dynamic: politics institutionalises expectation into the future, while the present remains captive to continuity.

Negotiation and the Illusion of Consensus

Negotiation is often hailed as the hallmark of governance, but it too reflects fear and desire. Rival groups negotiate not from perception of the whole but from bargaining among competing anxieties and ambitions. Peace agreements codify divisions by granting recognition to identities born of fear and desire rather than dissolving them. International negotiations around SDG 16 (Peace, Justice and Strong Institutions) reveal this paradox: frameworks promise peace but are driven by fragmented interests, ensuring that division is stabilised rather than ended.

Governance as Weaponisation of Division

The institutionalisation of fear and desire ensures that governance becomes less about coherence and more about managing division. Borders are drawn and defended as lines of fear and desire. National budgets justify massive military expenditure in the name of security, while citizens are promised prosperity through industrial and technological progress. In each case, governance reproduces fragmentation, presenting management of division as order.

True governance would mean responsibility for the whole, not manipulation of projections. It would not trade on fear or promise prosperity as desire but act from perception of what is. As long as governance remains bound to psychological time, it will perpetuate continuity of disorder while appearing to promise reform.

Sustainability: Fear and Desire in the Future of the Planet

Sustainability is often celebrated as humanity's noblest collective project. Yet when observed closely, it too is shaped by the dual movements of fear and desire projected into time. Climate agreements, biodiversity targets and the SDGs are not merely technical frameworks; they embody the same psychological structure that defers transformation into the future.

Fear of Collapse

Fear of ecological catastrophe drives much of the sustainability discourse. Images of melting glaciers, burning forests and displaced populations are invoked to generate urgency. SDG 13 (Climate Action) mobilises nations by warning of irreversible collapse if action is delayed. Yet fear often narrows perception, turning sustainability into crisis-management rather than holistic awareness. Policies framed by fear of collapse rarely dissolve the root causes of ecological violence; instead, they manage symptoms. Urgency requires perception of danger, not fear. If fear was working, the world would have changed, but what's missing is sensitivity which emerges when fear ends.

Desire for Growth and Security

Alongside fear, desire for prosperity and stability shapes sustainability frameworks. Nations pursue renewable energy transitions not to dissolve fragmentation but to secure competitive advantage. Corporations adopt sustainability branding to access markets and legitimacy. Citizens desire safety for their children and continuity of lifestyle. This dual movement, fear of loss, desire for advantage, translates ecological responsibility into bargaining over national interest. SDG 7 (Affordable and Clean Energy) and SDG 9 (Industry, Innovation and Infrastructure) illustrate this tension, where ecological goals are merged with desires for industrial growth (UNGA, 2015).

Timelines as Postponement

Timelines such as 'net zero by 2050' or 'halt biodiversity loss by 2030' are the structural language of sustainability. They provide the illusion of direction but often legitimise delay. Nations continue fossil fuel expansion even while pledging long-term reductions. Gupta (2024) emphasises that transformation is not in time but in the ending of time; by tying change to deadlines, institutions ensure continuity rather than immediate transformation and makes the same point at the systemic level: frameworks structured around future targets reproduce inertia because they are born of the same divided psyche that sustains disorder.

The Illusion of Paradise After Targets

Another assumption is that once targets are achieved, harmony will follow. Yet history shows otherwise. Even in wealthy nations with advanced climate policy and welfare systems, fear, alienation and division persist. Development in GDP, energy or technology may reflect certain improvements but cannot create transformation. To build models on the by-products of an 'evolved' society is to mistake surface for substance. Prosperity cannot coexist with fear and division; as long as borders and identities remain intact, sustainability will not be wholeness.

Climate Negotiations as Theatre

International summits dramatise this paradox. Negotiators argue over percentages, deadlines and financing while emissions rise. Leaders debate the colour of the bucket while the house burns. Fear of collapse and desire for advantage dictate positions: fear of economic loss, desire for geopolitical dominance. SDG dashboards showcase progress, yet ecosystems collapse in the present. Negotiation, when it treats truth as opinion, becomes a way of postponing rather than acting.

Sustainability frameworks, bound to fear and desire, mobilise activity but perpetuate delay. They create the appearance of progress while the underlying fragmentation deepens. True sustainability is not fear of collapse or desire for prosperity stretched into the future but the ending of division now. Without this, sustainability becomes another industry of reports and dashboards while the earth burns.

Implications

The wider implication is clear. As long as institutions are structured around psychological time, they cannot deliver transformation, only continuity. Fear and desire may mobilise activity, but that activity remains trapped in postponement, compromise and rivalry. Reform frameworks institutionalise insecurity when they act from fear and institutionalise dissatisfaction when they act from desire.

Transformation does not lie in the movement of becoming but in the ending of becoming. To see fear and desire as movements of time is already to question their authority. In that very questioning, the possibility of freedom opens, for individuals and for the systems they create.

7.5 Time as Postponement of Change

If fear and desire sustain psychological time at the emotional level, postponement sustains it at the level of action. Human beings rarely act directly. Instead, change is deferred into the future: 'tomorrow', 'next year', 'by 2050'. This displacement is not neutral. It is the very structure of psychological time: continuity disguised as progress.

The Illusion of Gradual Change

Psychological time creates the belief that change is a process, a gradual improvement from what is to what should be. The violent person says, '*I will practice non-violence*'. The addict says, '*I will overcome this habit*'. The divided society promises reconciliation through future reforms. Yet in each case, the fact of violence, addiction or division persists in the present, while thought sustains comfort by imagining eventual resolution.

Krishnamurti (1983) warned against this illusion, arguing that real change is never in time. To seek non-violence through gradual practice is to perpetuate violence itself, because becoming is still within the movement of violence. Bohm (1996) explained that thought sustains disorder by stringing fragments of the past into imagined futures, mistaking projection for transformation. Psychology confirms the same dynamic: Zimbardo and Boyd (2008) showed how individuals with 'future-positive' or 'future-anxious' orientations defer meaningful action, comforting themselves with the idea of eventual change while reinforcing inertia.

Postponement in Governance

The same illusion pervades governance. Peace agreements are drafted with timelines: 'ceasefire by 2025', 'elections by 2030'. Development agendas promise to 'eradicate poverty by 2030' or 'achieve equality by 2050'. While such targets provide structure, they also function as deferrals. The reality of violence, poverty or inequality is displaced into a horizon of gradual reform.

Koselleck (2004) captured this historical pattern through his distinction between a 'space of experience' (the past shaping the present) and a 'horizon of expectation' (the imagined future). Modern politics institutionalises this horizon through deadlines and targets. But when expectation is mistaken for transformation, the present remains captive to continuity.

This has grave consequences in crises that demand immediacy. Ecological tipping points, war and mass displacement are not matters of opinion but facts. When one's own family is in pain, relief is immediate; it is not postponed to a future date. Yet international frameworks rely on negotiation and compromise, entrenching delay even while conditions worsen.

Sustainability and Climate Commitments

Climate policy offers a stark example. Nations pledge 'net zero by 2050' while expanding fossil fuel infrastructure today. SDG dashboards track progress towards 2030, even as ecosystems collapse in the present. The language of timelines legitimises delay, providing the image of responsibility while deepening disorder. Singh and Bhatnagar (2023) note that sustainability policy often functions as a 'wicked problem' because temporal displacement prevents coherence in the present, entrenching rivalry and postponement (Rittel & Webber, 1973).

Personal Postponement

The same pattern operates at the personal level. The individual says, '*I will meditate tomorrow*', or '*I will be kind when I have more time*'. These promises soothe the mind but sustain the very habits they claim to end. Freud (2010/1930) described civilisation as a compromise between aggression and restraint, but compromise is itself continuity, not resolution. Trauma research makes this visceral: van der Kolk (2014) shows how the body holds on to pain in the hope of eventual release, yet healing occurs only in the immediacy of perception, never in imagined futures.

Implications

Postponement reveals how psychological time sustains inertia. By promising future transformation, it perpetuates present disorder. The violent person continues in violence, the divided society in division, the unsustainable system in exploitation. Time becomes a refuge for the self, a way of deferring responsibility while appearing to move forward.

The radical question is whether change requires time at all. Can transformation occur only in the immediacy of perception, not through projection? If so, change is not a matter of becoming, but of ending, an ending that opens the space for action grounded in truth, not in delay.

7.6 Inertia: The Weight of Psychological Time

If postponement sustains disorder through projection, inertia sustains it through continuity. Inertia is the weight of psychological time: the tendency of past patterns to carry forward, repeating themselves under new names. Violence becomes rebranded as security, domination as progress, conformity as education. Though the forms change, the underlying movement remains.

Inertia in the Psyche

At the personal level, inertia manifests as habit. The mind, conditioned by memory, repeats its responses: anger resurfaces when triggered, fear returns despite promises to overcome it, desire renews itself as soon as one goal is reached. The self resists change because continuity provides stability, even if that stability is painful.

Freud (1930/2010) described civilisation as the management of aggression, but management is compromise, not dissolution. Compromise is inertia, not transformation. Krishnamurti (1991) insisted that the self is itself inertia: built of memory and time, it is the repetition of what has been. The psyche seeks novelty, but it reconfigures the old into new forms. This is why reform so often fails: the reformer is the same as the reformed.

Bergson (1910/1991) offered a related insight with his notion of *durée*, or lived time. Consciousness endures as a fluid flow, yet thought freezes this flow into static forms, generating rigidity. Augustine, centuries earlier, described this as the 'distension of the soul' (*Confessions* XI): the stretching of consciousness between memory and anticipation, which prevents full presence. Both perspectives highlight that inertia is not just habit but the structural weight of time in the psyche.

Modern neuroscience corroborates this. McGilchrist (2021) shows how the left hemisphere, when dominant, imposes fixity on reality, recycling categories and resisting openness. Trauma research adds that unprocessed pain binds people to repetition: van der Kolk (2014) explains how trauma keeps the body 'living in the past', reactivating fear and pain in the present. Inertia is not metaphorical but embodied time refusing to end.

Inertia in Governance

Institutions mirror this pattern. Constitutions, laws and bureaucracies carry forward the assumptions of their founders. Even revolutions often reproduce the very structures they overthrow. Colonial administrations gave way to post-colonial states that frequently maintained the same hierarchies, surveillance systems and extractive logics (Chatterjee, 1993). What appeared as independence was, in many cases, the persistence of colonial rationalities under new banners.

Peace agreements institutionalise division in similar ways. Treaties that divide land, populations or resources often stabilise conflict rather than dissolve it, embedding hostilities into law. The 1993 Oslo Accords, for instance, promised a pathway to peace between Israel and Palestine, but by codifying territorial divisions and deferring resolution of core issues, they entrenched fragmentation rather than resolving it (Tilley, 2005).

Development frameworks display the same inertia. Structural adjustment in the 1980s, poverty reduction strategies in the 1990s, and today's SDGs all recycle metrics of growth, efficiency and state-centric planning. Acronyms change, but the psyche of measurement and control persists (Fukuda-Parr & McNeill, 2019). Reform loops appear innovative but sustain continuity.

Inertia in Sustainability

The climate crisis illustrates inertia starkly. Despite decades of agreements, global emissions continue to rise. Each new framework including Kyoto (1997), Paris (2015), COP26 (2021), COP28 (2023) restates urgency while underlying patterns of extraction and dependency persist (UNEP, 2023). Even as governments pledge 'net zero by 2050', subsidies for fossil fuels expand. New timelines provide legitimacy while deferring structural change.

Beck (1992) described modernity as a 'risk society', where efforts to manage crises generate new ones. Climate governance reflects this: carbon markets, offsetting schemes and 'loss and damage' funds reframe ecological destruction as managerial risk rather than transformation. Singh and Bhatnagar (2023) note that sustainability policy often functions as a 'wicked problem' because competing interests displace responsibility into the future. In this way, inertia ensures that present disorder continues beneath the rhetoric of targets.

The Weight of History

Inertia is also the burden of history carried forward. Nations sustain grievances for centuries, reproducing conflict in each generation. Communities transmit prejudice through culture and education. Individuals inherit traumas that shape responses long after the original event. Maté (2010) and Levine (2015, 2023) show how unresolved traumas resurface as illness, addiction or social violence, perpetuating cycles of suffering.

Implications

Inertia reveals that disorder is not sustained by accident but by continuity itself. As long as the past dictates the present, the future will be repetition. Reform, compromise and gradual improvement cannot break this cycle; they are its continuation.

The radical possibility lies in ending inertia at its root. This does not mean resisting habit through will, for will is itself continuity. It means observing continuity directly. When the mind sees that it is repeating, the illusion of stability collapses. What remains is openness to the present, where transformation is possible.

This possibility is not only for individuals. Groups, organisations and nations too carry the weight of their own continuity – repetition of structures and frameworks under new names. When collectives recognise that their reforms are often the past disguised as progress, the mask of novelty falls away. In such seeing, even at the institutional level, lies the space for action not bound by inertia but open to what is.

7.7 The Ending of Time

If psychological time sustains fear, desire, postponement and inertia, the pressing question is whether it can end. Not the ending of clocks or calendars, physical time must continue, but the ending of the inner movement of becoming: the self carried forward as memory, identity and projection.

Krishnamurti and Anderson (1991) stated this uncompromisingly: 'The ending of time is the ending of sorrow'. For him, psychological time is the structure of the self. To live in 'what was' and 'what will be' is to perpetuate conflict, division and suffering. Ending time is not a gradual process, for any process is already time. It is a direct perception of the movement of thought, in which the whole structure collapses. David Bohm (1996) pointed to dialogue as one way to glimpse this. When participants suspend assumptions and listen without defending identity, thought is seen in motion. The flow of meaning is no longer bound by past conclusions or projected outcomes. In such moments, awareness is present without the weight of continuity. Dialogue becomes a space where psychological time loosens its grip and relationship is no longer mediated by memory or projection.

Augustine, in *Confessions* XI, described human beings as 'distracted by times past and to come', distended between memory and anticipation. For him, only in eternity as timeless presence, was there true rest. Heidegger (1962), though coming from a different angle, insisted that authentic existence means confronting temporality without fleeing into distraction or projection. Levinas (1969) suggested that in ethical encounter with the other, time itself is disrupted: the immediacy of responsibility shatters the continuity of self-centred projection. Each of these perspectives gestures towards the possibility that the hold of psychological time is not inevitable.

McGilchrist (2021) suggested that the right hemisphere perceives reality in flow and context, unbound by the sequential logic of the left. This does not 'end' time in a metaphysical sense, but it illustrates that perception need not be chained to fragmentation. Awareness can perceive movement as a whole, not as fragments strung through psychological time. Trauma studies add another layer: van der Kolk (2014) describes how unresolved trauma traps people in 'the tyranny of the past'. Healing begins not by extending into future time but by integrating perception in the.

Eternal Movement (Gupta, 2024) makes the radical point that transformation is not in becoming but in ending. The idea of non-violence held in time is still part of violence, because the self sustains itself through projection. Change occurs only when the movement of becoming ceases, not when it is refined. The ending of time is the ending of the self's continuity. At the systemic scale, *Governance Without the Self* (Gupta, 2025) extends this insight. Frameworks structured around future targets perpetuate delay. To act from awareness, not from time, is to respond directly to the present. Governance in wholeness is not a process of reform but perception as action. Without this ending, institutions will continue to embody inertia, reforming endlessly while sustaining disorder.

The ending of time is not destruction but the cessation of projection. It does not require effort, for effort belongs to becoming. It requires seeing. When fear is observed as projection, its continuity ends. When desire is seen as imagination, its authority dissolves. When thought is recognised as fragment, it no longer masquerades as whole. The false falls away, leaving what is true.

Implications

The ending of time is the ending of the self as continuity. This does not mean the loss of memory or practical knowledge, but the dissolution of identification with them. It means meeting life without the residue of yesterday or the projection of tomorrow. In such ending lies freedom: freedom from sorrow, conflict and inertia for individuals, for organisations as actors, and for institutions as the rules and structures that shape collective life.

Individuals, organisations and institutions are not separate kinds of beings but expressions of a single movement of consciousness. That movement can be rigid, repeating the past under new names, or emergent, perceiving and acting without projection. When psychological time ends in the actor, its structures no longer reproduce inertia automatically; new forms of coordination arise from direct perception rather than from thought's continuity. This is not a prescription but a recognition: the same movement that binds the person binds the collective. Seeing this clearly is already the beginning of freedom.

7.8 Implications for Education, Governance and Sustainability

The distinction between physical and psychological time is not only philosophical. It has direct consequences for the challenges of education, governance and sustainability. All three domains operate on timelines, targets and projections. Yet when rooted in psychological time, they perpetuate the very disorder they claim to resolve.

Education Beyond Achievement Timelines

Education is structured around future achievement: exam preparation, career milestones, global rankings. Children are taught to become, to be more tomorrow than they are today. Learning, in this framework, is not inquiry in the present but preparation for a projected future. Fear of failure and desire for success dominate classrooms.

Freire (2000) critiqued this 'banking model' of education, in which students are treated as repositories of information. Dewey (1938) argued that education must be rooted in lived experience, because 'the present has its claims'. Zimbardo and Boyd (1999) showed that students oriented towards future achievement often experience anxiety and disengagement, while those rooted in present awareness show more creativity. These insights converge: the timeline-driven classroom sustains fear and conformity rather than awakening intelligence.

Governance Beyond Policy Timelines

Governance is dominated by targets and deadlines: 'eradicate poverty by 2030', 'achieve peace within five years', 'reach net zero by 2050'. While such goals create coordination, they also serve as escapes into the future, stabilising the present by projecting transformation outward.

Koselleck (2004) described this as 'futures past': horizons that rarely materialise but that regimes use to manage continuity. Arendt (1998) pointed instead to *natality* – the human capacity to begin anew in the present. Governance built on psychological time reduces politics to management; governance rooted in perception makes genuine action possible. Peace agreements in the Middle East and South Asia illustrate this: deadlines are set, yet conflict persists because its roots in identity and fear remain untouched.

Sustainability Beyond Roadmaps and Deadlines

Sustainability frameworks mirror the same postponement. Climate negotiations pledge 'net zero by 2050' or 'halt biodiversity loss by 2030', while extraction and emissions expand in the present. Timelines provide legitimacy but defer responsibility. Beck (1992) described modernity as a 'risk society', where attempts to manage crises generate new risks. Singh and Bhatnagar (2023) show how sustainability often functions as a 'wicked problem', entangled in competing interests and timelines. Bergson (1911/2022) noted that real life unfolds in *durée* – the lived time, not in measurements. By confusing lived transformation with dashboards and targets, sustainability frameworks risk perpetuating disorder rather than resolving it.

Implications

The lesson is clear. Timelines, targets and goals are not wrong in themselves, but when they substitute for direct perception, they reinforce disorder. Psychological time disguises inertia as progress. Transformation in education, governance and sustainability does not require better timelines but a different ground of action: awareness in the present. The ending of psychological time does not paralyse action; it liberates it. Action born of awareness does not wait for tomorrow or cling to yesterday. It is whole in the present and therefore coherent. Without love as its ground, even the most ambitious reforms remain projections. With love, education becomes inquiry, governance becomes responsibility without control and sustainability becomes care for the Earth rather than management of resources.

7.9 Conclusion

This chapter has traced the architecture of psychological time and its consequences. We distinguished between physical time, which is indispensable for practical life, and psychological time, which sustains the continuity of the self. In this inner time lie fear, desire, postponement and inertia. Together they form the movement of becoming – the effort to transform 'what is' into 'what should be'. The self is this continuity. It is not a permanent essence but a narrative constructed out of memory and projected into the future. Nations, communities and institutions mirror this same structure, sustaining identity through history, grievance and promise. Governance frameworks, development plans and sustainability targets all operate from this continuity. They provide the appearance of progress while entrenching disorder in subtler forms.

Krishnamurti (1983) called psychological time 'the enemy of man' because it displaces transformation into the future. Bohm (1996) explained how thought mistakes continuity for permanence, generating the illusion of an enduring self. Neuroscience echoes this, showing that the autobiographical self is continuously reconstructed, not fixed (Damasio, 2010). Gupta (2024, 2025) insists that change is not in becoming but in ending and extends this to systems: reform structured in time perpetuates continuity, while action born of awareness responds directly to the present.

The larger policy frameworks of our age reflect this trap. The SDGs, for instance, have set goals for the world to end poverty by 2030, achieve gender equality by 2050, or reach net zero in the second half of the century. These targets create legitimacy and order, yet they function as postponements. Dashboards may light up 'green' while ecosystems collapse, peace agreements coexist with militarisation and education enrolment deepens conformity. This is not failure of implementation but a deeper crisis: knowledge and time themselves become the frameworks of reform. When awareness is reduced to a skill or competency, as in many *Inner Development Goals* frameworks (Inner Development Goals, 2023), it becomes another instrument of time, performance and continuity. In this way, time disguises inertia as progress.

The radical implication is that transformation is not in time at all. It is not the product of reform, postponement or gradual improvement. Transformation occurs only when psychological time ends – when fear, desire, memory and projection are seen in their movement and their authority dissolves. What falls away is false; what remains is true.

This prepares the ground for the next chapter: if time ends, what remains? When relationship is no longer bound by image, comparison and becoming, what is the quality of the encounter? Chapter 8 will suggest that what arises in this ending is love – relationship beyond image, fear and time, the ground of wholeness.

References

Anderson, B. (2006). *Imagined communities: Reflections on the origin and spread of nationalism* Rev. ed.. Verso.

Arendt, H. (1998). *The human condition* (2nd ed.). University of Chicago Press. (Original work published 1958).

Beck, U. (1992). *Risk society: Towards a new modernity* M. Ritter, Trans.. SAGE.

Bergson, H. (1991). *Time and free will: An essay on the immediate data of consciousness* N. M. Paul & W. S. Palmer, Trans.. Dover. (Original work published 1910).

Bergson, H. (2022). *Creative evolution* A. Mitchell, Trans.. Dover. (Original work published 1911).

Bohm, D. (1996). *On dialogue*. Routledge.

Chatterjee, P. (1993). *The nation and its fragments: Colonial and postcolonial histories*. Princeton University Press.

Damasio, A. (2010). *Self comes to mind: Constructing the conscious brain*. Pantheon.

Dewey, J. (1938). *Experience and education*. Kappa Delta Pi.

Freire, P. (2000). *Pedagogy of the oppressed* 30th anniversary ed., M. B. Ramos, Trans.. Continuum. (Original work published 1970).

Freud, S. (2010). *Civilization and its discontents* J. Strachey, Trans.. W. W. Norton. (Original work published 1930).

Fukuda-Parr, S., & McNeill, D. (2019). *Knowledge and politics in setting and measuring the SDGs*. Routledge.

Girard, R. (1977). *Violence and the sacred* P. Gregory, Trans.. Johns Hopkins University Press. (Original work published 1972).

Gupta, D. (2024). *Eternal movement*. Emerge Publications. https://doi.org/10.65169/q7d2z6

Gupta, D. (2025). *Governance without the self: Consciousness and the end of fragmentation, systems, and complexity*. Emerge Publications. https://doi.org/10.65169/a9m3p4

Heidegger, M. (1962). *Being and time* J. Macquarrie & E. Robinson, Trans.. Harper & Row. (Original work published 1927).

Inner Development Goals. (2023). *IDG framework: Transformational skills for sustainable development*. IDG Initiative. https://www.innerdevelopmentgoals.org

Koselleck, R. (2004). *Futures past: On the semantics of historical time* K. Tribe, Trans.. Columbia University Press. (Original work published 1979).

Krishnamurti, J. (1983). *Krishnamurti to himself: His last journal*. Harper & Row.

Krishnamurti, J. (1991). *The awakening of intelligence*. Krishnamurti Publications of America.

Krishnamurti, J., & Anderson, A. W. (1991). *A wholly different way of living: Krishnamurti in dialogue with Professor Allan W. Anderson, San Diego 1974*. HarperSanFrancisco.

Levinas, E. (1969). *Totality and infinity: An essay on exteriority* A. Lingis, Trans.. Duquesne University Press.

Levine, P. A. (2015). *Trauma and memory: Brain and body in a search for the living past*. North Atlantic Books.

Levine, P. A. (2023). *An autobiography of trauma: A healing journey*. Sounds True.

Maté, G. (2010). *In the realm of hungry ghosts: Close encounters with addiction*. North Atlantic Books.

McGilchrist, I. (2021). *The matter with things: Our brains, our delusions, and the unmaking of the world*. Perspectiva Press.

Nietzsche, F. (1994). *On the genealogy of morality* K. Ansell-Pearson, Ed.; C. Diethe, Trans.. Cambridge University Press. (Original work published 1887).

Rittel, H. W. J., & Webber, M. M. (1973). Dilemmas in a general theory of planning. *Policy Sciences*, *4*(2), 155–169.

Singh, N., & Bhatnagar, D. (2023). *Applied spirituality and sustainable development policy*. Emerald Publishing.

Tilley, V. (2005). *The one-state solution: A breakthrough for peace in the Israeli–Palestinian deadlock*. University of Michigan Press.

United Nations Environment Programme. (2023). *Emissions gap report 2023*. Author.

United Nations General Assembly. (2015). *Transforming our world: The 2030 Agenda for Sustainable Development (A/RES/70/1)*. Author.

United Nations High Commissioner for Refugees. (2011). *The 1951 convention relating to the status of refugees and its 1967 protocol*. Author.

van der Kolk, B. A. (2014). *The body keeps the score: Brain, mind, and body in the healing of trauma*. Viking.

Whitehead, A. N. (1978). *Process and reality: An essay in cosmology* Corrected ed.; D. R. Griffin & D. W. Sherburne, Eds.. Free Press. (Original work published 1929).

Zimbardo, P. G., & Boyd, J. N. (1999). Putting time in perspective: A valid, reliable individual-differences metric. *Journal of Personality and Social Psychology*, *77*(6), 1271–1288.

Zimbardo, P. G., & Boyd, J. N. (2008). *The time paradox: The new psychology of time that will change your life*. Free Press.

Interlude: Crossing the Threshold

Part II of this inquiry has moved steadily inward, from observing the outer disorder of the world to uncovering the inner architecture of the self. We began with the proposition that the self is both the source and the effect of fragmentation, conflict and violence. Through each chapter, we have looked closely at its construction: the way thought sustains identity through memory, how knowledge divides, how belief creates psychological security, how relationships are shaped by images and how the movement of time sustains desire and fear.

Each step of this inquiry has not been analytical but observational. We have not sought to fix the self, improve it, or transcend it, but to see it as it is. And in seeing it clearly, without escape or judgement, something ends. That something is the very movement of the self, the endless becoming, comparison, control and continuation in time.

To come to this point is not to arrive at a new belief or conclusion. It is to stand in a different space altogether, a space where the known no longer guides perception, where the observer is not separate from the observed. If we have truly followed this movement with attention, we now find ourselves in a silence that is not constructed by effort. It is the silence that remains when thought has seen its limitation.

It is here, in this silence beyond the self, that something else may flower.

The next chapter begins from there.

Chapter 8

Love Without Image

Abstract

Chapter 8 unfolds the natural flowering of love when image ends. It investigates how fear, desire and memory distort love into attachment, dependence and control. This false love, visible in both personal life and public institutions, manifests as charity, duty or performance. The chapter contrasts emotional sentiment with the intelligence of love that arises when the self is quiet. Love without image is not an emotion but a state of awareness in which perception is whole. Such love is the essence of order, it acts without motive, comparison or reward. The chapter concludes that when the mind no longer seeks security through relationship, it discovers compassion that is inseparable from freedom. This marks the transition from seeing to being: the ground from which all right action flows.

Keywords: Love and awareness; ending image; fear vs love; dependency critique; natural compassion; perception in relationship; beyond attachment; intelligence of care

8.1 Introduction

Love is one of the most invoked words in human life, yet rarely understood in its essence. What is commonly called love is often memory, attachment or desire, shaped by images of the other, by fear of loss or by longing for fulfilment. These movements are not love. They are extensions of the self in time, the self that remembers, projects and seeks security.

Love cannot belong to memory. To say, '*I loved yesterday and I will love tomorrow*' is to reduce love to continuity, to thought carried over in time. Love is not the extension of the past into the present. It is alive only now.

Nor can love coexist with fear, desire or comparison. Where there is fear of abandonment, desire for possession or jealousy sustained by images, there is no love. These belong to psychological time, to the 'me' seeking comfort or control.

From Fragmentation to Wholeness, 165–176

doi:10.1108/978-1-80686-677-920261008

Love is not in becoming; it is not '*I will learn to love*' or '*I will practice love*'. The attempt to cultivate love is already its negation, because it turns love into an idea, a method, a goal.

Krishnamurti (1954) emphasised that love and freedom are inseparable. Where the self operates, with its ambitions, attachments and projections, there can be no love. Relationship under the dominance of the 'me' becomes exploitation, role or expectation. But when the self ends, when the movement of image, memory and becoming falls silent, what remains is care without condition. Love flowers naturally in the absence of the self.

This flowering is inseparable from the perception of oneness: the realisation that there is no division between self and other, human and nature, inner and outer. Yet this recognition cannot be approached through an idea of oneness. To begin from the thought '*we are all one*' is still the continuity of the fragment, the 'me' reaching towards what it imagines it should be. Such an idea becomes another projection in time, an escape from actuality and therefore a barrier to love. Only when the 'me' ends, not through effort, discipline or imitation, but through direct seeing of what is, can there be love that is not cultivated or performative.

Love is not created by thought, memory or practice. It cannot be possessed or secured. It is not an achievement. Love is the ending of image, time and fear. In that ending lies the possibility of relationship without division: a relationship not bound by the past or projected into the future, but whole in the present.

8.2 Love and Image

Human beings rarely meet each other directly. Between one person and another stands the image: the accumulated memory of words spoken, actions done, hopes projected and hurts remembered. This image becomes the lens through which relationship is filtered. A parent sees the child not as the living being before them, but as 'my son' or 'my daughter', shaped by expectation and ambition. A partner sees the other as husband, wife or role-bearer, rather than as a presence in the now.

Krishnamurti (1954) insisted that where there is image, there is no relationship. What is met is not the person but the past carried in thought. Bohm (1996) explained that thought creates categories that appear real and these categories then dictate interaction. Thus, relationship is mediated not by actuality but by memory, comparison and projection.

Poets have long intuited this fragility. Rahim (1997) wrote that bonds built on condition are as fragile as glass: once cracked, they cannot be restored to wholeness. Kabir (2003) sang that the thread of love breaks not with force but with subtle tension, and once broken, it cannot be joined again without a knot. Both point to the same truth: when relationship is built on images, it is always vulnerable, for images divide.

Premchand's literature shows the social consequences of this image-making. In Godan (1936/1957), Hori's life is governed not by love but by an image of religious duty, the idea that salvation requires donating a cow at death. His relationships are burdened by this illusion of duty; love of family collapses under ritual and fear. In Kafan (1936/2001), Ghisu and Madhav, trapped in hunger and despair, respond to death not with care but with apathy, their perception dulled by years of deprivation. In both cases, the image of duty, of survival, of societal expectation stands in place of direct relationship. What could have been love becomes continuity of fear, desire and conditioning.

Modern psychology confirms this subtle violence. Attachment theory shows how expectations, fears and projections from early childhood dominate adult relationships (Mikulincer & Shaver, 2007). Neuroscience suggests that the brain's predictive mechanisms constantly construct expectations about others, obscuring direct perception (Friston, 2010). These findings echo what poets, philosophers and novelists have long observed: where there is image, there is no meeting.

This dynamic extends into communities, institutions and nations. In education, students are seen through ranks and grades rather than as learners in the present moment. In governance, citizens are reduced to categories of voter blocs, migrants, minorities that serve political ends. Nations sustain themselves through myths of origin, enemy projections and symbols of identity. In each case, the image replaces relationship, sustaining conflict.

The question, then, is whether relationship can exist without image. To see another without the mediation of thought is not a matter of will or discipline. It happens only in awareness, when thought is seen as thought and memory as memory, without mistaking them for reality.

Such relationship is not fragile, for it does not rest on expectation. It does not depend on past or future, but flowers in the present. It is not the product of effort, but the natural movement of perception unclouded by image. In that seeing, love is not an ideal or a demand; it is the fact of direct encounter.

8.3 Love and Memory

Memory is indispensable in practical life. It allows us to recognise faces, speak language and navigate the world. Yet in relationship, memory often becomes a prison. When the mind relates to another through past pleasure or pain, it does not meet the person but an image sustained by memory. Love, in such a case, is replaced by recollection, expectation or resentment.

Krishnamurti (1983) observed that 'to carry the memory of pleasure or hurt into the present is to destroy relationship'. Love cannot coexist with continuity of grievance, nor with the repetition of pleasure projected into future demand. Bohm (1996) added that thought carries fragments of past experience as if they were present reality, sustaining distortion in dialogue.

Premchand's *Godan* captures this truth at the social scale. Hori lives under the weight of ancestral memory and ritual duty. The idea that salvation depends on performing *godan* is not born of present care but of inherited memory, sustained across generations (Premchand, 1936/1957). His love for family is distorted by this burden: providing for them is filtered through the imagined respectability of ritual. Love is obscured by the weight of memory masquerading as duty.

In *Kafan*, memory also operates, though negatively. Ghisu and Madhav's apathy is not only born of hunger; it is shaped by accumulated despair, a long memory of futility (Premchand, 1936/2001). Their refusal to act is conditioned by repeated patterns of failure. Thus even their grief is consumed by memory, what they carry of the past dictates their response in the present.

Nations operate similarly. Historical memory dictates collective identity. Wars, colonisation and partitions perpetuate division by repeating past hurt as present grievance. Anderson (1991) described nations as 'imagined communities', sustained not only by geography but by curated memory with rituals, textbooks and commemorations. In South Asia, the memory of Partition continues to shape hostility between India and Pakistan. Generations born long after 1947 inherit images of enemy and victim, ensuring that love between communities remains obstructed by inherited memory.

At the family level, memory equally distorts love. A parent who recalls a child's failure may unconsciously project that disappointment into every future encounter. The child is no longer seen freshly, but through the residue of past judgement. Love, in such cases, becomes conditional, bound to memory.

To see another without the weight of memory does not mean amnesia. It means perceiving when memory arises and not mistaking it for reality. Love can only exist when the past does not dictate the present. In this direct perception, memory finds its right place as a tool for life, not as a filter for relationship.

8.4 Love and Comparison

Comparison is another subtle violence that distorts relationship. To compare is to measure oneself against another in beauty, intelligence, wealth or virtue. Love, when mixed with comparison, becomes rivalry, envy or pride.

Krishnamurti (1983) argued that 'the moment you compare yourself with another, you destroy yourself'. In comparison, the fact of what is is denied in favour of an image of what should be. Love collapses under the pressure of measurement.

Modern psychology confirms this. Festinger's (1954) social comparison theory shows how self-worth is often derived from evaluating oneself against others, a process that produces chronic anxiety. Clark and Hogue (2020) demonstrate that comparison undermines trust, leading to reduced cooperation.

Institutions embody this same pattern. Education reduces students to ranks and grades, cultivating rivalry and fear. Workplaces measure performance through metrics that encourage competition over cooperation. Nations measure themselves through Gross Domestic Product (GDP), military strength or Sustainable Development Goal (SDG) rankings, perpetuating competitive nationalism.

Premchand's *Godan* again offers a poignant mirror. Hori's family is caught in endless comparison between classes, castes and religious duties. The desire to appear respectable in the eyes of others drives their suffering more than actual need (Premchand, 1936/1957). Similarly, in *Kafan*, Ghisu and Madhav's bitterness is amplified by comparison: their resentment towards wealthier villagers shapes their choices, as they measure dignity not in care but in fleeting indulgence (Premchand, 1936/2001). Comparison corrodes love, reducing life to rivalry and resentment.

In India's contemporary education system, millions of students prepare for high-stakes exams that determine careers. The result is not love of learning but a culture of fear, depression and even suicide (Deb et al., 2015). Globally, climate dashboards rank nations on performance, leading to blame-shifting rather than cooperation. Nations celebrate marginal improvements in rankings while ecosystems collapse.

In relationships, comparison erodes intimacy. A partner who measures affection against another's marriage no longer relates directly but through an image of 'better' or 'worse'. A parent who compares children fosters rivalry rather than care.

Love without comparison is possible only when measurement ends. To see another without asking '*how do they compare?*' is to meet them as they are. Such meeting is not conditioned by envy or pride, but by care and presence.

8.5 Love, Fear and Desire

Fear and desire are the strongest forces that masquerade as love. A parent fears losing a child's affection and clings; a partner desires security and confuses it with care; communities fear dissolution and enforce conformity as belonging. In each case, fear and desire sustain attachment, but attachment is not love.

Krishnamurti (1954) was unequivocal: love cannot exist where fear is. Fear narrows perception, distorts relationship and makes affection conditional. Desire, likewise, is not love. To desire another is to treat them as a means to fulfilment. As soon as desire is satisfied or frustrated, the 'love' turns into indifference, resentment or possession. Nietzsche (1887/1994) described desire as the will to power, a striving that never ends in completion. Girard (1977) showed how desire is mimetic: we desire what others desire, producing rivalry and scapegoating.

Neuroscience confirms the distinction. Damasio (2010) demonstrates how desire activates reward pathways in the brain, producing cycles of craving and dissatisfaction. Fear activates the amygdala, sustaining vigilance and defensiveness. Neither state allows the openness and presence that love requires.

Premchand's stories mirror this distortion. In *Godan*, Hori's devotion to his family is entangled with fear of dishonour and desire for social respectability (Premchand, 1936/1957). His sacrifices are not free acts of love but responses to insecurity. He gives, but his giving is conditioned by fear of exclusion and desire for recognition. In *Kafan*, Ghisu and Madhav's neglect of Budhiya is not simply cruelty, it is the victory of desire for food and drink over care. Fear of hunger and desire for fleeting indulgence suppress the possibility of love for the suffering woman (Premchand, 1936/2001). Both stories dramatise how fear and desire displace genuine relationship.

At systemic levels, the same holds. In education, students are driven by fear of failure and desire for success. In governance, leaders exploit fear of insecurity and desire for prosperity to consolidate power. In sustainability, climate negotiations are framed less by love of the earth than by fear of collapse and desire for economic advantage. Frameworks built on fear and desire perpetuate the very crises they aim to solve, for they mirror the psyche of becoming/

The US 'war on terror' after 9/11 exemplifies this. Policies were framed around fear of terrorism and desire for security, yet the result was endless war, surveillance and erosion of freedom. In climate policy, 'net zero by 2050' arises from fear of collapse and desire for stability, yet emissions continue to rise. Both fear and desire postpone transformation into the future, disguising inertia as action.

Love, by contrast, has no motive. It is not driven by fear of loss or desire for gain. It arises only when the mind is free of projection. In love, the other is not a means to security but an end in themselves.

8.6 Love and Dependency

Dependency is often mistaken for love. A partner may say, '*I cannot live without you'*, or a community may demand conformity as belonging. Yet dependency arises from fear and insecurity, not from freedom. It makes the other a crutch for the self, binding relationship in chains of need.

Krishnamurti and Anderson (1991) insisted that where there is dependency, there is no love. To depend is to use another as an escape from loneliness, emptiness or fear. Such dependency inevitably breeds resentment, because the one depended upon feels burdened, while the dependent feels perpetually insecure.

Psychology recognises this as attachment anxiety, where fear of abandonment drives clinging behaviour. Studies show that dependency reduces trust and increases conflict (Mikulincer & Shaver, 2007). What begins as affection turns into possession, jealousy and control.

Premchand's *Godan* illustrates this subtly. Hori's family relationships are shaped by economic dependency. Women and children depend on him for survival, while he depends on landlords and moneylenders. These chains of need suffocate genuine care. Love is obscured because relationship is mediated by survival (Premchand, 1936/1957). In *Kafan*, dependency appears differently: Ghisu and Madhav depend parasitically on villagers' charity. Their lack of responsibility reveals not love but manipulation, a bond sustained by need, not freedom (Premchand, 1936/2001). In both, dependency corrodes care, turning it into survival mechanism or exploitation.

At systemic levels, dependency operates in similar ways. Postcolonial states remain dependent on former colonial powers for aid and legitimacy. Economies are tied into dependency through debt and trade hierarchies. Development frameworks present dependency as 'partnership', but in reality, they reinforce asymmetry (Frank, 1967). In education, dependency arises when students are trained to rely on authority for answers rather than inquiry. Freire (1970) called this the 'banking model': students depend on teachers to deposit knowledge rather than engaging in co-discovery.

In households, economic dependency often traps women in abusive relationships, sustained by fear of survival without the provider. In governance, welfare schemes sometimes reinforce dependency by making support conditional on obedience. All these forms confirm that dependency is not care but bondage.

Love, by contrast, requires freedom. It is not clinging but meeting. It does not ask the other to fill a void but sees them as they are. In love, relationship is not born of psychological need but of wholeness. Only then can care, responsibility and joy flow without fear.

8.7 Love, Institutions and the Image of Care

Institutions, like individuals, often confuse the image of love with its reality. They create structures of 'care' through policies, programmes and symbolic gestures, but beneath the surface, these are frequently motivated by fear, desire or dependency. What appears as love is often management, performance or the pursuit of legitimacy.

The Image of Care in Education

Schools and universities frequently speak of nurturing students, but their practices are dominated by ranking, competition and discipline. Students are told that care lies in preparing them for success, but this 'care' is conditional on performance. Freire (1970) critiqued this as the hidden violence of the 'banking model', where students are treated as containers for knowledge, moulded for productivity rather than met as beings in relationship.

Premchand saw the same hollow care in *Godan*. Hori's sacrifices for family and society were framed as responsibility, yet they left him exploited, crushed under debt and robbed of dignity (Premchand, 1936/1957). His devotion was respected as 'duty', but not recognised as suffering. The system appeared to care, but only to the extent that he remained compliant. Such care is not love; it is the preservation of order under the guise of compassion.

The Image of Care in Governance

Governments present welfare and protection as love for citizens. Yet welfare programmes are often tied to compliance, dividing people into 'deserving' and 'undeserving'. Surveillance states justify control as protection, reducing citizens to regulated objects. Foucault (1977) described this as biopower: life itself is managed and normalised under the image of care.

Premchand's *Godan* again illuminates this dynamic. Hori's longing for a cow was not merely personal, it symbolised social legitimacy. Institutions as priests, landlords and caste elders profited from his sense of duty, while cloaking their exploitation in moral authority (Premchand, 1936/1957). The rhetoric of responsibility displaced love; rules of conformity became more important than the living man before them.

The Image of Care in Sustainability

The same pattern governs sustainability. Corporations market 'green initiatives', governments announce 'net zero' targets, and global institutions promote the SDGs as evidence of planetary compassion. Yet as Gupta (2025) observes, these frameworks are designed more for legitimacy than transformation. They produce dashboards and indicators that perform care while ecosystems collapse in the present. Beck (1992) called this the paradox of the risk society: efforts to manage crises often generate new risks, sustaining disorder under the language of responsibility.

Love Beyond Image

The difference between love and the image of love is radical. Love requires no proof, policy or performance. It is care without condition, responsibility without hierarchy, encounter without projection. Institutions can organise support, but unless their ground is awareness rather than fear, desire or dependency, their 'care' will remain only an image.

Krishnamurti (1954) pointed out that the image always divides. To care through an image, whether of the 'good student', the 'grateful citizen' or the 'sustainable consumer' is to reduce life to categories. Real care begins only when the image ends.

Premchand, through characters like Hori, revealed how societies use the image of duty and responsibility to perpetuate exploitation while silencing love (Premchand, 1936/1957). Krishnamurti showed that even in intimate relationships, the image stands in place of the living other. At both personal and institutional levels, the problem is the same: when care is mediated by image, it is control or conformity, not love.

Thus, the challenge for education, governance and sustainability is not to design better images of care, but to question whether care can exist at all without love. Without this, institutions will continue to perform compassion while perpetuating fragmentation.

8.8 Love and Freedom: Beyond Possession

Love and freedom are inseparable. Where there is possession, control or dependency, there is no freedom and therefore no love. The tragedy of human relationship is that we often mistake possession for care and dependency for love.

This distortion pervades not only families and communities, but also institutions and nations.

Possession in Personal Relationship

At the personal level, love is often entangled with attachment. A partner is seen as 'mine', a child as 'my extension', a friend as 'my ally'. Possession creates expectations: the other must conform to the image I hold of them. When they deviate, conflict arises. Krishnamurti (1954) observed that this possessiveness is fear masquerading as love. Fear of loneliness, fear of loss and fear of insignificance push the mind to cling to another. But this clinging is not care for the other as they are; it is a demand that they sustain my sense of security.

Premchand's *Kafan* dramatises this absence starkly. Ghisu and Madhav, numb with poverty and despair, cannot even grieve Budhiya's death. Their possession is not of her life, but of her death, the little money collected for her shroud is consumed in food and drink. What could have been dignity in ritual becomes possession in appetite. The story shocks precisely because it shows the collapse of love when survival, fear and desire dominate (Premchand, 1936/2001).

Possession in Communities

Communities, too, often mistake control for love. Religious groups claim to 'protect' their members, but this protection is conditional on conformity. Cultural traditions speak of 'care', yet enforce obedience through shame and exclusion. Possession here takes the form of collective control: belonging is granted only if one submits to the image of the group.

Possession in Governance

Governments frequently claim to act out of care for their citizens through security, welfare or development but their measures often embody control. Surveillance states insist that monitoring is for safety, yet it reduces citizens to objects of regulation. Welfare tied to compliance reduces care to dependency. Nations profess to love peace, yet cling to territory and dominance, equating possession with security. Foucault (1977) described this as the logic of discipline: institutions managing life itself under the guise of care.

Premchand's *Godan* echoes this critique. Hori's lifelong pursuit of respect through donating a cow is framed as duty, but in truth it is possession, his desperate attempt to own social legitimacy. The institutions around him reinforce this illusion, demanding conformity while profiting from his sacrifice (Premchand, 1936/1957).

Possession in Sustainability

Even sustainability discourse can fall into this trap. The planet is spoken of as a possession: *'our resources', 'our environment', 'our biodiversity'.* Care for the Earth is framed in terms of management, ownership and exploitation, even when described as 'protection'. Capra (1996) warned that this mechanistic relationship to nature sustains ecological destruction. Love for the Earth is not control, stewardship or management; it is freedom from the illusion of separateness.

Love and Freedom as One

The essence of love is freedom: freedom from fear, from image, from possession. To love is to allow the other to be, not to mould them into what one desires. In freedom, there is no dependency, for dependency is a subtle form of possession. In love, there is no domination, for domination is fear disguised as protection.

Premchand's stories reveal what happens in the absence of this freedom: relationships collapse into appetite (*Kafan*) or duty (*Godan*), leaving no space for love. Krishnamurti showed that the same distortion occurs in personal life and institutions: when relationship is mediated by fear, desire or control, it is not love but possession. Bohm (1996) added that dialogue collapses when participants defend identity rather than suspend it then freedom is lost, and with it, the possibility of genuine care.

Love without freedom is not love; it is attachment, dependency or control. Freedom without love is not freedom; it is isolation, indifference or neglect. Love and freedom are one movement: inseparable, indivisible and whole.

8.9 Conclusion

This chapter has explored love not as sentiment, attachment or ideology, but as the movement of freedom and wholeness. Love is not in time, for it is not the projection of desire or memory. Love is not in image, for it cannot coexist with comparison, possession or dependency. Love is not in ideology, for it cannot be contained by belief, duty or doctrine.

We saw how personal relationships are corrupted by image and memory, how fear and desire masquerade as care, how communities and nations mistake conformity and control for solidarity, and how institutions perform 'care' while enacting possession. Premchand's *Kafan* revealed how poverty and despair can erode even the possibility of love, reducing relationship to survival and appetite. His *Godan* showed how duty and ritual can imprison care within possession, turning love into obligation rather than freedom. These narratives expose the systemic absence of love when life is governed by fear, duty or possession.

Krishnamurti (1983) insisted that love cannot be cultivated, practised or secured by will. Where there is fear, image or dependency, there is no love. Bohm (1996) explained that dialogue collapses when identity is defended, and that only in freedom from possession can meaning flow between people. McGilchrist (2021) suggested that wholeness of perception requires a shift from control to openness, from left-hemispheric dominance to relational presence. Gupta (2024, 2025) reinforced that love flowers when the self ends, not as an idea or method, but as perception in the present.

At systemic levels, the same truth holds. Education organised around performance and comparison may claim to care for students, but in fact it trains ambition and fear. Governance framed around security and welfare may present itself as care, but in fact it manages through possession and dependency. Sustainability discourse may speak of protecting the Earth, but when framed in terms of ownership, *'our resources', 'our biodiversity'*, it sustains the same fragmentation that destroys ecosystems. These are not failures of implementation; they are distortions of care into possession. And just as Chapter 7 showed, without love, even the most well-intentioned sustainability targets and SDGs risk becoming instruments of postponement, timelines that defer transformation instead of ending disorder in the present.

Arendt (1958) described natality as the capacity to begin anew, to act without being bound by the inertia of the past or the projections of the future. Love has the same quality: it begins afresh in the present, unconditioned by memory, duty or possession. Just as natality makes politics possible, love makes relationship possible, not as management or reform, but as the ground of freedom. Without love, governance becomes control, sustainability becomes management and education becomes performance. With love, each becomes an expression of wholeness: care without dependency, responsibility without fear, belonging without exclusion.

The radical implication is that love is not separate from freedom, awareness or intelligence. It is the same movement. Where love is, action is whole. Where love is not, even noble frameworks reproduce fragmentation.

This prepares the ground for the next movement of this book. Part III turns from inward inquiry to outward life. It asks how intelligence and love express themselves as action in the world of governance, education and sustainability. If love is not in time, not in image, not in duty and not in possession, then how does action arise when life demands responsibility and decision?

Chapter 9 begins here, exploring choiceless action and intelligence, an action not driven by fear or desire, not chosen by will, but emerging from awareness and love as one indivisible movement.

References

Anderson, B. (1991). *Imagined communities: Reflections on the origin and spread of nationalism* Rev. ed. Verso.

Arendt, H. (1958). *The human condition*. University of Chicago Press.

Beck, U. (1992). *Risk society: Towards a new modernity*. Sage Publications.

Bohm, D. (1996). *On dialogue*. Routledge.

Capra, F. (1996). *The web of life: A new scientific understanding of living systems*. Anchor Books.

Clark, M. S., & Hogue, M. (2020). Social comparison in close relationships. *Current Opinion in Psychology*, *33*, 91–95. https://doi.org/10.1016/j.copsyc.2019.07.021

Damasio, A. (2010). *Self comes to mind: Constructing the conscious brain*. Pantheon.
Deb, S., Strodl, E., & Sun, J. (2015). Academic stress, parental pressure, anxiety and mental health among Indian high school students. *International Journal of Psychology and Behavioral Sciences*, *5*(1), 26–34. https://doi.org/10.5923/j.ijpbs.20150501.04
Festinger, L. (1954). A theory of social comparison processes. *Human Relations*, *7*(2), 117–140.
Foucault, M. (1977). *Discipline and punish: The birth of the prison* A. Sheridan, Trans. Pantheon.
Frank, A. G. (1967). *Capitalism and underdevelopment in Latin America: Historical studies of Chile and Brazil*. Monthly Review Press.
Freire, P. (1970). *Pedagogy of the oppressed*. Continuum.
Friston, K. (2010). The free-energy principle: A unified brain theory?. *Nature Reviews Neuroscience*, *11*(2), 127–138. https://doi.org/10.1038/nrn2787
Girard, R. (1977). *Violence and the sacred* P. Gregory, Trans. Johns Hopkins University Press. (Original work published 1972).
Gupta, D. (2024). *Eternal movement*. Emerge Publications. https://doi.org/10.65169/q7d2z6
Gupta, D. (2025). *Governance without the self: Consciousness and the end of fragmentation, systems, and complexity*. Emerge Publications. https://doi.org/10.65169/a9m3p4
Kabir. (2003). *Songs of Kabir* R. Tagore, Trans. Dover Publications. (Original work published ca. 15th century).
Krishnamurti, J., & Anderson, A. W. (1991). *A wholly different way of living: Krishnamurti in dialogue with Professor Allan W. Anderson, San Diego 1974*. HarperSanFrancisco.
Krishnamurti, J. (1954). *The first and last freedom*. Harper & Brothers.
Krishnamurti, J. (1983). *The wholeness of life*. Harper & Row.
McGilchrist, I. (2021). *The matter with things: Our brains, our delusions, and the unmaking of the world*. Perspectiva Press.
Mikulincer, M., & Shaver, P. R. (2007). *Attachment in adulthood: Structure, dynamics, and change*. Guilford Press.
Nietzsche, F. (1994). *On the genealogy of morality* K. Ansell-Pearson, Ed.; C. Diethe, Trans. Cambridge University Press. (Original work published 1887).
Premchand. (2001). The shroud (Kafan). In *The Oxford India Premchand* (T. Vijay, Trans) (pp. 457–465). Oxford University Press. (Original work published 1936).
Premchand. (1957). गोदान(*Godan*). Hind Pocket Books. (Original work published 1936).
Rahim, A. K. (1997). *Rahim ke dohe*(*Couplets of Rahim*). Rajkamal Prakashan. (Original work published ca. 17th century).

Turning Point: The Ending of 'Me' and the Emergence of Choiceless Action

Part I of this book began in the world: its crisis, fragmentation and search for sustainability. But the inquiry soon revealed that these external crises mirror the structure of human consciousness. Systems thinking, global goals and institutional reform cannot resolve fragmentation if they emerge from the same divided ground.

Part II turned inward. Chapter by chapter, it observed the self not as a fixed identity but as a movement of becoming, sustained by knowledge, image, fear and psychological time. We saw how thought divides perception, how relationships are corrupted by possession and how even love becomes entangled in duty, desire and conformity. But we also saw that when this movement is seen wholly, not resisted, not justified, it ends.

The ending of the 'me' is not despair. It is not silence imposed by will. It is the space in which love begins. Not love as sentiment, not as duty, not as possession, but as presence, attention, and freedom. The same movement that ends the self is the movement of intelligence and compassion. There is no division between love and awareness, between stillness and action.

From this wholeness, action arises, not as reaction, not as choice among alternatives, but as the natural unfolding of perception. This is choiceless action. It is not driven by fear or ambition, not shaped by identity or delayed by psychological time. It does not begin in thought; it begins in seeing. Where there is wholeness of perception, there is wholeness of action.

This marks the turning point of the book, and of the human journey. From fragmentation to wholeness. From becoming to being. From image to love. From choice to choicelessness. This is not a transition between methods but a movement beyond method. What ends here is the continuity of psychological separation. What begins is the flowering of intelligence.

Part III now explores how such intelligence moves in the world. If the self is no longer the centre, how does one act in education, governance and sustainability? If love and awareness are the same movement, how do they respond when the world demands responsibility, clarity and care?

We now step into this final arc, not as a return to reform, but as the unfolding of action born of insight.

Part III

Holistic Response

The journey so far has revealed that the crises we face in sustainability, education and governance are not separate from the fragmentation within. The Sustainable Development Goals, systems thinking and policy frameworks, though well-intentioned, are rooted in identity, knowledge and time. They seek transformation through methods that continue the same disorder.

In turning inward, Part II observed how this fragmentation arises from the structure of the self as memory, image, fear and becoming. That enquiry brought us to the ending of the 'me' and the emergence of a different movement altogether, one of presence, love and choiceless awareness.

Part III now explores how action flows from that ground. These chapters are not proposals for reform but enquiries into how awareness moves when it is not bound by identity, ideology or image.

Chapter nine explores choiceless action and intelligence, challenging the assumption that freedom lies in will or in choosing between alternatives. Chapter 10 examines leadership and governance as the outward extension of identity or intelligence. Chapter 11 considers how education and public systems can function not from correction but from coherence. Chapter 12 turns to institutions, asking if they can become mirrors of perception rather than instruments of control. Chapter 13 concludes with the insight that transformation is not deferred to the future but begins when attention is whole now.

This part is not about applying the previous insights, for love and intelligence are not applications. It is about seeing how life unfolds when perception is no longer fragmented. When the self is absent, wholeness is the ground. From that ground, action is choiceless, not determined by method but arising from clarity itself.

Chapter 9

Choiceless Action and Intelligence

Abstract

This chapter explores action beyond the dualities of choice and control. It shows that decision-making rooted in the self is always conditioned by fear, desire and memory, creating a contradiction between thought and perception. True freedom, the chapter argues, lies not in choosing between options but in perceiving without a chooser or a motive. Drawing parallels with insights from psychology and contemplative philosophy, it shows that intelligence acts instantly when the mind is silent. Choiceless awareness is not indifference; rather, it is clarity in which the right action is self-evident. This chapter also illustrates how this principle transforms leadership, education and interpersonal life: where perception replaces reaction, order arises naturally. Choiceless action thus becomes the practical expression of freedom.

Keywords: Choiceless awareness; freedom from choice; conditioned decisions; clarity in perception; psychological freedom; leadership through intelligence; direct action; awareness in life

9.1 Introduction: From Inward Inquiry to Response

The previous chapters have shown how fragmentation shapes human life. The self is a construct of memory and image; psychological time sustains fear, desire and postponement; and knowledge and belief solidify fragments into identity. In each case, the result is the same: an action that is divided, conditioned and incapable of resolving disorder.

Yet action is unavoidable. To live is to act in relationship, in society and in governance. The question, then, is not whether we act but from where action arises. Most action today is rooted in identity, choice and knowledge. It is propelled by fear of loss, desire for gain or the inertia of the past. Even well-intentioned reforms in education, sustainability or governance are often

From Fragmentation to Wholeness, 181–217

doi:10.1108/978-1-80686-677-920261009

extensions of this conditioned movement. They may alter appearances, but they rarely end the patterns of division beneath.

This brings us to a crucial turning point in our inquiry: is there a form of action that does not perpetuate disorder? If conditioned action inevitably reproduces fragmentation, what is the nature of right action? Can there be a movement that is not rooted in choice, calculation or identity but flows from a deeper clarity?

Jiddu Krishnamurti (1969/2009) described this possibility as choiceless awareness. To many, the phrase sounds paradoxical, as if awareness without choice implied passivity or indifference. Yet Krishnamurti insisted it is the opposite: when the mind sees directly, without distortion, action arises naturally. There is no need to choose, because clarity leaves no contradiction. In his words, 'When there is complete attention, there is no choice'.

Modern research echoes this challenge to the assumption that choice is freedom. Neuroscience shows that decisions often arise unconsciously before awareness registers them (Libet, 1983; Soon et al., 2008). Psychology demonstrates how behaviour is shaped by automatic processes and social cues (Bargh & Chartrand, 1999). Sociology reveals how institutions and structures embed dispositions, making certain 'choices' appear natural or inevitable (Bourdieu, 1977; Giddens, 1984). If both personal and collective decisions are conditioned, then freedom cannot lie in choice at all.

This chapter explores freedom as intelligence. Not intelligence as the accumulation of knowledge, nor as performance measured by systems, but intelligence as the clarity that perceives wholeness and responds directly. Action from such intelligence is not the continuation of the self but movement from awareness itself.

This movement is not limited to the inner life. The same structures of choice and identity that fragment the psyche reappear outwardly in leadership, governance and institutions. To understand choiceless action in relationship is, therefore, to prepare for the next step: seeing whether public life, too, can function from clarity rather than confusion, from perception rather than projection. Chapter 10 takes up this challenge.

9.2 The Illusion of Choice

In most societies, the ability to choose is celebrated as the hallmark of freedom. The consumer who selects between hundreds of products, the voter who casts a ballot, and the student who picks a career all are thought to embody autonomy. Policymakers and economists measure progress by expanding 'choice architectures', designing markets and governance systems to multiply alternatives. Having more choice is equated with being more free.

Yet beneath this celebration lies a paradox. The very act of choosing often hides the fact that alternatives are shaped by conditioning: by memory, desire, fear and social structures. What feels like freedom may actually be the repetition of patterns set long before the moment of decision.

Choice as Continuity of the Past

Every choice carries the weight of what came before. The food we prefer, the career we pursue, the political platform we support, are all guided by past experiences, cultural habits and emotional associations. Krishnamurti (1969/2009) expressed this with stark clarity: 'Choice exists only when the mind is confused'. Where there is clarity, he argued, there is direct action, not deliberation. Choice arises only when the mind is caught between competing desires or memories, and therefore is never free; it is continuity in disguise.

Contemporary neuroscience supports this observation. Antonio Damasio (2010) showed that decision-making relies on 'somatic markers', bodily signals linked to past emotional experiences. What seems like rational evaluation is, in fact, guided by memory, layered into the body itself. The chooser is never starting fresh; the past speaks through preference.

Unconscious Determinants of Choice

Research goes further. Benjamin Libet's (1983) pioneering experiments revealed that the brain initiates activity for voluntary actions milliseconds before subjects report the conscious intention to act. Later studies extended this temporal gap to several seconds (Soon et al., 2008). Conscious 'choice' appears as a narrative constructed after unconscious processes have already determined the direction of action.

Psychologists John Bargh and Tanya Chartrand (1999) call this 'the unbearable automaticity of being'. Much of human behaviour, they argue, is shaped by unconscious priming and social cues rather than conscious deliberation. Preferences emerge not in isolation but in relation to environmental triggers, many of which the individual does not notice.

Daniel Wegner (2002) pushed this idea further in *The Illusion of Conscious Will*, arguing that the sense of agency itself may be a post-hoc construction. What we call 'choosing' might be less an act of freedom than a story we tell ourselves after behaviour is already in motion.

Sheena Iyengar (2010) showed that multiplying options does not necessarily increase satisfaction or autonomy. In many cases, more choice leads to paralysis or anxiety, revealing that an abundance of alternatives is not the same as freedom.

Social and Institutional Conditioning

The illusion deepens at the collective scale. Pierre Bourdieu's (1977) notion of *habitus* shows how societies inscribe dispositions into individuals so deeply that certain actions feel 'natural' and others unthinkable. Choices appear free, but they are expressions of ingrained patterns. Anthony Giddens (1984) described this as structuration: institutions reproduce themselves by embedding practices that seem voluntary but in fact sustain continuity.

This is stark in governance. Nations 'choose' between economic models, but nearly all remain within growth-centric paradigms. Leaders 'decide' climate policy, yet fossil fuel subsidies persist. These contradictions are not accidents; they reveal how systemic inertia and identity shape what is even conceivable as an option. Ulrich Beck (1992) noted that modernity itself is a 'risk society', where institutions perpetuate crises while claiming to manage them.

Herbert Marcuse (1964) anticipated this critique in *One-Dimensional Man*: the supposed freedom to choose between products, parties or careers is already contained within a system of manufactured needs. Similarly, Byung-Chul Han (2017) argued that neoliberalism internalises control: individuals believe themselves free precisely when they conform most fully to invisible mechanisms of power.

The False Equation of Choice and Freedom

If choice is bound to memory, unconscious processing and institutional structures, then equating choice with freedom is a profound mistake. Krishnamurti (1969/2009) pointed out that freedom is not in choosing between conditioned alternatives but in seeing without choice. Where there is clarity, action is immediate; it does not pass through deliberation or preference.

This reframes freedom entirely. Freedom is not the multiplication of options but the ending of confusion. It is not the power to select but the absence of the chooser. In this light, the celebrated structures of modern life, the consumer markets, democratic elections and policy reforms are revealed as limited. They may expand options, but they rarely question the conditioning that shapes them.

Implications

The implications of this inquiry are profound. If choice is shaped by memory, unconscious processes and institutional structures, then transformation cannot be secured through 'better choices' alone. The individual who shops ethically, the citizen who votes for reform, or the policymaker who drafts new frameworks may act with good intentions, but if the movement of thought remains conditioned, the outcome will largely preserve the same patterns of division. Surface adjustments occur, but the root fragmentation remains untouched.

At the personal level, the assumption that freedom lies in choosing between desires sustains confusion. A person may deliberate endlessly between career paths, relationships or lifestyles, believing that the right choice will bring fulfilment. Yet when the chooser itself is bound by fear, ambition and memory, no choice can bring freedom. The cycle of uncertainty simply renews itself under a different form.

At the collective level, the celebration of democratic elections, consumer markets or global frameworks such as the Sustainable Development Goals (SDGs) often overlooks the deeper fact that these structures still operate within conditioned paradigms. Expanding the number of options does not dissolve the habits of competition, identity and control that shape them. As Marcuse (1964) and Han (2017) caution, systems of modern life often disguise conformity as freedom, offering the individual the experience of choice while leaving intact the very structures that perpetuate crisis.

At the level of governance, the illusion is even more visible. Nations appear to 'choose' between policy models, yet nearly all remain tethered to growth-centric and security-driven paradigms. Leaders debate climate policy, yet fossil fuel subsidies persist. Such examples reveal that the field of choice is itself limited by institutional inertia, historical identity and systemic fear. To respond to global crises with the same conditioned logic that created them is to reinforce disorder under new labels.

This leads to a radical conclusion. Freedom is not found in the multiplication of options, nor in the refinement of strategies, but in the ending of confusion. It is not the power to select but the absence of the chooser. Transformation requires a shift from asking 'What should I choose?' to asking 'Can there be perception without choice?'

This question opens the door to the possibility of choiceless awareness. It points to a different quality of action, one that is not driven by preference, calculation or identity, but arises from clarity itself. The next section explores this possibility and asks what it means for philosophy, psychology and governance when action flows from awareness rather than choice.

9.3 What Is Choiceless Awareness?

If choice is conditioned, then what does it mean to act without choice? The phrase *choiceless awareness*, used often by Jiddu Krishnamurti, describes a state of perception where the mind sees directly without the filter of preference, judgement or self-interest. To many, this idea sounds paradoxical: how can there be action without choice? Is not choice the essence of decision? Yet choiceless awareness reveals that true action arises not from selection among alternatives but from immediate perception of what is.

Seeing Without the Chooser

Krishnamurti (1969/2009) pointed out that whenever there is a chooser, the 'I' that deliberates between this and that, perception is already fragmented. The chooser looks through the lens of memory, desire or fear, and therefore never sees clearly. In choiceless awareness, there is no such filter. The mind observes without translating perception into preference. In that observation lies clarity, and from clarity flows natural action. He insisted, 'When there is complete attention, there is no choice'.

This does not imply passivity or indecision. Just as one withdraws a hand from fire without deliberation, so does choiceless awareness act directly in response to reality. The difference is that here, action is not born of calculation but of direct seeing.

Philosophical and Contemplative Parallels

Though articulated with singular force by Krishnamurti, the idea resonates across traditions. Heidegger's notion of *Gelassenheit* (releasement) describes a posture of letting beings be, in which one refrains from imposing will and allows truth to disclose itself (Heidegger, 1959/1966). Sartre (1943/1992) distinguished authentic existence from inauthenticity precisely in terms of fleeing from direct experience into roles or self-images. Both point to the possibility of freedom not as choosing, but as being present to what is.

Buddhist thought echoes this distinction. The *Satipatthāna Sutta* emphasises bare attention (*sati*), awareness without clinging or aversion. Thich Nhat Hanh (1999) called this 'the miracle of mindfulness', where awareness itself is the act of liberation, not through effort but through presence. Similarly, Varela et al. (1991) in *The Embodied Mind* describe enactive awareness: cognition not as representation but as direct engagement with the world. Awareness here is not about choosing between mental images but about participation in lived immediacy.

Simone Weil (2002a, 2002b) adds another dimension. For her, attention without demand is the highest form of openness: 'Attention, taken to its highest degree, is the same thing as prayer'. This echoes choiceless awareness as pure receptivity without the interference of will.

The Ending of Division

David Bohm (1996) emphasised that thought inevitably fragments perception, creating a division between observer and observed. Choiceless awareness is present when this division dissolves, when the observer is not separate from the observed but is the observation itself. In such moments, perception is whole, and action arises as an expression of that wholeness.

This is echoed in Merleau-Ponty's (1962) phenomenology of perception. For him, perception is not a detached representation but an embodied opening to the world. The body 'knows' before thought intervenes. Choiceless awareness is akin to this embodied immediacy: a seeing that is already participation.

Insights From Psychology and Neuroscience

Modern science provides converging insights. Antonio Damasio (2010) demonstrated that much of human action is guided by pre-reflective, embodied processes rather than deliberate reasoning. When not distorted by fear or conditioning, these processes orient behaviour in adaptive, coherent ways.

Neuroscience further distinguishes between the brain's hemispheres. McGilchrist (2019) argued that the left hemisphere isolates, categorises and manipulates fragments, while the right hemisphere perceives context, flow and wholeness. Choiceless awareness, in this sense, aligns with the right hemisphere's openness: perception not reduced to categories but attentive to the whole field.

Contemporary research on mindfulness and attention also provides parallels. Kabat-Zinn (1990) showed that non-judgemental awareness reduces reactivity and bias, enabling responses more attuned to the present. Davidson and Kaszniak (2015) note that such awareness fosters resilience, not by training the mind to prefer certain states but by suspending preference altogether. Ellen Langer (1989) highlighted a similar principle in her work on mindfulness in psychology: true awareness comes from noticing without clinging to categories, a direct engagement rather than habitual labelling.

Judson Brewer (2017) showed how awareness interrupts conditioned loops of craving. In his research, the simple act of observing craving without judgement dissolves its power. This reinforces Krishnamurti's point that observation without choice carries its own transformative action.

Not a Method, Not a Wilful Act

A central misunderstanding is to treat choiceless awareness as something to cultivate through method or effort. To attempt to produce choiceless awareness is already a movement of will, of the self trying to achieve a state. Krishnamurti, throughout his life, warned that this very effort reintroduces choice, comparison and becoming, precisely the movement that sustains disorder.

Choiceless awareness is not manufactured; it is uncovered. It is present when the mind is still enough to see without distortion. In that seeing, clarity itself organises action.

Direct Action, Not Passivity

Another misunderstanding is to confuse choiceless awareness with passivity. In fact, it is intensely active. When perception is whole, action flows directly, without conflict. There is no delay, no oscillation between alternatives and no contradiction between inner clarity and outer response. In this sense, choiceless awareness is the ground of right action, action that is neither reaction nor calculation but intelligence in motion.

Implications

If choiceless awareness is possible, it reshapes how freedom, responsibility and transformation are understood. Modern culture often equates freedom with the multiplication of options or the refinement of decision-making frameworks. Yet choiceless awareness reveals that freedom is not the ability to choose among conditioned alternatives, but the absence of the chooser itself.

This shift carries several implications.

At the personal level, choiceless awareness dissolves the restless pursuit of the 'right' decision as the path to fulfilment. Individuals often live in cycles of indecision, believing that the right choice of career, partner or ideology will bring stability. But when the chooser itself is seen as conditioned, the cycle of becoming loses its authority. Freedom, then, is not found in managing choices but in perceiving without the division of the self.

At the relational level, choiceless awareness transforms communication. Much of human conflict arises because individuals respond to one another through the filters of memory, hurt or expectation. Choiceless awareness interrupts this cycle. When perception is direct, without translating the other into an image, the relationship becomes a meeting in the present. Compassion, then, is not cultivated as an ideal but arises naturally when division ends.

At the societal level, the implications are profound. Governance systems operate largely by designing choices: policy alternatives, market incentives and institutional reforms. Yet these choices remain bound by identity, fear and the inertia of tradition. If awareness itself is fragmented, then the structures it creates will reproduce fragmentation. Choiceless awareness suggests that true change in governance cannot come merely from better systems of decision but from perception that is not rooted in identity or ideology.

In education, choiceless awareness challenges the dominant model of achievement and preference. Students are trained to choose careers, ranks or skills as if these were markers of freedom. In reality, this sustains conditioning. An education rooted in awareness would not condition the mind to make 'better' choices but would open the mind to see without distortion. Such education frees intelligence rather than reinforcing identity.

In sustainability, the same pattern appears. Current strategies multiply choices: carbon offsets, green technologies or new consumer options. Yet they often remain within the same paradigm of control and exploitation. Choiceless awareness points to a different foundation, where action arises not from managing resources but from perceiving the indivisibility of life. When perception is whole, exploitation loses its basis, and sustainability becomes the natural outcome of intelligence.

Taken together, these implications reveal that choiceless awareness is not a private state of mind but a ground that reshapes how action appears in every domain. It dissolves the assumption that progress lies in expanding options or refining strategies. Instead, it shows that transformation begins in seeing, in ending the division between observer and observed.

This recognition prepares us for the next step. If choiceless awareness is the ground, then what moves from that ground? The answer is not knowledge or belief, but intelligence: a clarity that perceives the whole and acts directly from it. Section 9.3 will turn to this question and explore how intelligence differs fundamentally from knowledge.

9.4 Intelligence Beyond the Self

If choiceless awareness is perception without the distortion of preference or identity, then what is the quality that acts from such perception? That quality may be called intelligence, not the intelligence measured by IQ or accumulated through knowledge, but a living clarity that perceives wholeness and acts accordingly. To grasp this distinction is crucial, for much of modern civilisation has equated intelligence with knowledge and, in doing so, confused memory with insight.

Knowledge as the Past

Knowledge, as we have seen, is indispensable. It allows bridges to be built, illnesses to be diagnosed and cultural memory to be preserved. Yet knowledge is always the past, the residue of experience stored in memory and applied to the present. Its strength is also its limitation. It can describe, analyse and compare, but it cannot perceive directly. Krishnamurti (1969/2009) emphasised that knowledge is never complete and therefore cannot be the ground of freedom. If action rests on knowledge alone, it will always be partial and conditioned.

This limitation is visible in global policy frameworks. Economic forecasts project growth based on historical data. Climate models extrapolate from past emissions to future scenarios. These tools are valuable, but they remain maps, not the territory. When treated as reality itself, they bind action to abstraction. The world is then approached not as it is, but as thought has already structured it.

Intelligence as Direct Perception

Intelligence, in contrast, is not an accumulation but a movement of seeing. It is the responsiveness of awareness to the fact of the moment, unmediated by the self. When awareness is choiceless, intelligence comes alive. Bohm (1980, 1994) described this as a perception in which the observer and observed are not divided, allowing meaning to flow holistically.

Philosophers of process have expressed this in related ways. Whitehead (1929/1978) argued that reality is a continuous process of becoming, not a collection of static things. To perceive reality is therefore to perceive movement. Merleau-Ponty (1962) similarly described perception as embodied and relational: we do not stand apart from the world but are already enmeshed in it. Intelligence is not detached calculation but participation in this inseparable flow.

In Buddhist psychology, this distinction appears between vijñāna (conceptual, analytical knowledge) and prajñā (direct wisdom). Prajñā is not learnt; it is awakened in awareness. It is insight into the impermanence and interdependence of all phenomena, inseparable from compassion.

Science and Complexity

Contemporary science offers further analogies. Mitchell (2009) shows how intelligence in complex systems arises not from central control but from distributed interactions. Ant colonies, immune systems and ecosystems exhibit coherence without a single controller. Intelligence emerges from relationship, not possession. Capra (1996) makes a similar point: life is organised through networks of interdependence, and true ecological intelligence perceives systems as wholes rather than as aggregates of parts.

Systems theorists extend this insight into practice. Meadows (1999) demonstrated that effective change requires perceiving leverage points in systems, not multiplying information. Ostrom (2010) revealed that resilient governance arises when communities perceive interdependence and organise polycentrically, rather than relying on centralised control. Intelligence here is not imposed but discovered in lived relationship.

Neuroscience also supports this view. Damasio (2010) showed that much adaptive behaviour arises not from conscious reasoning but from embodied processes integrating feeling, memory and perception. McGilchrist (2019) distinguished between the brain's hemispheres: the left isolates and manipulates fragments, while the right perceives the whole field. Intelligence arises not from the dominance of either, but from balance in which the whole is not sacrificed to the part. Friston's (2010) free energy principle and Kirchhoff and Kiverstein's (2019) work on extended active inference likewise suggest that cognition is not a closed system of representation but a continuous participation with the world.

Social Structures and Intelligence

Sociology helps clarify what intelligence is not. Bourdieu's (1977) concept of habitus and Giddens' (1984) theory of structuration show how human action is shaped by culture and institutions. What feels like free choice often reflects embedded norms and historical inertia. Yet to see habitus in operation, to perceive conditioning as it arises, is already to loosen its hold. Awareness of conditioning is the beginning of intelligence.

Intelligence Beyond the Self

The key is that intelligence does not belong to the self. When intelligence acts, it is not 'my' intelligence or 'your' intelligence. To claim it is already to reduce it to knowledge, a possession. Krishnamurti all his life warned against this subtle trap: the moment intelligence is attributed to the 'I', it becomes memory, a fragment and ceases to be alive.

Intelligence, then, is a movement of wholeness responding appropriately to what is. It is universal, though it expresses itself uniquely in particular contexts. One might say it is like light: it illuminates everything, but the colours revealed depend on what is present.

Lived Examples

Personal Relationships

Research shows that relationships are most resilient when partners act not from rigid expectations but from responsiveness to present needs. *Perceived partner responsiveness*, the sense that one is seen, understood and cared for, predicts well-being, intimacy and long-term stability (Reis et al., 2004; Reis & Gable, 2015). This is not the result of calculation or contract but of attunement in the moment. John Gottman (1999) found that couples who thrive are not those who avoid conflict, but those who consistently 'turn toward' their partner's small bids for connection. Even subtle gestures of attention, like responding to a glance or a sigh, sustain emotional bonds. These findings suggest that intelligence in relationships is not strategic negotiation but direct sensitivity. In such moments, awareness organises action without deliberation.

Leadership

True leadership reveals the difference between identity-driven decisions and intelligence beyond the self. History offers many examples where leaders acted not by aligning with ideological camps but by perceiving the whole situation. Nelson Mandela, after his release from prison, resisted pressures for retribution and instead chose reconciliation, guided by the clarity that the survival of South Africa required dissolving division rather than deepening it (Mandela, 1994). Václav Havel, in leading Czechoslovakia through transition, emphasised what he called 'living in truth' (Havel, 1985), pointing to a kind of authenticity grounded not in political manoeuvring but in direct perception of moral reality. Leadership grounded in intelligence acts not from ambition or expediency but from seeing clearly what sustains life as a whole.

Ecology

Indigenous governance systems embody intelligence in action by relating to the land as kin rather than a commodity. Community forestry in the Himalayas, for instance, regulates forest use not only through abstract rules but also by observing ecological rhythms and responding collectively (Agrawal & Gibson, 1999). Practices such as limiting grazing during certain seasons or designating sacred groves are not derived from distant policy models but from lived attention to the ecosystem's needs.Robin Wall Kimmerer (2013), drawing from Potawatomi traditions, writes that true ecological intelligence is born of reciprocity: 'All flourishing is mutual'. Similarly, Elinor Ostrom's (2010) studies of commons governance showed that communities often manage resources more sustainably than centralised states or private markets because decisions are guided by awareness of interdependence, trust and accountability. Intelligence here is not abstract management but lived responsiveness, a movement of attention that aligns human action with ecological coherence.

Implications

The implications are profound. If intelligence is not knowledge, not belief and not identity, then transformation requires another ground. Policies, institutions and education must shift from accumulation to perception, from abstraction to wholeness. This does not mean discarding knowledge but placing it in its right place: indispensable in the technical sphere, yet inadequate in the psychological and existential.

When intelligence is reduced to information or performance, reform inevitably cycles back into fragmentation. Climate frameworks expand metrics, but they often miss the perceptual roots of exploitation. Education systems multiply curricula, but they still reward conformity over understanding. Governance designs more accountability systems, but these often reinforce mistrust rather than dissolve it. Each intervention, when rooted in knowledge alone, addresses the surface but leaves the underlying division untouched.

Concrete examples illustrate this pattern. In climate negotiations, nations debate carbon credits and deadlines, but the underlying assumption remains: the earth is a resource to be managed. As Bruno Latour (2017) observed, modern politics often treats the earth as a backdrop for human projects rather than as a living participant in collective life. The result is a sustainability discourse that manages symptoms while perpetuating separation.

In education, reforms emphasise 'skills for the future', yet the competitive structure persists. Paulo Freire (1970/2000) warned that education that only deposits knowledge into students, what he called the 'banking model', reproduces oppression rather than awakening freedom. Without a perceptual shift, reforms remain another version of the same conditioning.

Humanitarian governance shows similar patterns. International aid systems mobilise resources in response to crises, but often through bureaucratic frameworks that reproduce dependency. James Ferguson (1990) described this in his study of development in Lesotho, showing how well-meaning programmes can depoliticise poverty and reinforce control. Compassion becomes management, and disorder deepens.

Intelligence beyond the self suggests a different orientation. In governance, Elinor Ostrom (2010) demonstrated how communities manage common-pool resources sustainably when decision-making emerges from lived awareness of interdependence rather than from centralised enforcement. Her research revealed that intelligence in practice is not imposed from above but arises from participation, trust and local perception. Hence, frameworks built on identity and control perpetuate fragmentation, while governance that begins from perception can embody coherence and wholeness.

In sustainability, Indigenous practices offer living examples of intelligence in action. Community forestry in the Himalayas, as studied by Agrawal and Gibson (1999), regulates forest use not by abstract quotas alone but by observing ecological rhythms and responding collectively. Robin Wall Kimmerer (2013) emphasises in *Braiding Sweetgrass* that true ecological intelligence arises from reciprocity, seeing the land as kin rather than a commodity. These cases show intelligence as responsiveness to life, not the application of method.

In relationships, psychological research underscores the same principle. John Gottman (1999) found that resilient couples are not defined by conflict avoidance but by their capacity to 'turn toward' one another in everyday moments. Responsiveness to a partner's needs, even in small gestures, sustains connection more than negotiation or strategy. Reis and Gable (2015) framed this as perceived partner responsiveness: the sense that one is seen, understood and cared for. This is not calculation but attunement, a form of intelligence in daily life.

Economics also reveals the tension between knowledge and intelligence. Conventional growth models multiply data points yet rarely question their underlying assumptions. Barry Schwartz (2004) observed that when freedom is equated with choice, societies often produce anxiety rather than well-being. Alternative approaches, like Bhutan's Gross National Happiness index, attempt to reframe value, yet even these risk becoming abstract frameworks unless rooted in a deeper perception of interdependence. The true question is whether economic life can move from accumulation to coherence with life itself.

This perspective reframes responsibility. Responsibility is not simply adhering to external rules or pursuing socially sanctioned goals. It is the capacity to see clearly and respond from that clarity. Simone Weil (1952/2002a, 2002b) described attention as the purest form of generosity: 'Attention is the rarest and purest form of generosity'. Responsibility in this sense is not born of obligation but of perception, a seeing that naturally brings care.

Finally, this shift reveals the limits of reform as it is currently practiced. Reforms driven by ideology or identity may temporarily patch symptoms, but they cannot end the cycle of disorder. Ulrich Beck (1992) noted that modern societies become 'risk societies', producing crises while claiming to manage them. Hence, reform that arises from the same conditioned structures reproduces division, whereas intelligence as awareness in action can respond coherently. Only intelligence, arising from choiceless awareness, has the capacity to act without reproducing disorder.

In this sense, the implications of intelligence beyond the self are not abstract. They touch every domain of life, including climate, education, governance, economics, relationship and daily living. They suggest that the deepest transformation is not the creation of better systems but the presence of a mind that sees without distortion and acts from wholeness.

9.5 Action Without Identity

If intelligence is beyond the self, then action that flows from it cannot be anchored in identity. Yet most human action, whether in personal life, politics or global governance, is inseparable from identity. We act as Hindus or Muslims, liberals or conservatives, men or women, citizens or professionals. Identity provides stability and belonging, but it also fragments life. It places action within the boundaries of 'us' and 'them', 'my interest' and 'your interest'. To act from identity is to act from division.

The Trap of Identity

Identity offers continuity, yet the same continuity sustains violence and inertia. At the national scale, Benedict Anderson (1991) described nations as 'imagined communities', story-worlds that bind strangers into a collective 'we'. That very binding also creates a 'they'. At the micro level, social psychology shows how little is needed to trigger division: Henri Tajfel and John Turner (1979) found that even arbitrary group assignments produced favouritism towards one's own group and distrust of others. What appears as cohesion within simultaneously generates separation without.

When tensions rise, groups often maintain stability by locating a culprit. René Girard (1977) described this as the scapegoat mechanism, in which conflict is displaced onto an 'other' so the community can preserve its unity. Modern institutions reinforce these patterns by classifying people in ways that determine what can be done with them. Michel Foucault (1977) analysed how labels such as 'delinquent', 'patient' or 'immigrant' are not neutral descriptors but technologies of power that shape life and policy alike.

Psychoanalysis adds another layer. Freud (1930/2010) argued that drives constrained at the individual level frequently re-emerge collectively inside group identities. Belonging protects the self, but it also mobilises hostility against outsiders. Postcolonial critique widens the lens further. Edward Said (1978) showed how the 'Orient' was fabricated as an inferior identity to justify domination and control. Across these perspectives, identity emerges not as a neutral marker but as machinery that manufactures division.

Reform Through Identity

Much reform proceeds within this machinery. Political activism often takes the form of opposing camps, as environmentalists versus industrialists and progressives versus conservatives, achieving gains but leaving the deeper pattern of contention intact.

Climate negotiations provide a stark example. COP summits are often staged as contests among nations defending growth or security. Pledges such as 'net zero by 2050' are framed less as shared commitments than as competitive positioning. The SDGs institutionalise a similar tension. SDG 8 calls for sustained economic growth, while SDG 13 demands urgent climate action. These are not merely technical targets but expressions of competing identities: the industrial development paradigm on one side and the ecological preservation paradigm on the other. The result is perpetual bargaining rather than transformation.

Gender policy reveals parallel traps. Mainstreaming initiatives have expanded access for women but often by reinforcing binary categories. Judith Butler (1990) warned that stabilising categories can reproduce the very hierarchies reform seeks to dismantle, and Nancy Fraser (2013) argued that recognition without redistribution risks leaving material inequalities intact. Development discourse amplifies this caution. Arturo Escobar (1995) showed how the 'Third World' was constructed as an identity needing rescue, while James Ferguson (1990) documented how aid bureaucracies depoliticise poverty, embedding control under the banner of help. In all these cases, identity-led reform yields continuity disguised as change.

The Nature of Action Beyond Identity

To act without identity does not mean abandoning culture, community or relationship. Nor does it mean erasing difference. It means that action is not determined by identification. In choiceless awareness, the mind perceives a situation directly, not through the filter of 'me' or 'us'. From such perception, action arises naturally, free of distortion.

Krishnamurti (1969/2009) described this as action without motive, where the mind is not seeking reward or avoiding punishment but simply perceiving and responding. David Bohm (1994) emphasised that such action is holistic because it arises from seeing the interconnectedness of the entire field rather than privileging a fragment. Emmanuel Levinas (1969) added that responsibility begins prior to identity, in the immediate face-to-face encounter with the other. Genuine action, then, is not the extension of identity but the expression of direct perception.

Personal Relationships

In daily life, identities often act as filters. Parents may insist on the role of authority and miss the child's actual need. Partners may cling to the identity of the wronged one and perpetuate cycles of blame. To act without identity is to meet the other as they are, not as one imagines them to be.

Empirical studies support this view. Harry Reis and Gabrielle Gable (2015) found that perceived partner responsiveness: the sense that one's needs are noticed, understood and valued is the strongest predictor of relationship quality. John Gottman (1999) showed that resilient couples are not defined by the absence of conflict but by their ability to 'turn towards' one another in small, everyday bids for connection. These responses are not calculated but emerge from attunement. Carol Gilligan (1982) introduced the ethic of care, emphasising responsiveness to lived relationships over abstract principles. Clark and Lemay (2010) demonstrated that relationships thrive when partners express concern without defensiveness or role-playing. In this light, action beyond identity is not about performing the role of the 'good partner' or 'good parent', but about listening without the distortion of self-image. Intelligence here is the capacity to attend rather than to defend.

Governance

Most political systems are organised around identities: nations, parties and interest groups. Yet history offers glimpses of leaders who acted beyond these constraints. Nelson Mandela oriented South Africa's transition not around vengeance but reconciliation. His willingness to negotiate with former adversaries was not allegiance to identity but clarity about the need for wholeness (Mandela, 1994). Václav Havel (1985) similarly argued for 'living in truth', grounding politics in authenticity rather than partisan image.

These examples show that even within institutions structured by identity, leaders can act from another ground. What distinguished Mandela and Havel was not strategy alone but perception of the whole. Leadership rooted in perception rather than image does not weaken political responsibility. Instead, it allows decisions to dissolve cycles of antagonism where identity would only deepen them.

Ecology

Ecological stewardship reveals perhaps the clearest expression of action beyond identity. Modern economies define people as 'resource users' or 'economic agents', reducing life to categories of utility. Indigenous traditions often refuse this division. Robin Wall Kimmerer (2013) describes reciprocity with the land as kinship rather than extraction: 'All flourishing is mutual'. Vandana Shiva (1989) critiques industrial agriculture for uprooting relational ways of knowing and imposing extractive identities on communities and ecosystems.

Agrawal and Gibson (1999) documented Himalayan community forestry systems where forests are regulated not by abstract quotas but through observation of ecological rhythms and collective responsibility. Elinor Ostrom (2010) confirmed that communities across the world sustain common-pool resources without centralised enforcement when governance emerges from lived awareness of interdependence. More recently, Ashish Kothari et al. (2019) have argued for alternatives to development that build on such practices, where identity as a consumer or producer gives way to participation in ecological wholeness. These examples embody intelligence in practice: responsiveness to life rather than allegiance to ideology.

Nationalism and Collective Identity

At the national scale, identity can mobilise belonging but also perpetuate enmity. Anderson's (1991) 'imagined communities' provide continuity, yet they are bounded by exclusion. Amartya Sen (2006) warned against the solitarist fallacy that reduces people to a single identity (national, religious or cultural), at the expense of plural belonging.

Conflict zones make this especially vivid. National histories that glorify past victories or traumas often trap societies in cycles of hostility. The persistence of disputes in Kashmir, the Middle East or the Balkans shows how narratives of identity, once solidified, create generational conflict. Identity-driven politics narrows perception, reducing complex realities to binary narratives. To act beyond identity in such contexts would mean perceiving the shared needs of people and ecosystems directly, not through the lens of national pride or grievance. This shift is rare but it is the only path out of repeated antagonism.

Love as the Ground of Action

When action is not filtered through identity, what remains is love. This is not sentimental or selective attachment but what Krishnamurti (1969/2009) described as love without motive, a flame without smoke. It is the quality of care that arises when division ends.

bell hooks (2000) framed love as an ethic of freedom, a commitment to nurture growth in self and other beyond domination. Thich Nhat Hanh (1991) called this interbeing, the recognition that one's life is inseparable from all others, and that compassion flows naturally from that recognition. Martin Luther King Jr. (1963/2010) argued that love in action is not weakness but the deepest form of strength, the only force capable of transforming enemies into friends. Action from such love is not commanded by morality but expressed in perception itself.

Freedom in Action

Action without identity is free because it is not bound by memory, ideology or belonging. It responds to the present rather than reacting to the past. Hannah Arendt (1958) defined action as the capacity to begin anew, a quality that withers when politics is consumed by antagonisms of identity. Chantal Mouffe (2000) and Iris Marion Young (1990) both showed how identity-based politics can lock democracies into perpetual contestation.

Action beyond identity dissolves this trap. It is free not because it has endless options but because it is no longer tethered to the chooser. Freedom here is not the power to impose but the absence of compulsion. Simone Weil (1952/2002a, 2002b) observed that attention itself is a form of generosity. In seeing without self, one is already acting beyond division. Only when the self does not act is there true action. Only when identity is absent can intelligence move freely.

Implications of Action Beyond Identity

The implications of action beyond identity are profound. If most reforms and legal frameworks continue to operate from the standpoint of identity, whether national, ideological or institutional, they risk reproducing the very divisions they seek to heal. Moving beyond identity does not mean discarding frameworks, treaties or policies. It means enlivening them with a spirit that transcends selective application. Without such a perceptual shift, even universal principles risk collapsing into instruments of contest.

The SDGs illustrate this tension clearly. While they embody a global commitment to end poverty, protect the planet and ensure dignity, they also contain contradictions that reflect identity-driven compromises. SDG 8's emphasis on 'sustained economic growth' sits uneasily beside SDG 13's urgent call for climate action. This is not a mere technical conflict but a clash of identities: the industrial growth paradigm versus the ecological survival imperative (United Nations, 2015). Unless these contradictions are seen at the level of worldview, the framework risks becoming a fragmented checklist rather than a coherent expression of responsibility.

The Geneva Conventions provide another striking case. These treaties codified universal protections for civilians and combatants, yet their spirit is undermined in practice. Amnesty International (2019) documented a 'catastrophic failure' of states to uphold humanitarian protections, while the UN Security Council's 2024 anniversary report noted that violations remain widespread, especially in deliberate attacks on civilians and essential infrastructure. Legal debates such as the Diakité case (C-285/12) at the European Court of Justice further expose how classifications of conflict, whether 'international' or 'domestic', determine the scope of protections (Court of Justice of the European Union, 2014). When civilians' safety depends on such labels, universality erodes and humanitarian law becomes entangled in the very identity politics it was designed to overcome.

The same distortion is evident in the application of jus cogens norms and the Responsibility to Protect (RtoP). These principles, which prohibit genocide, slavery and torture, are binding on all states. Yet enforcement often reflects political interest rather than universality. The UN Secretary-General's 2024 report acknowledged that responses to atrocity crimes are frequently delayed or inadequate because they are subordinated to geopolitical alliances (United Nations, 2024). The contrast between NATO's rapid intervention in Libya in 2011 and the paralysis in Syria illustrates how identity politics compromise the promise of universality. More recently, reports of mass graves discovered in Gaza hospitals and schools have reignited debates on selective enforcement, showing how even obligations towards the dead, among the clearest in humanitarian law, can be disregarded when filtered through national and strategic identity (Reuters, 2024).

International agreements reveal a similar tension. The principle of pacta sunt servanda, that treaties must be kept, is undermined when accords are routinely suspended, reinterpreted or ignored in the name of national interest. Climate accords, arms reduction treaties and refugee conventions are often honoured selectively, exposing how the letter of the law survives while its spirit disintegrates.

The Universal Declaration of Human Rights (UDHR) perhaps best reflects the aspiration to transcend identity. Yet even here, implementation is uneven. Rights are championed for allies while muted for adversaries. As Zook (2020) notes, the invocation of rights often mirrors political alliances, turning universals into instruments of strategy rather than reflections of shared humanity. What was meant as a mirror of dignity for all becomes another tool of contest.

These failures are not inevitable but symptomatic of perception rooted in identity. To reanimate global frameworks requires a different ground of action. Elinor Ostrom's (2010) research on common-pool resource governance shows that communities sustain shared goods not through rigid enforcement but through trust, reciprocity and lived awareness of interdependence. These examples suggest that law and policy can be infused with intelligence beyond identity if they are rooted in direct perception of interdependence.

If such perception were applied globally, existing frameworks could take on new meaning. The Geneva Conventions would not only regulate war but also embody a universal refusal to treat human beings as expendable. RtoP would cease to be a tool of selective intervention and instead become an immediate recognition of shared humanity. Pacta sunt servanda would mean honouring the deeper intention of agreements, fidelity to trust across nations, rather than clinging only to their letter. The UDHR would no longer be an aspirational text but a lived principle, expressing relationship rather than rhetoric.

Ulrich Beck's (1992) reminder that modern societies are 'risk societies' remains pertinent here: institutions often generate the very crises they claim to manage. Unless action shifts from identity to intelligence, frameworks like the SDGs, Geneva Conventions, RtoP and the UDHR will remain caught in cycles of fragmentation. But with a shift in perception, these same structures could be animated by intelligence beyond the self, becoming expressions of coherence rather than instruments of division.

In this sense, the implication is not that existing mechanisms are irrelevant but that their renewal depends on transformation at the root. Without a shift in perception, even the most carefully crafted reforms will repeat the past. With such a shift, however, these frameworks could become channels for a different movement, one grounded not in identity but in intelligence, capable of acting for the whole.

9.6 Holistic Perception in Practice

If action without identity arises from intelligence, how does it appear in lived experience? For many, the idea of choiceless awareness or action without self may seem abstract. Yet glimpses of this quality exist across diverse domains of life. These are not applications of a method or the result of cultivated practice, but expressions of perception when it is whole.

Psychological Dimension: The Flow of Attention

In psychology, the closest analogue to choiceless action is often described as 'flow'. Mihaly Csikszentmihalyi (1990) documented how athletes, musicians and scientists enter states where action unfolds effortlessly, without deliberation or self-consciousness. Flow shows that action can emerge from direct engagement rather than reflective choice.

Yet there is a crucial distinction. Flow is typically tied to performance, hitting the right note, making the right move and solving the problem. Choiceless awareness, by contrast, is not bound to achievement. It is available in everyday living, in listening, walking or observing, where perception is unclouded by thought.

Neuroscience reinforces this. Iain McGilchrist (2019) demonstrated that the brain's right hemisphere perceives context, flow and relationship, while the left isolates fragments for manipulation. When the analytic mode dominates, perception narrows into categories and choices. When the holistic mode is active, action flows more naturally, less distorted by self-conscious calculation. Research on mindfulness supports this distinction. Davidson and Kaszniak (2015) found that non-judgemental awareness reduces bias and reactivity, making responses more attuned to the present.

Leadership and Integrity

History provides rare instances of leaders whose action reflected wholeness rather than identity. Dag Hammarskjöld, Secretary-General of the United Nations from 1953 to 1961, exemplified this. His private reflections in *Markings* reveal a search for silence and clarity that grounded his leadership. He often defied national pressures, acting not as a representative of one bloc but from a perception of the whole. His death in 1961, while mediating conflict in the Congo, underlined his refusal to compromise integrity for expediency (Hammarskjöld, 1964).

More recent cases echo this spirit. Jacinda Ardern's leadership after the Christchurch mosque shootings in 2019 similarly displayed holistic perception. She refused to frame the tragedy in terms of 'us versus them' and emphasised compassion as the ground of political response (New Zealand Government, 2019). These examples suggest that intelligence in leadership does not mean possessing more strategies but perceiving without distortion, allowing action to arise from clarity.

Indigenous Ecological Practices

Indigenous ecological traditions often embody holistic perception. Their governance is not organised primarily through abstract frameworks but through attentiveness to the rhythms of the natural world. Community forestry in the Indian Himalayas has shown that villagers regulate use not by market incentives or legal penalties but by observing ecological limits and collective responsibility (Agrawal & Gibson, 1999).

This orientation is echoed globally. Robin Wall Kimmerer (2013) emphasises reciprocity, describing ecological responsibility as kinship rather than management: 'All flourishing is mutual'. Vandana Shiva (1989) critiques industrial agriculture for imposing extractive categories on life, severing relationships that once embodied reciprocity. Elinor Ostrom's (2010) research confirmed that communities across the world sustain common-pool resources without centralised enforcement when governance emerges from lived awareness of interdependence.

At COP28 in Dubai, Indigenous delegations again urged negotiators to recognise Indigenous stewardship as a model for climate governance (United Nations Framework Convention on Climate Change [UNFCCC], 2023). Their interventions highlighted that ecological intelligence arises not from 'managing carbon' but from perceiving inseparability with the land. These voices remind us that holistic perception is not an abstract philosophy but a living practice at the frontlines of the climate crisis.

Everyday Relationship

Holistic perception is also present in daily life. In dialogue, when one listens without preparing a response, defending a position or projecting an image, communication changes. David Bohm (1996) described dialogue as a 'stream of meaning flowing among and through us', possible only when participants suspend assumptions. In such listening, conflict often subsides, not because arguments are won, but because the very structure of division is absent.

In families, the same dynamic holds. A parent who listens to a child without projecting ambition or expectation meets the child as a human being rather than as an extension of self-image. In partnerships, when one listens without defensiveness, care becomes the natural ground of action. Studies confirm this: Reis and Gable (2015) found that perceived partner responsiveness is the strongest predictor of relationship satisfaction, while Gottman (1999) showed that resilience depends on attentiveness in small daily moments rather than dramatic gestures. These findings affirm that intelligence in action is not calculation but attunement.

Education as Inquiry

Education offers a powerful arena to see the difference between fragmented and holistic perception. Paulo Freire (1970/2000) critiqued the 'banking model' where knowledge is deposited into students, reproducing conformity and fear. By contrast, learning as inquiry, where teacher and student observe together without hierarchy, awakens intelligence.

This was made visible during the COVID-19 pandemic, when many schools adopted rigid digital models that deepened inequities (UNESCO, 2021). Yet some classrooms experimented with dialogic or contemplative approaches, allowing students to process the uncertainty of the pandemic together. Research in dialogic pedagogy (Wegerif, 2011) and contemplative education (Zajonc, 2013) shows that when learning is rooted in inquiry, students demonstrate greater openness, creativity and ethical awareness. Holistic education does not condition identity but cultivates awareness.

Governance as Perception, Not Framework

Modern governance largely operates through frameworks: treaties, policies, indicators and reforms. These are designed from knowledge and identity and therefore reproduce contradiction. Critical theorists like Ulrich Beck (1992) and Arturo Escobar (1995) have shown that frameworks describe crises but often fail to dissolve them.

Recent global events highlight this limitation. COP28 negotiations in 2023 ended with a historic call to 'transition away from fossil fuels', yet fossil fuel subsidies remain entrenched in nearly every major economy (International Energy Agency [IEA], 2023). The framework describes the crisis but does not end it, because the perception that growth and extraction are necessary remains unchallenged. SDG 8's call for 'sustained economic growth' continues to contradict SDG 13's demand for urgent climate action (United Nations, 2015). Without a shift in perception, even well-meaning frameworks perpetuate the very contradictions they seek to resolve.

International humanitarian law faces the same distortion. The Geneva Conventions were designed to protect civilians in war, yet conflicts in Syria, Yemen and more recently Gaza have shown repeated violations. Reports by Amnesty International (2019) and the International Committee of the Red Cross (2023) document systematic attacks on hospitals, schools and water systems. When laws are applied selectively, enforced against adversaries but overlooked for allies, their universality collapses.

The RtoP demonstrates a similar trap. The principle was affirmed unanimously by the UN in 2005, but its application has been inconsistent. In Libya 2011, NATO intervened swiftly; in Syria, paralysis allowed mass atrocities to unfold. The UN Secretary-General's 2024 report acknowledged that RtoP responses remain subordinated to geopolitical alliances, leaving many populations unprotected (United Nations, 2024). When identity dictates which lives matter, law loses coherence.

Even refugee protection reveals the limits of identity-led governance. The 1951 Refugee Convention commits states to protect those fleeing persecution, yet responses vary depending on geography, religion and politics. The open-armed welcome of Ukrainian refugees in Europe in 2022 contrasted sharply with the restrictive policies applied to Syrians or Afghans just years earlier (UNHCR, 2023). Such double standards expose how identity shapes the 'worthiness' of protection.

The economic and legal domains reflect the same pattern. The World Trade Organization (WTO) was created to enforce rules-based trade, yet disputes are often settled along power hierarchies. The paralysis of the WTO Appellate Body since 2019, due to the United States' refusal to appoint judges, shows how legal mechanisms collapse when national identity overrides collective commitment (WTO, 2020). Similarly, climate accords like the Paris Agreement enshrine 'common but differentiated responsibilities', yet wealthier states often evade their obligations while pressing poorer nations to comply. This selective burden-sharing reinforces historical inequities rather than dissolving them (IPCC, 2022).

International financial systems also embody the tension between identity and universality. Debt relief frameworks, such as the Heavily Indebted Poor Countries (HIPC) initiative, were intended to support development but were frequently tied to structural adjustment programs that deepened inequality. Critics argue these programs served creditor interests more than debtor societies, perpetuating dependency through the language of reform (Stiglitz, 2002). In practice, 'conditionality' became an identity filter, classifying nations as 'disciplined' or 'irresponsible', rather than perceiving their lived realities.

Yet there are glimpses of governance rooted in a different perception. In Bolivia, the 2010 Law of Mother Earth recognised nature as a living subject with rights. This framework, inspired by Indigenous worldviews, attempted to reframe governance from managing 'resources' to honouring interdependence (Gudynas, 2011). While implementation has faced political and economic resistance, the law reflects a shift in perception: ecosystems as kin, not commodities.

In Ecuador, the 2008 Constitution incorporated the principle of *Buen Vivir* (good living), drawn from Andean traditions. Buen Vivir does not separate human well-being from ecological integrity but frames them as inseparable. This represents a radical departure from identity-led development models, offering a relational understanding of governance (Acosta, 2013).

South Africa's Truth and Reconciliation Commission (TRC) (1996–2003) offers another example. Rather than reducing post-apartheid justice to retribution, the TRC emphasised testimony, acknowledgement and forgiveness. Desmond Tutu described its goal as restoring the fabric of humanity rather than perpetuating cycles of vengeance (Tutu, 1999). Though imperfect and contested, the TRC signalled a governance experiment rooted less in identity antagonism than in the attempt to perceive shared humanity.

In Rwanda, after the 1994 genocide, the Gacaca courts drew from traditional practices to process mass atrocity cases. While criticised for uneven fairness, they emphasised community participation and reconciliation over retribution, embodying a form of justice rooted in lived perception rather than solely in institutional categories (Clark, 2010).

These cases, though partial and contested, reveal that action without identity is not merely a philosophical ideal but a lived possibility. When Bolivia and Ecuador inscribed Indigenous cosmologies into law, when South Africa chose reconciliation over vengeance, or when Rwanda turned to community justice, governance shifted, even if briefly, from identity antagonisms toward perception of the whole. These moments did not emerge from new frameworks alone but from a willingness to see beyond the categories of 'us' and 'them'. They embody what Krishnamurti (1969/2009) described as action without motive, Bohm's (1994) call for holistic perception and Levinas's (1969) insistence that responsibility begins in the encounter with the other. Though imperfect, these examples show that when perception shifts, even institutions long shaped by division can act as mirrors of intelligence rather than machinery of identity.

These global experiments, though partial and often compromised, highlight that action beyond identity is not confined to individual psychology but can shape collective life. Bolivia's Law of the Rights of Mother Earth and Ecuador's recognition of Buen Vivir challenged the extractive paradigm by reframing humans not as resource-users but as part of a larger living whole. South Africa's TRC, despite its shortcomings, created a space where the cycle of vengeance could be interrupted by acknowledging shared humanity. Rwanda's Gacaca courts, with their emphasis on face-to-face accountability, demonstrated that reconciliation requires more than legal categories; it demands direct encounter. In each case, the possibility of wholeness appeared not because identity was affirmed, but because, however briefly, it was set aside.

What unites these examples is that transformation arose not from the invention of new methods, but from a different quality of perception. These were not purely technical achievements but moments when leaders, communities and citizens glimpsed a reality larger than their own affiliations. In this sense, they confirm that frameworks alone cannot bring coherence; coherence comes when perception itself is whole. As Bohm (1996) noted, dialogue becomes meaningful only when participants suspend assumptions and meet one another without defence. Likewise, governance becomes transformative only when action flows from awareness rather than allegiance.

Yet the fragility of these examples is equally instructive. Each was quickly constrained by the return of political interest, economic pressures or entrenched narratives of identity. This suggests that while frameworks can provide openings, only a shift in perception can sustain transformation. Holistic perception in practice is therefore not a blueprint to be replicated, but a way of being that must be continually rediscovered in leadership, in communities and in everyday relationship. It is not about designing systems that guarantee wholeness, but about seeing directly in each moment so that action naturally arises from intelligence rather than identity.

Not a Method, But a Way of Being

The temptation, after seeing such examples, is to ask how they can be repeated or institutionalised. Yet to turn holistic perception into a method is to miss its essence. The moment it becomes a strategy, a practice or a programme, it falls back into the movement of identity and control. What gave those moments of transformation their power was not design but perception, an attentiveness in which division momentarily ceased.

Krishnamurti (1969/2009) warned against this trap: awareness cannot be cultivated as one cultivates skill. It is not the outcome of effort or will, but the natural clarity that emerges when the mind is quiet. To attempt to 'practice' choiceless awareness is already to reintroduce motive, the desire to become something, to achieve a result. In that very movement, the self returns.

This does not make holistic perception rare or esoteric. On the contrary, it is present in ordinary life. It is there when one listens to another without preparing a defence. It is there when a community observes the needs of a forest directly, rather than abstracting them into quotas. It is there when a leader acts from clarity rather than allegiance. It is there when a child learns through discovery rather than comparison.

Contemporary life continues to offer glimpses. When Jacinda Ardern resigned as New Zealand's Prime Minister in 2023, she framed her decision not as defeat but as recognition that she no longer had the clarity to serve with full presence. Her departure, rare in modern politics, was not an act of identity preservation but of attentiveness to what the moment required. In climate response, when Pacific island communities relocate together from rising seas, their choice reflects more than technical adaptation: it shows recognition that survival depends not on defending land as property but on preserving relationships as wholeness. In classrooms around the world, educators experimenting with dialogical inquiry, where teachers and students explore questions together rather than imposing predefined answers, show that intelligence can emerge whenever authority is replaced by shared attention.

The same contrast is visible at the level of global governance. International treaties and humanitarian mechanisms often falter because they are interpreted through the lens of national or ideological interest. Climate accords are debated in terms of burden-sharing rather than recognition of planetary interdependence. The Geneva Conventions are invoked selectively, their universality compromised by strategic calculations. The RtoP is applied in some crises while ignored in others.

Legal principles meant to embody universality collapse under the weight of identity. *Jus cogens* norms, which prohibit genocide, slavery and torture, are binding on all states. Yet in practice, their enforcement depends on political alliances: atrocities in some regions are condemned, while similar acts elsewhere are met with silence. The principle of *pacta sunt servanda*, that treaties must be kept in good faith, is equally fragile. States reinterpret, suspend or abandon agreements when they conflict with perceived national interests. Climate accords, arms reduction treaties and refugee conventions all suffer from selective compliance, where the letter of law survives but its spirit disintegrates. Even the Universal Declaration of Human Rights (United Nations, 1948), perhaps the clearest attempt to transcend identity, is often invoked strategically, championed for allies and disregarded for adversaries.

A holistic perception would cut through these distortions. It would not ask whether upholding human rights serves national image, whether enforcing humanitarian law advances strategic interests or whether climate commitments can be traded for economic advantage. It would see directly: that torture is always a violation of humanity, that civilians are never expendable, that the atmosphere is not property to be bargained but the shared condition of life. From such perception, action would not be compliance with treaties for the sake of reputation, but fidelity to life itself.

Holistic perception is therefore not a technique to be applied but a way of being to be recognised. It cannot be transmitted through curricula or enforced by law, yet it can be seen whenever thought does not interfere with perception. Its presence transforms action, not because it follows a plan, but because in the absence of distortion, life responds coherently to itself. In this sense, every moment of genuine attention is already a glimpse of intelligence beyond the self, already the seed of transformation.

To see without identity is not an abstraction but a possibility present in every sphere of life. It begins in the smallest gestures of attention of a parent listening to a child without projection, a friend responding without defence. It extends into communities that care for forests as kin, and into classrooms where learning unfolds as inquiry rather than comparison. It can guide leaders who act from clarity rather than allegiance, and it can breathe spirit into global frameworks that too often collapse into bargaining and selectivity. The same perception that ends division in a conversation can also reanimate the Geneva Conventions, the RtoP and the Universal Declaration of Human Rights, not as instruments of power but as living mirrors of our shared humanity. Whether in personal relationship or international law, the ground is the same: when perception is whole, action flows without distortion, and in that movement lies the possibility of coherence, care and transformation.

9.7 Leadership as Perception, Not Identity

The inquiry into choiceless action has profound implications for how societies organise themselves. Education, governance, sustainability and leadership are the very fields where humanity claims to pursue order, yet they are also where fragmentation is most deeply institutionalised. Choices are celebrated as freedom, frameworks as transformation and targets as solutions, but beneath these appearances lie the same movements of identity, fear, desire and continuity. The result is incoherence: institutions stabilise disorder while claiming to resolve it. Only leadership that arises from perception, not identity, can respond coherently to the crises of our time.

The Limits of Fragmented Action

Governance today is defined by trade-offs and negotiations. Policymakers speak of balancing growth with environment, security with freedom or equality with efficiency. These dilemmas are treated as if they were the essence of responsibility. In reality, they reveal the fragmentation of perception. As Krishnamurti (1969/2009) observed, choice exists only where there is confusion. When there is clarity, action is direct, not deliberative.

Climate policy illustrates this contradiction sharply. At COP28, nations renewed commitments to 'phasing down' fossil fuels while simultaneously expanding drilling projects and subsidies. The European Union pledged ambitious reductions in emissions, yet many member states extended coal power in the name of energy security. These are not isolated oversights but expressions of divided consciousness: the pull of national identity and economic allegiance overpowering perception of planetary interdependence.

The SDGs reveal the same paradox. They aspire to universality but fragment crises into separate targets – poverty (SDG 1), education (SDG 4), climate (SDG 13) and peace (SDG 16). Dashboards may 'light up green' even as ecosystems collapse and militarisation accelerates. Targets conceal continuity by displacing transformation into the future. Frameworks designed from accumulated knowledge reform symptoms but leave the roots of disorder untouched.

Even 'inner' frameworks risk this trap. The Inner Development Goals (IDGs) attempt to complement the SDGs by adding competencies such as awareness and empathy. Yet awareness cannot be reduced to a competency. The moment it is instrumentalised for performance, it collapses into knowledge and time: another technique of the self. Awareness is not a skill to be trained; it is perception unmediated by identity.

Leadership From Perception

Most leaders act from identity – national, ideological or corporate – rather than from perception of the whole. This produces incoherence disguised as pragmatism.

Identity-driven collapse is visible in Myanmar, where the military clings to power under the banner of 'national security', unleashing violence that fractures communities. Fragmentation dominates climate politics as well: at COP28, leaders issued ambitious pledges only to retreat under pressure from fossil-fuel lobbies. Corporate leadership often follows the same pattern. ExxonMobil suppressed climate science for decades to defend shareholder interests, while Big Tech profits from surveillance capitalism even as it promotes 'ethical AI' (Zuboff, 2019). These are not neutral failures but examples of leadership bound to identity and continuity.

By contrast, glimpses of leadership from perception arise when action flows from seeing rather than from allegiance. Māori ecological governance in New Zealand recognises rivers as legal persons, reflecting the perception of interdependence rather than resource identity. Citizens' climate assemblies in France and the United Kingdom have produced proposals more coherent than those of elected officials, precisely because they emerge from dialogue and shared perception. After the 2011 Japanese tsunami, community-led disaster relief showed how spontaneous action rooted in attentiveness to need can bypass bureaucratic inertia.

These contrasts prepare the ground for a closer examination of leadership in practice. The following case studies – spanning education, governance, ecology and global politics – reveal in concrete terms how identity produces incoherence and how perception can generate coherence.

Case Studies: Incoherence and Coherence

In contexts of protracted conflict, education has been deeply shaped by identity. In Israel and Palestine, school textbooks construct mutually exclusive narratives that legitimate one side while erasing the other (Podeh, 2002). Children are taught to inherit antagonism before they can inquire for themselves, ensuring that conflict is transmitted across generations. UNESCO's 2021 global curriculum review warns that such approaches undermine SDG 4's aspiration of fostering peace and critical thinking. Afghanistan provides a similar pattern in a different register. From the Soviet era through Taliban rule and the US.-backed period, curricula were rewritten repeatedly to secure political legitimacy. Heroes, values and even access to education shifted with each regime. The Taliban's latest decrees banning girls from schools collapse SDG 4 and SDG 5 in one stroke (Human Rights Watch, 2023). In both contexts, education becomes an instrument of continuity rather than a mirror of perception, reducing students to vehicles of political identity rather than enabling them to see directly.

By contrast, South Africa's Truth and Reconciliation Commission (TRC) created rare spaces where perception momentarily interrupted identity. Between 1996 and 2003, more than 21,000 victim statements were collected, and perpetrators testified before survivors. In some hearings, cycles of vengeance gave way to recognition, as victim and perpetrator encountered one another as human beings rather than images (TRC, 1998). Archbishop Desmond Tutu described the TRC's task as restoring humanity itself, though critics warn that reconciliation risks becoming procedural when the frame is mistaken for the seeing (Wilson, 2001). The lesson is clear: reconciliation processes serve SDG 16 not when they are perfected as frameworks, but when they preserve the vitality of perception.

Ecology also reveals the divergence. In Nepal and India, community forestry groups sustain ecosystems by observing rhythms and adjusting practices collectively rather than enforcing distant abstractions. Studies confirm that forests governed this way are healthier and more resilient (Agrawal & Gibson, 1999; Chhetri et al., 2013). These practices dissolve the binary of 'livelihood versus conservation' and embody SDG 13 and SDG 15 as inseparable. By contrast, global climate governance remains mired in identity-led fragmentation. The IPCC's 2023 synthesis warned of cascading tipping points, yet COP28's Global Stocktake created loopholes even as fossil-fuel subsidies and drilling approvals expanded (UNFCCC, 2023; IEA, 2023). Dashboards light up green, but ecosystems collapse. Climate policy framed as trade-offs between 'growth' and 'action' is not merely inadequate; it is a symptom of fractured perception.

The humanitarian field shows similar tensions. In Gaza, the ongoing atrocities by the IDF produced mass civilian casualties and destruction of essential systems, documented by UNOCHA (2024). Here, identity-driven narratives converted human beings into symbols of cause or resistance, foreclosing perception of suffering itself. By contrast, South Africa's TRC or Rwanda's Gacaca courts, though contested, demonstrated that even in the aftermath of atrocity, perception of the other as human can interrupt cycles of vengeance (Clark, 2010).

Economic and ecological governance reinforce the point. Brazil's Amazon illustrates both regression and possibility. When perception aligned with accountability, fines on companies like JBS and new herd-tracking rules curbed illegal deforestation (AP, 2023). Yet when economic identity regained dominance, the soy moratorium was weakened, reversing ecological gains. Research even shows that reducing deforestation lowers violence in Amazonian municipalities (Arxiv, 2025), suggesting that care for ecosystems and social stability are inseparable. Where leadership acts from perception, SDG 13 and SDG 15 reinforce each other; where it acts from identity, they collapse.

Emerging technologies now magnify the stakes. Artificial intelligence is framed by corporations as competition for market dominance and by states as a race for national security. This identity-driven framing repeats the same distortions visible in climate and conflict. Yet UNESCO's 2021 Recommendation on the Ethics of AI placed dignity, sustainability and inclusivity at the centre, while citizens' assemblies in Europe demonstrated that ordinary people, freed from partisan allegiance, can deliberate more coherently than politicians (Ada Lovelace Institute, 2022). If AI governance remains identity-bound, SDG 8, SDG 10 and SDG 16 will be undermined. If rooted in perception of shared humanity, AI could instead reflect the UDHR's founding principle of equal dignity for all.

Education in the United States further illustrates how identity can colonise perception. Recent state-level restrictions on teaching systemic racism, colonialism or gender identity illustrate curricular violence (UNESCO, 2023). By narrowing what can be taught, legislatures enforce identity as the lens of learning. Research confirms that racial and class segregation persists in US schools, reinforcing inequality (UCLA Civil Rights Project, 2024). Rather than expanding perception, education here reinforces division, undermining SDG 4 and SDG 10.

Migration policy in Europe shows the same oscillation. Since 2015, Europe has alternated between humanitarian gestures and restrictive border controls. The 2024 EU Pact on Migration and Asylum tightened asylum processing and outsourced enforcement to third countries, a move widely criticised by NGOs as sacrificing rights for political expediency (Le Monde, 2024). Refugees were treated not as vulnerable human beings but as risks to sovereignty and electoral legitimacy. The result is incoherence: policies designed to protect life end up reinforcing identity-driven exclusion. SDG 10, SDG 16 and SDG 17 are all compromised.

Across these cases, the pattern is unmistakable: where leadership is rooted in identity, frameworks become weapons, knowledge becomes propaganda and universals collapse into selective bargains. Where leadership arises from perception, trade-offs dissolve and action aligns without coercion. The contrast between incoherence and coherence is not abstract but lived in classrooms, forests, commissions, treaties and even algorithms.

Seeing the Whole: From Identity to Perception

Taken together, these cases reveal the structural difference between action driven by identity and action arising from perception. Where identity governs, institutions serve continuity rather than intelligence. In Israel and Palestine, curricula transmit antagonism before children can inquire for themselves. In Afghanistan, textbooks are rewritten to secure legitimacy rather than awaken understanding. In climate negotiations, leaders pledge reductions while expanding subsidies, as national image outweighs planetary care. In the Amazon, economic identity unravels ecological progress, and in Europe's asylum systems, rights fluctuate with political convenience. Identity weaponises knowledge, converts frameworks into instruments of contest and transforms universals into conditional bargains.

The result is systemic incoherence: institutions stabilise disorder while appearing to manage it.

When perception surfaces, coherence arises naturally. South Africa's Truth and Reconciliation Commission allowed victims and perpetrators to meet as human beings rather than categories. Himalayan forest communities sustain ecosystems by observing ecological rhythms rather than enforcing distant quotas. Citizens' climate assemblies in Europe reveal that when ordinary people deliberate without partisan allegiance, their conclusions are often more coherent than those of political leaders. Māori recognition of rivers as legal persons reframes law as relationship rather than ownership, and UNESCO's ethical framework for AI places human dignity at its centre, showing that even technology can reflect awareness of interdependence.

These examples are fragile and often temporary, yet they demonstrate that coherence does not emerge from perfected systems but from perception that is whole. Where identity dominates, commitments fracture; where perception leads, action flows coherently, dissolving false dilemmas and revealing inseparability. Frameworks, treaties and policies, therefore, find coherence not in design but in the quality of seeing that animates them. When perception is whole, even imperfect mechanisms become vessels of intelligence. Leadership as perception is thus not a philosophical ideal but a lived possibility, glimpsed whenever allegiance gives way to awareness.

Practical Implications

The contrast between incoherence from identity and coherence from perception offers three practical directions for governance and leadership.

Beyond Trade-offs

Identity-led governance thrives on dilemmas. Policymakers frame decisions as contests – growth versus climate, freedom versus security, equality versus efficiency. These binaries create the appearance of responsibility while masking fragmentation. COP28 illustrates this: leaders promise both decarbonisation and new fossil expansion, as if the economy could be preserved apart from ecology. In reality, the economy is a subsystem of ecology; its separation is artificial. Similarly, security without freedom is coercion, and equality without justice is tokenism. Whole perception dissolves such oppositions. It does not negotiate between fragments but sees their inseparability. Once the whole is perceived, action is direct. Climate becomes survival, not a bargaining chip. Education becomes inquiry, not propaganda. Migration becomes a shared human condition, not a threat.

Beyond Frameworks

Frameworks, treaties and dashboards are tools, not transformation. They provide coordination but cannot substitute for perception. When mistaken for change, they perpetuate delay. Climate dashboards 'light up green' even as ecosystems collapse. The Geneva Conventions are cited selectively according to alliances. SDG targets improve statistics while leaving structures of antagonism intact. Without perception, frameworks hollow out universality by making rights and duties contingent on identity. With perception, even imperfect frameworks can function as vessels of coherence. The Māori recognition of rivers, or the TRC's hearings in South Africa, show how frameworks can embody wholeness when animated by awareness. The challenge is to treat frameworks as mirrors of perception rather than as ends in themselves.

Towards Coherence

If identity produces systemic incoherence, leadership must embed conditions that allow perception to surface. Three conditions are essential. First, proximity: suffering must be encountered directly, not abstracted into categories of 'ally', 'enemy', or 'collateral'. Proximity explains why community-led disaster relief often succeeds where bureaucratic systems falter: people act from what they see, not from ideology. Second, dialogue: genuine dialogue suspends assumptions and allows the relationship to precede identity. Citizens' assemblies show that when dialogue is prioritised, coherence emerges more easily than in partisan negotiation. Third, accountability: institutions must remain tethered to life rather than image. The Amazon cases reveal how accountability aligned with perception, herd-tracking and fines tied to ecological rhythms can reduce destruction, while a lack of accountability enables regression.

Together, these conditions make it possible for institutions to respond to what is, rather than to what identity demands. They do not guarantee wholeness, but they open the space where perception can replace allegiance as the ground of action.

Bridging to Holistic Perception

These lessons confirm the argument of the previous section: governance and leadership cannot be transformed by methods or frameworks alone. The coherence glimpsed in South Africa, the Himalayas or Māori ecological governance did not emerge from the invention of new strategies but from a momentary shift in perception. In each case, identity was set aside long enough for the relationship to be seen directly.

The fragility of these examples is equally instructive. Each was quickly constrained by the return of political interest, economic pressures or entrenched narratives of identity. This suggests that while frameworks can provide openings, only a shift in perception can sustain transformation. Holistic perception in practice is therefore not a blueprint to be replicated, but a way of being that must be continually rediscovered in leadership, in communities and in everyday relationship. It is not about designing systems that guarantee wholeness, but about seeing directly in each moment so that action naturally arises from intelligence rather than identity.

9.8 Conclusion: From Choice to Choiceless Intelligence

This chapter has traced the movement from the illusion of choice to the possibility of choiceless action. We began by examining how choice, so often equated with freedom, is in fact conditioned by memory, unconscious processes and institutional inertia. Neuroscience demonstrates that neural activity often precedes conscious intention, while social theory shows how habitus, culture and structural power inscribe patterns that masquerade as autonomy. To call such conditioned decisions freedom is to mistake continuity for transformation.

Against this backdrop, we explored the meaning of choiceless awareness. Far from being passive, it is a state of direct perception in which the chooser, the self-shaped by memory, fear and desire, is absent. In such perception lies clarity, and from clarity flows natural action. Choiceless awareness is not cultivated as a method, for the moment it becomes a practice or goal, motive re-enters and the self returns. It is present when the mind is still enough to see without distortion, when attention is whole rather than fragmented.

From this ground arises intelligence. This is not intelligence as accumulation of knowledge or refinement of strategies, but intelligence as responsiveness to what is. Intelligence is not personal property. It is not 'my' intelligence or 'yours'. It is a movement of wholeness, beyond identity, which expresses itself in action that does not fracture life into competing parts. This is why action from intelligence is coherent, while action from identity perpetuates contradiction even when it wears the language of progress or reform.

The contrast is visible across the crises of our time. In climate governance, nations negotiate as blocs with competing interests, producing pledges that coexist with new drilling and subsidies. This is identity disguised as pragmatism. When communities perceive directly the inseparability of ecology and survival, as in indigenous stewardship, action arises without the trade-off of growth versus conservation. In migration, the same fragmentation dictates who is welcomed and who is excluded, not on the basis of vulnerability but on geography, race or religion. Choiceless intelligence would see migration not as a threat to identity but as a shared human reality. In inequality, frameworks proliferate while structures of privilege remain intact, because identity defends its continuity. In war, human beings are turned into images of adversaries, collateral damage, strategic assets. Perception without identity sees only the fact of suffering, and from that clarity arises the refusal to treat any life as expendable. Even in emerging technologies such as artificial intelligence, governance framed by national rivalry or corporate identity collapses into competition. But when dignity and shared vulnerability are perceived directly, ethics ceases to be an external framework and becomes the natural ground of action.

The implication is clear. Frameworks, reforms and competencies cannot substitute for awareness. They may coordinate action, but when mistaken for transformation, they become instruments of continuity. When awareness is reduced to a skill, as in contemporary models such as the Inner Development Goals or regenerative leadership checklists, it collapses into another possession of the self, another strategy of identity. Its transformative essence is lost. By contrast, when awareness is whole, action is choiceless. It is not bound to identity, not driven by fear or desire and not postponed into time. It does not turn awareness into a framework or competency but remains alive as direct perception. It is free because it arises directly from seeing what is, here and now.

This quality of action dissolves false dilemmas between growth and climate, between equality and efficiency, between security and freedom. It shows that the very structure of our global crises is not external but a mirror of our inner fragmentation. Climate breakdown, war, inequality and technological disruption cannot be resolved by more sophisticated reforms if those reforms themselves are born of the same divided consciousness. Only action that flows from choiceless intelligence can respond coherently because it does not carry forward the division that created the crisis.

The ground of holistic response is therefore clear. True action is choiceless action, intelligence moving without identity. This alone can meet the crises of our time, because it arises from wholeness and not from fragmentation.

The next chapter turns to governance and leadership. If choiceless intelligence is the ground of right action, how can those entrusted with collective responsibility embody it? What does leadership look like when it arises not from identity or ideology but from wholeness? Chapter 10 explores these questions under the theme of conscious governance and leadership.

In the end, freedom is not the power to choose, but the ending of the chooser.

References

Acosta, A. (2013). *Buen vivir: Sumak kawsay*. https://www.researchgate.net/publication/260373466

Ada Lovelace Institute. (2022). *AI governance report*. https://www.adalovelaceinstitute.org/report/

Agrawal, A., & Gibson, C. C. (1999). Enchantment and disenchantment: The role of community in natural resource conservation. *World Development*, *27*(4), 629–649. https://doi.org/10.1016/S0305-750X(98)00161-2

Amnesty International. (2019). *Human rights in the digital age*. Amnesty International.

Anderson, B. (1991). *Imagined communities: Reflections on the origin and spread of nationalism* Rev. ed.. Verso.

Arendt, H. (1958). *The human condition*. University of Chicago Press.

Associated Press. (2023, July 6). Brazil fines meatpacking giant JBS for buying cattle from illegally deforested land. *AP News*. https://apnews.com/article/amazon-deforestation-cattle-jbs-nyse-brazil-76bd1fe0c0f3cd77f0e5a4be8e366ab5

Bargh, J. A., & Chartrand, T. L. (1999). The unbearable automaticity of being. *American Psychologist*, *54*(7), 462–479. https://doi.org/10.1037/0003-066X.54.7.462

Beck, U. (1992). *Risk society: Towards a new modernity* M. Ritter, Trans.. SAGE.

Bohm, D. (1980). *Wholeness and the implicate order*. Routledge.

Bohm, D. (1994). *Thought as a system*. Routledge.

Bohm, D. (1996). *On dialogue*. Routledge.

Bourdieu, P. (1977). *Outline of a theory of practice* R. Nice, Trans.. Cambridge University Press.

Brewer, J. (2017). *The craving mind: From cigarettes to smartphones to love—Why we get hooked and how we can break bad habits*. Yale University Press.

Butler, J. (1990). *Gender trouble: Feminism and the subversion of identity*. Routledge.

Capra, F. (1996). *The web of life: A new scientific understanding of living systems*. Anchor Books.

Clark, P. (2010). *The Gacaca courts, post-genocide justice and reconciliation in Rwanda: Justice without lawyers*. Cambridge University Press.

Chhetri, N., Chaudhary, P., Tiwari, P. R., & Yadaw, R. B. (2013). Institutional and technological innovation: Understanding agricultural adaptation to climate change in Nepal. In J. B. Smith, R. J. T. Klein, & S. Huq (Eds.), *Climate change, adaptive capacity and development* (pp. 147–161). Imperial College Press.

Court of Justice of the European Union. (2014). *Google Spain SL and Google Inc. v Agencia Española de Protección de Datos (AEPD) and Mario Costeja González (C-131/12)*. https://curia.europa.eu/juris/liste.jsf?num=C-131/12

Csikszentmihalyi, M. (1990). *Flow: The psychology of optimal experience*. Harper & Row. https://www.harpercollins.com/products/flow-mihaly-csikszentmihalyi

Damasio, A. (2010). *Self comes to mind: Constructing the conscious brain*. Pantheon.

Davidson, R. J., & Kaszniak, A. W. (2015). Conceptual and methodological issues in research on mindfulness and meditation. *American Psychologist*, *70*(7), 581–592. https://doi.org/10.1037/a0039512

Escobar, A. (1995). *Encountering development: The making and unmaking of the third world*. Princeton University Press.

Ferguson, J. (1990). *The anti-politics machine: Development, depoliticization, and bureaucratic power in Lesotho*. Cambridge University Press.

Foucault, M. (1977). *Discipline and punish: The birth of the prison* A. Sheridan, Trans.. Pantheon.

Fraser, N. (2013). *Fortunes of feminism: From state-managed capitalism to neoliberal crisis*. Verso.

Freire, P. (2000). *Pedagogy of the oppressed* 30th anniversary ed.; M. B. Ramos, Trans.. Continuum. (Original work published 1970).

Freud, S. (2010). *Civilization and its discontents* J. Strachey, Trans.. W. W. Norton. (Original work published 1930).

Friston, K. (2010). The free-energy principle: A unified brain theory?. *Nature Reviews Neuroscience*, *11*(2), 127–138. https://doi.org/10.1038/nrn2787

Giddens, A. (1984). *The constitution of society: Outline of the theory of structuration*. University of California Press.

Gilligan, C. (1982). *In a different voice: Psychological theory and women's development*. Harvard University Press. https://www.hup.harvard.edu/books/9780674445444

Girard, R. (1977). *Violence and the sacred* P. Gregory, Trans.. Johns Hopkins University Press. (Original work published 1972).
Gudynas, E. (2011). Buen vivir: Today's tomorrow. *Development*, *54*(4), 441–447. https://doi.org/10.1057/dev.2011.86
Gottman, J. (1999). *The seven principles for making marriage work*. Crown.
Hammarskjöld, D. (1964). *Markings*. Knopf. https://www.penguinrandomhouse.com/books/175023/markings-by-dag-hammarskjold/
Han, B.-C. (2017). *Psychopolitics: Neoliberalism and new technologies of power* E. Butler, Trans. Verso.
Havel, V. (1985). *The power of the powerless*. Routledge.
Heidegger, M. (1966). *Discourse on thinking* J. M. Anderson & E. H. Freund, Trans.. Harper & Row. (Original work published 1959).
hooks, b. (2000). *All about love: New visions*. William Morrow.
Human Rights Watch. (2023). *World report 2023*. https://www.hrw.org/world-report/2023
Intergovernmental Panel on Climate Change. (2022). *Climate change 2022: Impacts, adaptation and vulnerability*. https://www.ipcc.ch/report/ar6/wg2/
International Committee of the Red Cross. (2023). *Annual report 2023*. https://www.icrc.org/en/publication/annual-report-2023
International Energy Agency. (2023). *World energy outlook 2023*. https://www.iea.org/reports/world-energy-outlook-2023
Iyengar, S. (2010). *The art of choosing*. Twelve.
Kabat-Zinn, J. (1990). *Full catastrophe living: Using the wisdom of your body and mind to face stress, pain, and illness*. Delacorte.
Kimmerer, R. W. (2013). *Braiding sweetgrass: Indigenous wisdom, scientific knowledge, and the teachings of plants*. Milkweed Editions.
Kirchhoff, M. D., & Kiverstein, J. (2019). Extended active inference: Constructing predictive cognition beyond skulls. *Mind & Language*, *34*(1), 97–119. https://doi.org/10.1111/mila.12268
Kothari, A., Salleh, A., Escobar, A., Demaria, F., & Acosta, A. (Eds.). (2019). *Pluriverse: A post-development dictionary*. Tulika Books. https://www.tulikabooks.com/books/pluriverse
Krishnamurti, J. (2009). *Freedom from the known*. HarperCollins. (Original work published 1969).
Langer, E. J. (1989). *Mindfulness*. Addison-Wesley.
Latour, B., & Porter, C. (2017). *Facing Gaia: Eight lectures on the new climatic regime* (Vol. Trans.). Polity Press.
Levinas, E. (1969). *Totality and infinity: An essay on exteriority* A. Lingis, Trans.. Duquesne University Press.
Libet, B. (1983). Time of conscious intention to act in relation to onset of cerebral activity (readiness-potential): The unconscious initiation of a freely voluntary act. *Brain*, *106*(3), 623–642. https://doi.org/10.1093/brain/106.3.623
Mandela, N. (1994). *Long walk to freedom: The autobiography of Nelson Mandela*. Little, Brown and Company.
Marcuse, H. (1964). *One-dimensional man: Studies in the ideology of advanced industrial society*. Beacon Press.
McGilchrist, I. (2019). *The master and his emissary: The divided brain and the making of the Western world* (2nd ed.). Yale University Press.

Meadows, D. H. (1999). *Leverage points: Places to intervene in a system.* The Sustainability Institute.
Merleau-Ponty, M. (1962). *Phenomenology of perception* C. Smith, Trans.. Routledge & Kegan Paul.
Mitchell, M. (2009). *Complexity: A guided tour.* Oxford University Press.
Mouffe, C. (2000). *The democratic paradox.* Verso.
New Zealand Government. (2019). *The wellbeing budget 2019.* https://www.treasury.govt.nz/publications/wellbeing-budget/wellbeing-budget-2019
Ostrom, E. (2010). *Governing the commons: The evolution of institutions for collective action* (New ed.). Cambridge University Press.
Podeh, E. (2002). *The Arab-Israeli conflict in Israeli history textbooks.* Bergin & Garvey.
Reis, H. T., Clark, M. S., & Holmes, J. G. (2004). Perceived partner responsiveness as an organizing construct in the study of intimacy and closeness. In D. J. Mashek & A. Aron (Eds.), *Handbook of closeness and intimacy* (pp. 201–225). Lawrence Erlbaum.
Reis, H. T., & Gable, S. L. (2015). Responsiveness. *Current Opinion in Psychology, 1*, 67–71. https://doi.org/10.1016/j.copsyc.2015.01.001
Reuters. (2024, April 25). Mass graves in Gaza: What do we know? *Reuters.* https://www.reuters.com/world/middle-east/mass-graves-gaza-what-do-we-know-2024-04-25/
Said, E. W. (1978). *Orientalism.* Pantheon Books.
Sartre, J.-P. (1992). *Being and nothingness* H. E. Barnes, Trans.. Washington Square Press. (Original work published 1943).
Schwartz, B. (2004). The paradox of choice: Why more is less. *Ecco.* https://www.harpercollins.com/products/the-paradox-of-choice-barry-schwartz
Sen, A. (2006). *Identity and violence: The illusion of destiny.* W. W. Norton & Company. https://wwnorton.com/books/9780393329296
Shiva, V. (1989). *Staying alive: Women, ecology and development.* Zed Books. https://www.zedbooks.net/shop/book/staying-alive/
Soon, C. S., Brass, M., Heinze, H. J., & Haynes, J.-D. (2008). Unconscious determinants of free decisions in the human brain. *Nature Neuroscience, 11*(5), 543–545. https://doi.org/10.1038/nn.2112
Stiglitz, J. E. (2002). *Globalization and its discontents.* W. W. Norton & Company. https://wwnorton.com/books/9780393324390
Tajfel, H., & Turner, J. C. (1979). An integrative theory of intergroup conflict. In W. G. Austin & S. Worchel (Eds.), *The social psychology of intergroup relations* (pp. 33–47). Brooks/Cole.
Thich, N. H. (1991). *Peace is every step: The path of mindfulness in everyday life.* Bantam.
Thich, N. H. (1999). *The miracle of mindfulness: An introduction to the practice of meditation* M. Ho, Trans.. Beacon Press.
Truth and Reconciliation Commission. (1998). *Final report.* https://www.justice.gov.za/trc/report/
Tutu, D. (1999). *No future without forgiveness.* Doubleday. https://www.penguinrandomhouse.com/books/117353/no-future-without-forgiveness-by-desmond-tutu/
UCLA Civil Rights Project. (2024). *Report.* https://civilrightsproject.ucla.edu/

UNESCO. (2021). *Reimagining our futures together: A new social contract for education.* https://unesdoc.unesco.org/ark:/48223/pf0000379707

UNESCO. (2023). *Recommendation on education for peace, human rights and sustainable development.* https://unesdoc.unesco.org/ark:/48223/pf0000397418

United Nations. (1948). *Universal declaration of human rights.* https://www.un.org/en/about-us/universal-declaration-of-human-rights

United Nations. (2015). *Transforming our world: The 2030 agenda for sustainable development.* https://sdgs.un.org/2030agenda

United Nations. (2024). *World social report 2024.* https://www.un.org/development/desa/dspd/wp-content/uploads/sites/22/2024/01/World-Social-Report-2024.pdf

United Nations Framework Convention on Climate Change. (2023). *Global stocktake.* https://unfccc.int/topics/global-stocktake

United Nations High Commissioner for Refugees. (2023). *Global trends report 2023.* https://www.unhcr.org/global-trends-report-2023

United Nations Office for the Coordination of Humanitarian Affairs. (2024). *Global humanitarian overview 2024.* https://gho.unocha.org/

Varela, F. J., Thompson, E., & Rosch, E. (1991). *The embodied mind: Cognitive science and human experience.* MIT Press.

Wegerif, R. (2011). *Towards a dialogic theory of education* (Education, Dialogue and Argumentation, Vol. 1). Springer.

Wegner, D. M. (2002). *The illusion of conscious will.* MIT Press.

Weil, S. (2002a). *Gravity and grace* E. Craufurd, Trans.. Routledge. (Original work published 1952).

Weil, S. (2002b). *Waiting for god* E. Craufurd, Trans.. Harper Perennial.

Whitehead, A. N. (1978). *Process and reality: An essay in cosmology* Corrected ed.; D. R. Griffin & D. W. Sherburne, (Eds.). Free Press. (Original work published 1929).

Wilson, R. A. (2001). *The politics of truth and reconciliation in South Africa: Legitimizing the post-apartheid state.* Cambridge University Press. https://doi.org/10.1017/CBO9780511522291

World Trade Organization. (2020). *World trade report 2020.* https://www.wto.org/english/res_e/booksp_e/wtr20_e/wtr20_e.pdf

Young, I. M. (1990). *Justice and the politics of difference.* Princeton University Press. https://press.princeton.edu/books/paperback/9780691023151

Zajonc, A. (2013). Contemplative pedagogy: A quiet revolution in higher education. In J. Simmer-Brown & F. Grace (Eds.), *Meditation and the classroom: Contemplative pedagogy for religious studies* (pp. 83–94). State University of New York Press.

Zook, M. (2020). *The data revolution and its consequences.*

Zuboff, S. (2019). *The age of surveillance capitalism.* PublicAffairs. https://www.publicaffairsbooks.com/titles/shoshana-zuboff/the-age-of-surveillance-capitalism/9781610395694/

Chapter 10

Conscious Governance and Leadership

Abstract

Chapter 10 extends the inquiry from individual transformation to public life. It contrasts leadership rooted in identity, fear and image with leadership grounded in awareness and intelligence. Governance, traditionally organised through control and enforcement, is reimagined as consciousness in action. This chapter explores responsibility as perception: the capacity to respond wholly, not react partially. When awareness replaces ideology, governance ceases to manage disorder and begins to reflect harmony. Institutions designed without a centre of identity act as mirrors of collective intelligence rather than instruments of authority. This chapter concludes that governance, rightly understood, is not a framework of rules but a field of attention where action flows from clarity. This movement marks the integration of spirituality and public policy: awareness as the essence of good governance.

Keywords: Conscious governance; leadership awareness; identity dissolution; responsibility as perception; intelligence in action; order from wholeness; institutional reflection; beyond control

10.1 Introduction: From the Illusion of Choice to Leadership as Seeing

Chapter 9 concluded with a simple but radical finding: choice belongs to confusion. When the mind moves as the chooser, it compares fragments in memory, projects into time and mistakes this movement for freedom (Krishnamurti & Anderson, 1991). In that field, action is delayed; reform becomes adjustment within the same pattern. We saw how psychological choice is sustained by identity and desire, and how clarity appears only when the chooser is absent.

This inquiry cannot remain confined to the personal. The same structures of identity, choice and projection operate in public life. What moves inwardly as the

From Fragmentation to Wholeness, 219–232

doi:10.1108/978-1-80686-677-920261010

'me' appears outwardly as leadership, governance and institutions. The names change: nation, party, market, brand, community, but the structure remains the same. Choice becomes policy menus, negotiations and trade-offs; knowledge becomes frameworks, dashboards and treaties. When action starts from identity, order is created in a fragment, while disorder spreads across the whole (Bohm, 1980/1996).

The philosopher Jean-Paul Sartre (1943/1992) described how much of human action is carried out in *bad faith*: clinging to roles, categories and masks to avoid the radical responsibility of freedom. Leaders, too, often act in bad faith: invoking identity and institutional continuity rather than perceiving directly. Nietzsche (1887/1994) traced the will to power in politics as the reassertion of identity in new forms (domination, pride and survival) cloaked as necessity. These insights converge with Krishnamurti's insistence that clarity begins only when the chooser is absent.

Our task is not to add another governance method but to look precisely at how identity organises leadership and how frameworks carry identity forward in time. We then ask: if choice marks confusion, what is decision when there is clarity? If frameworks stabilise continuity, what is governance when perception is the ground? These questions matter because frameworks that mistake continuity for transformation become instruments of postponement (Meadows, 2008; Morin, 2008).

Two threads from Chapter 9 remain central as we proceed. The first is the continuity of the self over time: in public life, this appears as targets, pledges and projections that defer action while preserving identity. The second is the role of attention: where attention is whole, action does not begin from a category. A river is not a 'resource', a child not a 'unit', a civilian not a 'symbol'. From direct seeing, action is simple and coherent (Bohm, 1992).

The argument is plain. Leadership that starts from identity cannot deliver wholeness. Governance that starts from frameworks cannot end fragmentation. Intelligence is not more information; it is the ending of the chooser so that perception can act.

10.2 Leadership as Identity and Its Limits

Leadership is usually celebrated as strategy, decision and vision. Weber (1947/2019) described legitimacy in terms of authority types, but in practice legitimacy is nearly always mediated by identity. A president speaks as the embodiment of a nation, a CEO as the guardian of shareholder value and a religious leader as the custodian of faith. Identity stabilises authority, but it also constrains it. To lead 'us' requires defining 'them'. Boundaries are managed, division is reproduced and what appears as decisive action secures only short-term stability while reproducing long-term crises (Arendt, 1958/1998).

From empires to modern nation-states, leadership has been inseparable from identity formation. Colonial administrations consolidated power by inventing categories of race and tribe, ruling through division (Mamdani, 1996).

Nationalist movements mobilised counter-identities to overthrow empire but often left new states governed by inverted versions of the same logic. Democracies continue the cycle through party politics, where leaders represent fragments of class, religion, ethnicity or region, and leadership becomes a competition among identities rather than a perception of the whole (Anderson, 1983). Authoritarian systems intensify this dynamic by fusing the leader and the nation, dissolving dissent into treason.

The pattern continues today. Populist leaders in the United States, Brazil, India and Hungary rise by presenting themselves as protectors of a threatened identity (Mounk, 2018). Climate summits perform responsibility while governments expand drilling and subsidies in allegiance to their economic identity (International Energy Agency [IEA], 2023). Corporations pledge 'net zero' while investing in new extraction, preserving their corporate identity as competitive actors in global markets (Oreskes & Conway, 2010). Big Tech frames surveillance as innovation, masking exploitation as progress (Zuboff, 2019). In Myanmar and Ukraine, military and political leaders justify violence in the name of national security, reducing human beings to instruments of identity narratives. Even democratic systems incentivise appeals to partisanship rather than perception, rewarding loyalty to constituencies rather than responsiveness to the whole (Stiglitz, 2002).

From a psychological perspective, identity-based leadership is a conditioned action. Decisions are shaped by loyalty to memory, institutional roles and affective residues (Bourdieu, 1977; Damasio, 2010). What appears as decisiveness is often repetition of inherited patterns. As Krishnamurti (1969/2009) observed, action based on an image breeds contradiction. Leaders tethered to identity are structurally incapable of perceiving the whole. National leaders treat climate breakdown as a trade-off with growth. Corporate leaders treat ecological limits as risks to competitiveness. Community leaders treat outsiders as threats to cohesion. In each case, perception is filtered through 'our interest' versus 'theirs'.

The consequences are visible in global commitments. Climate pledges under SDG 13 are neutralised by fossil subsidies defended as national growth. SDG 16 collapses when security identity fuels cycles of violence. SDG 4 is compromised when education systems become vehicles of propaganda. SDG 5 falters when cultural identity restricts women's rights. Reform may bring tactical victories, but the deeper disorder persists (Sachs, 2015; Stiglitz, 2002).

Conventional leadership, therefore, is not the solution to fragmentation but its embodiment. It unites followers by dividing humanity, producing order within fragments while sustaining disorder in the whole. Leaders who act from identity may inspire loyalty, but they cannot act from wholeness. The task ahead is radical: to reconceive leadership not as the performance of identity but as the expression of perception.

This distinction becomes clearer when we separate the *leader* from *leadership*. The leader is often reduced to a role or identity, the figure who embodies 'us'. Leadership, however, need not depend on that figure. It can signify a quality of perception and responsibility. Singh and Bhatnagar (2023) articulate *spiritual leadership* in the context of sustainable development policy, arguing that leadership grounded in inner awareness and values beyond ego can guide public action with integrity. Alongside this, Rajneesh's *conscious systemic leadership (CSL)* shifts attention from the person to coherence in systems. These perspectives gesture toward the possibility that leadership may outlast the leader. Yet, if they are crystallised into leadership styles, methods or tools, we are back to the same fragmentation from which we intend to emerge (Gupta, 2025).

10.3 Conscious Governance: Beyond Frameworks and Reform Loops

If leadership rooted in identity perpetuates fragmentation, governance built on identity's instruments cannot yield coherence. Contemporary governance is dominated by frameworks including constitutions, treaties, agencies, policies and metrics. These provide stability but rarely resolution. By translating living crises into targets and dashboards, they produce administrable fragments. What is gained in tractability is lost in wholeness (Capra, 1997).

Frameworks promise transformation yet often result in delay. 'Net zero by 2050' pledges and COP28's ambiguous 'transition away' language exemplify temporal deferral (IEA, 2023). SDG 8's growth imperative contradicts SDG 13's climate demands, institutionalising category deferral. Responsibility deferral is seen in the Paris Agreement's 'common but differentiated responsibilities', where responsiveness is reduced to bargaining (UNFCCC, 2015).

Here the SDGs deserve particular attention. Agenda 2030 was presented as a universal pact, yet its structure fragments crises into 17 silos with 169 indicators (United Nations, 2015). Progress is reported in dashboards, but the contradictions remain systemic: economic growth is pursued as an independent goal, directly undercutting climate stability. Foucault (1978/1991) would describe this as governmentality, a regime of measurement and surveillance that produces governable subjects rather than transformation. Bourdieu (1977, 1990) would add that such frameworks inscribe dispositions into the field of governance, ensuring continuity through the very categories that claim to reform. What appears as universality is often a codification of identity into metrics.

Universality also collapses under pressure. The Geneva Conventions, intended to universalise the protection of civilians, have been repeatedly violated in Syria, Yemen and Gaza, with enforcement applied selectively depending on alliances (Amnesty International, 2019; ICRC, 2023). COVAX, designed to guarantee equitable vaccine access during COVID-19, collapsed into nationalism, with high-income countries securing over 70% of doses while many African states vaccinated less than 10% of their populations by late 2021 (WHO, 2022). Ukrainian refugees in 2022 were embraced in Europe while Syrians and Afghans were rejected (UNHCR, 2023). The Responsibility to Protect justified intervention in Libya but was paralysed in Syria and Gaza (United Nations, 2024). Even inward-facing frameworks, like the Inner Development Goals (IDGs), risk performativity, turning awareness into a measurable skill rather than perception (Krishnamurti, 1969/2009).

Conscious governance does not reject frameworks but rather relocates them. They can coordinate and signal intent, but they cannot supply perception. Placed rightly, they act as mirrors, enabling collaboration across boundaries; placed wrongly, they substitute design for intelligence (Bohm, 1992). Governance becomes conscious only when perception precedes instrumentation.

10.4 Intelligence as the Ground of Leadership

Knowledge organises, but intelligence resolves. Knowledge is retrospective, bound to categories, frameworks and memory. It models, forecasts and optimises. Intelligence, by contrast, is present-seeing – attention unclouded by identity that discerns inseparability. Knowledge generates reform loops, multiplying dashboards and KPIs. Intelligence dissolves contradiction itself, recognising that economy is a subsystem of ecology and that survival cannot be traded against growth (Capra, 1997; Meadows, 2008).

When knowledge operates under identity, it becomes bargaining material. Climate models become instruments of minimal negotiation, and statistical thresholds transform into ceilings rather than baselines (IPCC, 2023). Complexity and expertise often legitimise postponement, as leaders invoke uncertainty to delay action while defending identity (Beck, 1992). Intelligence cuts through by refusing separability. From perception, action aligns without compulsion because it arises directly from seeing.

Examples clarify this distinction. Angela Merkel's response to the 2015 refugee crisis, 'Wir schaffen das' ('We can manage this'), expressed recognition of vulnerability rather than partisan calculus, despite political backlash (Triadafilopoulos, 2019). Uruguay's José Mujica governed with simplicity, donating most of his salary and rooting reforms in dignity rather than ideology (Lissardy, 2013). Finland's education system prioritised equity and well-being over competition, achieving high outcomes (Sahlberg, 2011). Bhutan reframed development through Gross National Happiness, integrating ecological and cultural integrity with material welfare (Ura et al., 2012). During COVID-19, governments in New Zealand, Taiwan and Kerala acted swiftly in solidarity, contrasting with delays where leaders framed the crisis as a trade-off between lives and the economy (Hale et al., 2020).

These cases confirm that intelligence in leadership is not prediction or technical expertise but responsiveness to the whole. As Bohm (1994) noted, thought belongs to memory, while intelligence arises from wholeness. Krishnamurti (1969/2009) insisted that intelligence is present when the self is absent. When leadership acts from identity, policies stabilise disorder. When it acts from intelligence, coherence flows naturally because perception and action are indivisible.

10.5 Action From Perception, Not Projection

Contemporary governance is dominated by projections, including targets, scenarios and long-term roadmaps. These instruments are not neutral; they relocate responsibility into the future while preserving present identity. COP28 pledged to 'transition away' from fossil fuels even as drilling expanded, exemplifying projection as delay (IEA, 2023; UNFCCC, 2023). SDGs project improvements into 2030 while entrenching contradictions between endless growth (SDG 8) and urgent climate limits (SDG 13) (United Nations, 2015). The Responsibility to Protect projected universality but collapsed when filtered through alliances (United Nations, 2024). Projection multiplies options where clarity is required.

Perception, by contrast, begins with inseparability. A forest is a living system, not a resource; health and economy are one tissue; migration is a shared human condition. When inseparability is seen, false dilemmas dissolve. The question 'How do we balance?' becomes 'What coherence is required now?'

Examples show perception in action. Ecuador's 2008 constitution enshrined Buen Vivir, an Andean philosophy integrating ecology and well-being (Acosta, 2013). New Zealand recognised the Whanganui River as a legal person, institutionalising Māori cosmology that perceives rivers as kin (Charpleix, 2018). Vanuatu secured a 2023 ICJ advisory opinion on climate obligations, reframing justice as survival rather than negotiation (United Nations, 2023). Costa Rica abolished its military in 1948, redirecting resources to education and health because militarisation was seen as a threat, not security (Biesanz et al., 1999). Citizens' assemblies in France, Ireland and the United Kingdom revealed that when ordinary participants suspend partisan identity, they advance more coherent proposals than politicians (Smith, 2021).

The lesson is clear: projection sustains delay by displacing responsibility, while perception responds directly to what is. As Bohm (1987) emphasised, the word is not the thing; description cannot substitute for direct seeing. Conscious governance does not compute acceptable futures but refuses to obscure the present.

10.6 Responsibility as Perception, Not Burden

Public discourse treats responsibility as a burden to be distributed. The Paris Agreement institutionalised 'common but differentiated responsibilities', which led to bargaining over quotas and timelines (UNFCCC, 2015). The Responsibility to Protect, unanimously adopted in 2005, has been inconsistently applied, invoked in Libya, paralysed in Syria and Gaza, and filtered through alliances (United Nations, 2024). Even at the level of citizens, responsibility is reduced to compliance tasks such as recycling or paying taxes.

Conscious responsibility is different. It is not allocated but arises directly from perception. To see the suffering of a child, the bombing of civilians or the destruction of a forest is already to be responsible. No appeal to duty or role is needed; response flows from the encounter itself (Levinas, 1969). Krishnamurti (1969/2009) made the same point: responsibility is care where the self is absent.

Examples illustrate this quality. After Japan's 2011 tsunami, citizens organised spontaneously to share food and shelter before official relief arrived, demonstrating solidarity rooted in perception (Aldrich, 2012). In Kenya, pastoralist and farming communities manage drought through committees that integrate traditional ecological knowledge with climate science, acting through shared responsiveness rather than compliance with external quotas (Nyong et al., 2007).

Responsibility as a burden fragments into bargaining. Responsibility as perception unites through immediacy and care. Institutions can support this by enabling direct encounters such as community monitoring, response rooms or universal legal mirrors. Responsibility becomes lighter, not heavier, because it arises from care, not compulsion.

10.7 Decision-Making Beyond Choice

Choice appears as freedom but often conceals confusion. Krishnamurti (1969/2009) observed that choice exists only where there is division; when perception is clear, there is no conflict between options. Neuroscience reinforces this: Libet's (1985) and Soon et al.'s (2008) experiments suggest that conscious decisions often rationalise actions already initiated unconsciously.

Damasio's (1994) *somatic marker hypothesis* adds another dimension. He argued that decision-making is guided by bodily signals tied to past experiences, often before conscious thought intervenes. This means that even when leaders believe they are making rational choices, they are frequently acting from embodied residues of history with fear, reward and memory rather than fresh perception. Institutions mirror this process: they are collective 'bodies of memory', codifying habits and logics that guide decisions long before new awareness can intervene.

Policy science confirms the illusion. Simon (1957) described bounded rationality as satisficing rather than clarity. Lindblom (1959) characterised governance as 'muddling through'. March and Olsen (1989) showed how institutions encode logics of appropriateness, shaping decisions based on identity and role. Decision-making framed as choice stabilises disorder because it negotiates fragments without dissolving contradiction.

Contemporary crises make this plain. Climate negotiations oscillate between pledges and subsidies, treating growth and climate as trade-offs even as thresholds are breached (IEA, 2023; IPCC, 2023). COVID-19 was framed as a choice between lives and the economy, resulting in delays and heavier losses, while those perceiving inseparability acted swiftly (OECD, 2021). Migration policy in Europe oscillates between openness and restriction, applying principles selectively to Ukrainians versus Syrians (Le Monde, 2024; UNHCR, 2023).

Choiceless awareness does not abolish deliberation but transforms it. Decisions grounded in perception begin from inseparability, filter through irreversibility and dissolve when identity reasserts itself. Many 'hard choices' are revealed as false dilemmas once perception is whole.

The same illusion shapes institutions. As Weber (1922/1978), North (1990) and Pierson (2000) observed, institutions reproduce path dependence, resisting transformation. Bourdieu (1990) showed how institutions encode habitus, ensuring the continuity of power. What begins as an encounter ossifies into identity. Education systems transmit conflict narratives, as in Israel and Palestine (Podeh, 2002), or rewrite curricula with each regime, as in Afghanistan (Human Rights Watch, 2023; Senzai, 2011). Even awareness-based initiatives risk ossification, as with the IDGs reducing perception to performance. Legal frameworks also falter, with the Geneva Conventions selectively enforced depending on alliances (Amnesty International, 2019; ICRC, 2023).

Yet glimpses exist of institutions without identity. South Africa's Truth and Reconciliation Commission briefly allowed encounters beyond roles, where victims and perpetrators recognised one another as human (Tutu, 1999). The Whanganui River's legal recognition reframed law as a relationship, not ownership (Charpleix, 2018). Community forestry in Nepal and India aligned governance with ecological rhythms, sustaining ecosystems and livelihoods (Agrawal & Gibson, 1999). Citizens' assemblies on climate showed that institutions can host spaces where identity is suspended and coherence arises through shared perception (Smith, 2021).

Institutions anchored in identity stabilise fragmentation; institutions arising from perception serve as mirrors. Their coherence lies not in design but in the quality of attention sustaining them.

10.8 Institutions Without Identity

Institutions are often treated as the backbone of order. Yet as Weber (1922/1978), North (1990) and Pierson (2000) showed, institutions reproduce path dependence and resist transformation. Bourdieu (1990) emphasised how they encode *habitus*, ensuring the continuity of power. What begins as an encounter with reality ossifies into procedures and platforms of identity. Institutions become guardians of memory rather than mirrors of perception.

Education systems illustrate this ossification. In Israel and Palestine, textbooks transmit antagonistic narratives, embedding conflict from childhood (Podeh, 2002). In Afghanistan, successive regimes rewrote curricula to secure legitimacy (Human Rights Watch, 2023; Senzai, 2011). Awareness-based initiatives also risk ossification: the IDGs reduce awareness to competency checklists, turning perception into performance.

Legal and humanitarian frameworks follow similar patterns. The Geneva Conventions are cited selectively depending on alliances, collapsing universality into strategy (Amnesty International, 2019; ICRC, 2023). In Gaza, Syria and Yemen, hospitals and schools have been repeatedly attacked while enforcement falters (UNOCHA, 2024; Human Rights Watch, 2024). The SDGs fragment crises into targets, masking the continuity of disorder even as dashboards report progress (United Nations, 2015).

Yet glimpses of institutions without identity exist. South Africa's Truth and Reconciliation Commission briefly allowed encounters beyond roles, where victims and perpetrators recognised one another as human (Tutu, 1999). The Whanganui River's legal recognition reframed law as a relationship, not ownership (Charpleix, 2018). Community forestry in Nepal and India has shown that governance can align with ecological rhythms, sustaining ecosystems and livelihoods together (Agrawal & Gibson, 1999). Citizens' assemblies on climate further reveal that institutions can host spaces where identity is suspended, and coherence arises through shared perception (Smith, 2021).

Institutions anchored in identity stabilise fragmentation; institutions arising from perception can function as mirrors rather than monuments. Their coherence lies not in design but in the quality of attention that sustains them.

10.9 The Arc of Seeing: Governance, Intelligence and the Death of the Framework

Our inquiry has moved from the illusion of choice, through the incoherence of identity, to the possibility of perception as the ground of governance. Frameworks multiply when insecurity rises. Dashboards, models and metrics soothe with the promise of control, yet when perception is absent they disguise the continuity of disorder (Bohm, 1987; Foucault, 1977).

Here we can speak of the 'death of the framework' in a sense reminiscent of Nietzsche's (1882/1974) 'death of God'. Nietzsche did not mean nihilism but the collapse of a false ground that clears space for a deeper orientation. Similarly, the death of the framework does not imply anarchy. It means the form is provisional, transparent and dissolvable. Governance ceases to ask which framework will solve crises and begins to ask what quality of perception sustains action.

David Bohm's (1980/1996) implicate order resonates here: frameworks are explicate unfoldings within a deeper wholeness. They are not false in themselves, but false when mistaken for the ground. Ulrich Beck (1992) argued that modernity manufactures risks while promising institutions to manage them. Frameworks designed to manage climate risk, migration or technological disruption often amplify insecurity because they treat fragments as separable. The arc of seeing reveals that wholeness cannot be systematised; it must be perceived.

This arc unfolds as follows: from choice (oscillation), to identity (consolidation), to perception (dissolution of false dilemmas), to intelligence (action that is simple because it is whole). Along this arc, the framework dies, not because form disappears but because form is no longer mistaken for substance. A treaty or law remains a vessel, but when intelligence withdraws, it is reshaped without attachment.

Examples that illustrate this arc include South Africa's TRC, Māori ecological law, Himalayan community forestry and citizens' assemblies in Europe. These were not perfect systems but openings where perception briefly interrupted identity (Smith, 2021). Their coherence lay not in design but in attention.

The 'death of the framework' is therefore liberation, not loss. It clears the space for governance as the movement of intelligence itself.

10.10 Implications for Global Transformation

The implications of this inquiry extend far beyond theory. If governance continues as the negotiation of identities and the multiplication of frameworks, disorder will deepen. Climate breakdown will accelerate, pandemics will reappear, inequality will widen and violence will recur. The fragmentation of consciousness will continue to appear as the fragmentation of systems.

But if governance begins with perception, a different arc becomes possible. Institutions can act as mirrors rather than monuments. Leaders can act from intelligence rather than identity. Citizens can recognise responsibility as immediate care rather than a burden. In such a shift, universals like human dignity, ecological integrity and freedom cease to be abstract ideals and become direct expressions of perception.

The transformation required is not technical but perceptual. This is why reforms often fail despite resources and expertise: they seek to manage fragments without dissolving the fragmentation. As Bohm (1992) noted, thought operating on its own produces confusion; only intelligence arising from wholeness can resolve it.

Policy and governance studies must therefore expand their field. Complexity theory, systems thinking and institutional economics offer valuable insights, but they remain trapped when they treat perception as just another variable. What is needed is not a new framework but a different ground: the recognition that perception itself is the beginning of coherence.

10.11 Conclusion: Leadership as Seeing

This chapter began with the illusion of choice and extended the inquiry into leadership and governance. We saw how identity organises leadership, how frameworks stabilise disorder and how knowledge alone produces reform loops. We traced how perception, intelligence and choiceless awareness offer a different beginning.

The argument is radical but simple: leadership that acts from identity cannot embody wholeness, and governance that begins from frameworks cannot end fragmentation. Only when perception precedes identity, and intelligence precedes framework, does coherence flow naturally.

The arc of seeing culminates in the death of the framework, not as a collapse into chaos but as liberation from false ground. Governance becomes the movement of intelligence itself, provisional in form but whole in action.

As we turn to Chapter 11, the inquiry deepens further: if governance is to move without the self, how can institutions, leaders and communities embody this wholeness in practice? The following chapter explores the transition from reform and continuity to perception and transformation, where intelligence begins to take root in the fabric of governance itself.

References

Acosta, A. (2013). Extractivism and neoextractivism: Two sides of the same curse. In M. Lang & D. Mokrani (Eds.), *Beyond development: Alternative visions from Latin America* (pp. 61–86). Rosa Luxemburg Stiftung.

Agrawal, A., & Gibson, C. C. (1999). Enchantment and disenchantment: The role of community in natural resource conservation. *World Development*, *27*(4), 629–649. https://doi.org/10.1016/S0305-750X(98)00161-2

Aldrich, D. P. (2012). *Building resilience: Social capital in post-disaster recovery*. University of Chicago Press.

Amnesty International. (2019). *Human rights in the digital age*. Human rights in the digital age.

Anderson, B. (1983). *Imagined communities: Reflections on the origin and spread of nationalism*. Verso.

Arendt, H. (1958/1998). *The human condition* (2nd ed.). University of Chicago Press. (Original work published 1958).

Beck, U. (1992). *Risk society: Towards a new modernity* M. Ritter, Trans. Sage.

Biesanz, M. H., Biesanz, R., & Biesanz, K. Z. (1999). *The Ticos: Culture and social change in Costa Rica* (2nd ed.). Lynne Rienner Publishers.

Bohm, D. (1992). *Thought as a system*. Routledge.

Bohm, D. (1994). *Thought as a system*. Routledge.

Bohm, D. (1987). *Unfolding meaning: A weekend of dialogue with David Bohm*. Routledge.

Bohm, D. (1980/1996). *Wholeness and the implicate order*. Routledge. (Original work published 1980).

Bourdieu, P. (1977). *Outline of a theory of practice* R. Nice, Trans. Cambridge University Press.

Bourdieu, P. (1990). *The logic of practice* R. Nice, Trans. Stanford University Press.

Capra, F. (1997). *The web of life: A new scientific understanding of living systems*. Anchor Books.

Charpleix, L. (2018). The Whanganui river as Te Awa Tupua: Place-based law in a legally pluralistic society. *The Geographical Journal*, *184*(1), 19–30. https://doi.org/10.1111/geoj.12238

Damasio, A. R. (1994). *Descartes' error: Emotion, reason, and the human brain*. Putnam.

Damasio, A. R. (2010). *Self comes to mind: Constructing the conscious brain*. Pantheon Books.

Foucault, M. (1977). *Discipline and punish: The birth of the prison* A. Sheridan, Trans. Pantheon Books.

Foucault, M. (1978/1991). Governmentality. In G. Burchell, C. Gordon, & P. Miller (Eds.), *The Foucault effect: Studies in governmentality* (pp. 87–104). University of Chicago Press.

Gupta, D. (2025). *Governance without the self: Consciousness and the end of fragmentation, systems, and complexity*. Emerge Publications. https://doi.org/10.65169/a9m3p4

Hale, T., Angrist, N., Goldszmidt, R., Kira, B., Petherick, A., Phillips, T., Webster, S., Cameron-Blake, E., Hallas, L., Majumdar, S., & Tatlow, H. (2020). Variation in government responses to COVID-19. *Nature Human Behaviour*, *4*(5), 529–538. https://doi.org/10.1038/s41562-020-0909-7

Human Rights Watch. (2023). *Education denied: Afghanistan's Taliban deprive girls of secondary education*. Human Rights Watch.

Human Rights Watch. (2024). *World report 2024: Events of 2023*. Seven Stories Press.

Intergovernmental Panel on Climate Change (IPCC). (2023). *AR6 synthesis report: Climate change 2023*. IPCC.

International Committee of the Red Cross (ICRC). (2023). *Annual report 2022*. ICRC.

International Energy Agency (IEA). (2023). *World energy outlook 2023*. IEA.

Krishnamurti, J., & Anderson, A. W. (1991). *A wholly different way of living: San Diego, California 1974—Conversations with Professor Allan W. Anderson*. HarperSanFrancisco.

Krishnamurti, J. (1969/2009). *Freedom from the known*. HarperCollins. (Original work published 1969).

Le Monde. (2024, April 14). *EU migration pact criticized for double standards*. Le Monde. https://www.lemonde.fr

Levinas, E. (1969). *Totality and infinity: An essay on exteriority* A. Lingis, Trans. Duquesne University Press.

Libet, B. (1985). Unconscious cerebral initiative and the role of conscious will in voluntary action. *Behavioral and Brain Sciences*, *8*(4), 529–566. https://doi.org/10.1017/S0140525X00044903

Lindblom, C. E. (1959). The science of "muddling through." *Public Administration Review*, *19*(2), 79–88. https://doi.org/10.2307/973677

Lissardy, G. (2013, March 1). *José Mujica: The world's "poorest" president*. BBC News. https://www.bbc.com/news/magazine-20243493

Mamdani, M. (1996). *Citizen and subject: Contemporary Africa and the legacy of late colonialism*. Princeton University Press.

March, J. G., & Olsen, J. P. (1989). *Rediscovering institutions: The organizational basis of politics*. Free Press.

Meadows, D. H. (2008). *Thinking in systems: A primer*. Chelsea Green Publishing.

Morin, E. (2008). *On complexity*. Hampton Press.

Mounk, Y. (2018). *The people vs. democracy: Why our freedom is in danger and how to save it*. Harvard University Press.

Nietzsche, F. (1887/1994). *On the genealogy of morals* W. Kaufmann & R. J. Hollingdale, Trans. Vintage. (Original work published 1887).

Nietzsche, F. (1882/1974). *The gay science* W. Kaufmann, Trans. Vintage. (Original work published 1882).

North, D. C. (1990). *Institutions, institutional change and economic performance*. Cambridge University Press.

Nyong, A., Adesina, F., & Osman Elasha, B. (2007). The value of indigenous knowledge in climate change mitigation and adaptation strategies in the African

Sahel. *Mitigation and Adaptation Strategies for Global Change*, *12*(5), 787–797. https://doi.org/10.1007/s11027-007-9099-0

Oreskes, N., & Conway, E. M. (2010). *Merchants of doubt*. Bloomsbury Press.

Organisation for Economic Co-operation and Development (OECD). (2021). *OECD policy responses to coronavirus (COVID-19): The territorial impact of COVID-19: Managing the crisis across levels of government*. OECD.

Pierson, P. (2000). Increasing returns, path dependence, and the study of politics. *American Political Science Review*, *94*(2), 251–267. https://doi.org/10.2307/2586011

Podeh, E. (2002). *The Arab-Israeli conflict in Israeli history textbooks, 1948–2000*. Bergin & Garvey.

Sachs, J. D. (2015). *The age of sustainable development*. Columbia University Press.

Sahlberg, P. (2011). *Finnish lessons: What can the world learn from educational change in Finland?*. Teachers College Press.

Sartre, J.-P. (1992). *Being and nothingness* H. E. Barnes, Trans. Washington Square Press. (Original work published 1943).

Senzai, F. (2011). Education reform in Afghanistan: A missed opportunity. *Journal of International Affairs*, *64*(1), 193–208.

Simon, H. A. (1957). *Administrative behavior* (2nd ed.). Macmillan.

Singh, N., & Bhatnagar, D. (2023). *Applied spirituality and sustainable development policy*. Emerald Publishing.

Smith, G. (2021). *Can democracy safeguard the future?*. Polity Press.

Soon, C. S., Brass, M., Heinze, H.-J., & Haynes, J.-D. (2008). Unconscious determinants of free decisions in the human brain. *Nature Neuroscience*, *11*(5), 543–545. https://doi.org/10.1038/nn.2112

Stiglitz, J. E. (2002). *Globalization and its discontents*. W. W. Norton & Company.

Triadafilopoulos, T. (2019). Merkel's refugee policy and the limits of cosmopolitanism. *German Politics*, *28*(3), 412–429.

Tutu, D. (1999). *No future without forgiveness*. Doubleday.

United Nations Framework Convention on Climate Change (UNFCCC). (2023). *Conference of the parties serving as the meeting of the parties to the Paris agreement (CMA 5): Decision on the outcome of COP28*. UNFCCC.

United Nations Framework Convention on Climate Change (UNFCCC). (2015). *Paris agreement*. UNFCCC.

United Nations High Commissioner for Refugees (UNHCR). (2023). *Global trends: Forced displacement in 2022*. UNHCR.

United Nations Office for the Coordination of Humanitarian Affairs (UNOCHA). (2024). *Global humanitarian overview 2024*. UNOCHA.

United Nations. (2023). *Request for an advisory opinion of the International Court of Justice on the obligations of states in respect of climate change*. United Nations.

United Nations. (2024). *Responsibility to protect (R2P): Annual report 2024*. United Nations.

United Nations. (2015). *Transforming our world: The 2030 Agenda for sustainable development*. United Nations.

Ura, K., Alkire, S., Zangmo, T., & Wangdi, K. (2012). *A short guide to Gross National Happiness index*. Centre for Bhutan Studies.

Weber, M. (1978). *Economy and society: An outline of interpretive sociology* G. Roth & C. Wittich, Eds. University of California Press. (Original work published 1922).

Weber, M. (2019). *The theory of social and economic organization* A. M. Henderson & T. Parsons, Trans. Martino Fine Books. (Original work published 1947).
World Health Organization (WHO). (2022). *COVAX: Working for global equitable access to COVID-19 vaccines*. WHO.
Zuboff, S. (2019). *The age of surveillance capitalism: The fight for a human future at the new frontier of power*. PublicAffairs.

Chapter 11

Policy, Education, Systems; Acting From Coherence

Abstract

This chapter translates insight into the sphere of practice. It contrasts 'policy as continuity', which organises fragments through frameworks, with 'policy as perception', which acts from coherence in the present. Education and systems thinking are examined as extensions of the same movement of thought that seeks order through method. Drawing on cases in leadership and pedagogy, this chapter identifies guiding principles for transformation: perception before prescription, dialogue before decision, proximity to life, accountability to wholeness and simplicity in action. These are not tools but qualities of awareness operating in a relationship. When institutions act from this coherence, sustainability and compassion cease to be goals; they become the character of perception itself. This chapter situates this as the bridge between insight and structure, where living intelligence organises form without control.

Keywords: Coherent policy; educational perception; systems awareness; dialogue in action; simplicity principles; wholeness in governance; proximity to life; transformation beyond tools

11.1 Introduction: Beyond Reform Towards Wholeness

The inquiry of the previous chapters has revealed a decisive insight. What is commonly celebrated as freedom, reform or progress is in fact continuity: the self in movement, institutions in negotiation, frameworks in delay. Chapter 9 showed that choice, when rooted in memory and identity, is not freedom but a conditioned response. Chapter 10 demonstrated how governance and leadership, when tethered to frameworks and reform loops, perpetuate fragmentation under the guise of responsibility. In both cases, what appears as transformation is continuity in disguise.

From Fragmentation to Wholeness, 233–265

doi:10.1108/978-1-80686-677-920261011

This chapter turns to the practical domains where such continuity is most evident: policy, education and systemic reform. These domains have historically been seen as the levers for societal transformation, yet they frequently stabilise fragmentation. Policy frameworks, from the Sustainable Development Goals (SDGs) to the Paris Agreement, attempt to order crises by projecting solutions into the future. Ulrich Beck (1992) described modern societies as 'risk societies', constantly producing frameworks to manage crises that their own logic generates. Educational systems promise empowerment but often reduce learning to performance, competition and identity formation. Paulo Freire (1970, 2005) critiqued this as the 'banking model of education', where students are treated as receptacles of knowledge rather than active participants in inquiry. Deborah Stone (2012), in *Policy Paradox*, pointed out that the modern tendency to treat policy-making as a science strips it of the very messiness of politics and human values. Attempts to make policy purely technical end up masking the contest of meanings, interests and identities beneath the surface.

Complexity and systems thinking represent an important response to such reductionism. Edgar Morin (2008) emphasised that complexity requires not only mapping interdependence but also reintegrating the subject into knowledge. His 'dialogic logic' resists simplification by holding together apparent opposites: order and disorder, unity and multiplicity, without collapsing one into the other. In this, Morin cultivates what he calls an 'epistemology of humility', a recognition that knowledge must live with contradiction. Yet, as Krishnamurti and Bohm observed in *The Ending of Time* (1985/2008), even this movement of thought cannot touch the deeper root of fragmentation: the very structure of the self that observes, compares and seeks order. So long as the observer remains separate from the observed, contradiction persists. As Krishnamurti put it, 'The observer is the observed' (Krishnamurti & Bohm, 1985/2008, *p.* 23).

This distinction is crucial. Morin shows that thought can be reflexive and dialogic, but Krishnamurti shows that true order does not come through the reconciliation of opposites but through the ending of conflict in perception itself. As he said in *The Future of Humanity*: 'Order comes only when there is no conflict at all' (Krishnamurti & Bohm, 1986/2009, *p.* 43). Perception that is free of such division is not dialogic but whole. In that stillness, disorder is met without resistance and action flows from clarity rather than from the management of opposites. Krishnamurti reminds us in *Wholeness of Life* (1992) that 'perception itself is action' (p. 32). When there is direct seeing, no gap remains between understanding and doing. In that immediacy, wholeness acts without the distortion of desire, fear or conceptual delay.

This is also why reform so often fails to liberate. As Krishnamurti observed in *A Wholly Different Way of Living* (1992), the conditioned pursuit of improvement, whether through religious practice, political reform or personal effort, inevitably strengthens the very structure it seeks to transcend. Reform becomes continuity in disguise because it begins from a centre that is itself fragmented. A framework, however refined, is only as clear as the consciousness that employs it. A tool reflects the mind of the user; if the mind is clouded by division, even the most sophisticated framework will reproduce disorder.

Arturo Escobar (1995) highlighted how development discourse, even when framed in terms of participation or integration, often reproduces division by inscribing identities and hierarchies into its very frameworks. Across these domains, good intentions are undermined by the same root problem: governance and action that begin from fragmentation rather than from a perception of the whole. Frameworks compartmentalise, targets defer and reforms negotiate contradictions without dissolving them. The result is what Hannah Arendt (1958) called the reduction of politics to administration: action is replaced by behaviour, and institutions reproduce continuity instead of opening to transformation.

The crises of climate, inequality and conflict mirror the structure of the fragmented self and cannot be resolved by technical interventions alone; policy and institutional reform, when based on identity or frameworks, stabilise disorder under the language of progress.

The central question, then, is simple yet radical: How can policy, education and systemic efforts act from coherence rather than identity or continuity? This requires a shift in starting point. Instead of designing ever more refined frameworks, we must begin with perception. Instead of imagining transformation as the outcome of negotiation, reform or scaling, we must see that transformation is already present when awareness is whole. From that seeing, institutions may still exist, but they serve life rather than categories.

The sections that follow explore this possibility in detail. Section 11.2 examines how policy reproduces continuity through dashboards and trade-offs and what it might mean for policy to arise from perception. Section 11.3 explores education, contrasting conditioning and achievement with the possibility of awakening intelligence. Section 11.4 turns to systems thinking and complexity, distinguishing between integration of fragments and direct perception of wholeness. Case studies in Section 11.5 illustrate glimpses of coherence in action, while Section 11.6 distils practical principles for acting from coherence. The chapter concludes in Section 11.7 by showing that institutions cannot replace awareness but can mirror it, and that the measure of their legitimacy lies in whether they embody perception or perpetuate identity.

11.2 Policy as Continuity Versus Policy as Perception

Policy has long been imagined as the primary tool through which societies respond to crisis. Constitutions, treaties and programmes are crafted to offer direction, stability and reform. Yet much of contemporary policy functions less as transformation than as continuity. It organises fragments, manages contradictions and postpones change into the future.

Policy as Continuity

Modern governance exemplifies what Roberts and Parks (2007) describe as the 'architecture of delay'. The Paris Agreement is a celebrated example: hailed as historic cooperation, it rests on voluntary, nationally determined contributions. Each state interprets commitments through the lens of national interest, while global emissions continue to rise (IPCC, 2023). The agreement produces legitimacy but defers enforcement, institutionalising postponement as policy.

This logic extends across domains. Debt relief frameworks such as the Heavily Indebted Poor Countries (HIPC) programme promised liberation from financial dependency. Yet they were tied to structural adjustment programmes that dismantled public services and deepened inequality. As Stiglitz (2002) noted, assistance was often structured to serve creditors more than debtors. Similarly, humanitarian conventions like the Geneva Conventions (1949) proclaim universality but collapse under selective enforcement, as in Syria, Yemen and Gaza (Amnesty International, 2019; ICRC, 2023).

Policy dashboards deepen the illusion of transformation. The SDGs project a universal vision of justice, equality and sustainability. Yet their structure fragments life into silos, poverty in SDG 1, climate in SDG 13, peace in SDG 16, while embedding contradictions within the framework itself. SDG 8 calls for indefinite economic growth, directly undermining ecological limits affirmed in SDG 13 (United Nations, 2015). Dashboards light up green even as ecosystems collapse.

This tendency reflects what Arendt (1958) warned of politics reduced to administration. Policy becomes an exercise in balancing trade-offs, legitimising delay and projecting change into the future, while avoiding transformation in the present.

Illustrations of Fragmentation

- *Pandemic Governance*: COVID-19 exposed the limits of policy framed by national interest. The COVAX initiative promised equitable vaccine distribution, yet by 2021, high-income countries had secured over 70% of global doses, while low-income nations remained below 10% coverage (WHO, 2022). Policy language affirmed solidarity, but in practice, vaccine nationalism prevailed.
- *EU Migration Pact (2024)*: The European Union's New Pact on Migration and Asylum aimed to balance solidarity with border control. In practice, it outsourced responsibilities to transit countries, tightened detention procedures and reinforced selective welcome. Ukrainian refugees were embraced with expedited rights in 2022, while Syrians and Africans faced deterrence and offshore containment (Le Monde, 2024). Universality disintegrated once filtered through geography and political identity, leaving SDG 10 and SDG 16 hollow.
- *UNGA Exclusion of Palestine (2025)*: In September 2025, the United States, as host country, denied visas to Palestinian President Mahmoud Abbas and his delegation, preventing them from attending the United Nations General Assembly in New York. This act not only violated the United Nations (1947), which obliges the host nation to permit access for all accredited UN participants, but also undermined the UN Charter's principles of sovereign equality and universal participation (United Nations, 1945). Scholars note that it further contravenes the Vienna Convention on Diplomatic Relations (1961), which protects the movement of diplomatic representatives. UN experts

publicly urged the United States to honour its obligations, stressing that barring a member's delegation erodes the integrity of international law (Office of the United Nations High Commissioner for Human Rights, 2025). This episode reveals the fragility of institutions: policy, tools and organisations have no power apart from the actors who animate them. When ego and identity dominate, institutions lose their authority and become psychological covers offering only temporary legitimacy. In moments of conflict, their weakness is laid bare.

These cases illustrate that policy frameworks, however noble in intent, often collapse when tested against identity, power and fear. International law has no sovereign outside of society itself; without transformation in consciousness, law becomes a mask of order that dissolves under stress. As John Austin (1832/1995) famously argued, law is 'the command of the sovereign backed by a sanction'. But this logic depends on fear, reward or punishment. Where individuals and societies remain fragmented, law functions as external compulsion. Yet when there is wholeness, awareness itself is sovereign. In that state, law, compassion and right action are not imposed but arise naturally, without separation, without fear.

Glimpses of Partial Coherence

Alongside fragmentation, there are policies that hint at coherence, though they remain fragile.

- *South Africa's Just Energy Transition Partnership (JETP)*: In 2021, South Africa secured $8.5 billion from international donors to accelerate its shift away from coal. Unlike conventional climate finance, the JETP explicitly linked emissions reduction with social justice, seeking to protect workers and communities during the transition (Baker, 2023). While implementation has been contested, the attempt to align ecological urgency with human livelihood reflects a perception of inseparability rather than simple trade-offs.
- *Bhutan's Gross National Happiness (GNH)*: Since the 1970s, Bhutan has measured success not by GDP but by holistic well-being, integrating cultural, ecological and spiritual dimensions into governance. While implementation has faced pressures from modernisation and external trade, the principle itself represents a shift from fragmented economic growth to a more integrated perception of life (Ura et al., 2012).
- *Citizens' Climate Assemblies*: In France and the United Kingdom, deliberative assemblies composed of randomly selected citizens produced recommendations on climate policy that reflected ecological urgency more clearly than parliamentary negotiations (Courant, 2020).
- *Legal Personhood of Nature*: New Zealand's recognition of the Whanganui River as a legal person reframed environmental law from resource management to relationship (Charpleix, 2018).

- Gaza *Freedom Flotilla (2025): Action Beyond Outcome* – In September 2025, hundreds of civilians from Europe, supported by Spain, Italy and Turkey, sailed towards Gaza in a humanitarian flotilla to break the blockade. Unlike conventional diplomacy, this movement transcended nationality: people and states acted from a sense of shared humanity, risking confrontation with overwhelming force. Their action was not mediated by dashboards or frameworks but by the immediacy of compassion. As Krishnamurti (1996) reminded, '*Perception itself is action*' (p. 32). This was an action born of love, which carried its own strength. Perhaps they did not fully assess what it would mean to confront the Israeli Defence Forces, or perhaps they knew and accepted it. Either way, the action was not conditioned by the outcome. Aid might or might not be delivered, but the act itself revealed something more vital: that love, when unbound by calculation, becomes its own force. In choiceless action, success and failure lose their grip; what remains is light and strength. The flotilla did not only carry aid; it carried a mirror to the world. In facing it, the IDF revealed its own insecurity and intent: the fear that even unarmed compassion must be suppressed. Governments that failed to protect their citizens exposed the limits of their solidarity. Nations, and their people, were given a chance to see directly how far state power will go, and where it will not go, when confronted with love. The flotilla, therefore, bore a message deeper than its cargo. It showed that when people act as one, not for ideology, not for identity, but for life itself, the logic of weapons and deterrence is exposed as brittle. Against overwhelming force, they placed not resistance but love. And in that simple act, they revealed another ground of strength, one that no blockade can contain, and also how structures and international laws have no strength, and it's the actor who acts, either from love or separation.

These glimpses show that policy can move closer to coherence when it arises from a perception of interdependence rather than a calculation of interests. They also reveal fragility: coherence falters when initiatives are absorbed back into the machinery of negotiation, growth and political identity.

Policy as Perception

Conscious policy begins from a different ground. It does not ask how to balance competing interests but sees directly what is inseparable. In ecological governance, this means perceiving that the economy is not external to nature but dependent on it. The question 'growth versus climate' dissolves into the recognition that without ecological integrity, no economy can survive in the long term. In education, policy shaped by perception would not calculate labour-market outcomes but recognise that learning is the awakening of intelligence, not the accumulation of credentials. In justice, policy rooted in perception would not merely legislate rights but ensure conditions where human beings encounter one another beyond categories.

Examples like citizens' assemblies or legal personhood of rivers demonstrate that such shifts are possible, but they remain fragile as long as policy is approached as structure rather than perception. Reform calculates; perception responds. Reform negotiates contradictions; perception dissolves them. Reform defers; perception acts in the immediacy of what is.

The illustrations reveal the fragile foundations of policy when anchored in ego and identity. Climate agreements defer, migration pacts fracture, humanitarian conventions collapse into selectivity and even the United Nations itself fails when the host nation denies entry to a member delegation. The 2025 exclusion of Palestinian President Mahmoud Abbas from the UNGA in New York violated the United Nations (1945), the United Nations (1947) and the Vienna Convention on Diplomatic Relations (1961), yet institutions proved powerless to enforce their own principles (Office of the United Nations High Commissioner for Human Rights, 2025). In contrast, the 2025 Gaza flotilla demonstrated how ordinary citizens, joined by Spain, Italy and Turkey in offering naval protection, transcended national boundaries to act from compassion and solidarity. Where institutions faltered, people embodied oneness.

Legal philosophy has long wrestled with this tension between law, legitimacy and action. John Austin (1832/1995) reduced law to the sovereign's command, backed by sanctions. H. L. A. Hart (1961/2012) advanced this by showing that law rests on a 'rule of recognition': a shared acceptance of what counts as valid law. John Rawls (1971/1999) sought fairness through the veil of ignorance, imagining principles chosen without knowledge of personal advantage. Immanuel Kant (1785/2012) went further, grounding morality in the categorical imperative: act only according to that maxim that could be universal law. Each contribution refines the rational and structural dimensions of law, yet all remain within the sphere of thought and abstraction. Austin depends on fear and reward; Hart on collective recognition; Rawls on hypothetical neutrality and Kant on rational duty. None dissolves the self that sustains division.

This is where Krishnamurti and Gupta offer a radical departure. Krishnamurti (1983/1992) insisted that 'perception itself is action'. Justice, order and compassion do not emerge from sovereign command, rational recognition or abstract duty, but from the ending of division in the perceiver. Gupta (2024, 2025) extends this insight to governance: 'Law and order are not imposed from above. They are lived from within. Justice is not a code. It is a movement in which there is no division between right and being'. Here, sovereignty is not external authority but awareness itself, a movement beyond fear, desire and identity.

Seen through this lens, the Gaza flotilla was not merely a political act but a message to the ego: a demonstration that love, expressed collectively, has strength greater than coercion. When individuals and communities act holistically, their transformation reshapes institutions from within, since those transformed individuals occupy every layer of society. The UNGA exclusion, by contrast, shows that institutions collapse into irrelevance when actors remain rooted in ego and power. Policy, tools and organisations are not autonomous forces; they are animated by the quality of perception in those who enact them.

Without transformation of the self, they become psychological covers, offering legitimacy while concealing disorder.

Thus, Austin's command, Hart's recognition, Rawls' fairness and Kant's duty all point towards frameworks of order, but they remain within the shadow of the ego. Krishnamurti and Gupta show that true order arises only when love and awareness act without a separate centre. In such action, law is not enforced but lived, not imposed but embodied. Where identity rules, law collapses into contradiction; where perception is whole, love itself becomes the sovereign.

11.3 Education as Conditioning Versus Education as Awakening

As shown in Chapter 7, education often mirrors the movements of fear and desire that permeate society. Schools, universities and military academies, rather than liberating intelligence, reproduce ambition, comparison, obedience and allegiance to insecure structures. Model United Nations simulations, debating circuits and classrooms frequently become theatres of projection rather than fields of inquiry: students debate war crimes and atrocities while competing for recognition, their performance driven by fear of failure and desire for prestige. Teachers, conditioned by their own anxieties and ambitions, cling to authority rather than entering genuine dialogue. Curricula glorify nations, obscure inconvenient histories and train children to serve national or economic agendas rather than to question them. Even the military frames obedience, discipline and sacrifice as virtues, instilling fear through punishment and desire through rewards of medals, ranks and imagined permanence.

This diagnosis is not limited to authoritarian or conflict-ridden contexts. As Freire (1970, 2005) argued, even seemingly neutral systems adopt the 'banking model' of education, treating students as receptacles for information rather than participants in inquiry. In Israel and Palestine, curricula encode antagonistic national narratives, producing identity-bound citizens who normalise division (Human Rights Watch, 2021). In Afghanistan, governments have rewritten curricula repeatedly to secure legitimacy, reducing education to propaganda (UNICEF, 2023). In the United States, cultural battles over curricula, from bans on critical race theory to politicised inclusion mandates, demonstrate how classrooms mirror political polarisations (The Civil Rights Project, 2024). Even in 'advanced' systems, competition and grading foster insecurity rather than learning (Kohn, 1992), while Fromm (1941) warned that societies emphasising conformity and achievement produce alienated individuals. In all these cases, education functions as conditioning, training students to survive within fragmentation rather than perceive wholeness.

The Principles of Right Education

Against this backdrop, the question arises: what would education look like if it were freed from fear and desire? The answer is not a new pedagogy or framework, but a radical shift in purpose. Education must begin with the child not as a future worker, soldier or citizen, but as a living human being. Its purpose is not to prepare for becoming but to open the possibility of seeing.

From this ground, several principles of right education can be discerned:

- *The educator as mirror, not transmitter*: The teacher's role is not to mould but to invite inquiry. A true relationship between educator and student is communion, not authority. As Krishnamurti (2001) reminded, 'The description is not the described'. Talking about love is not love; repeating ideals is not presence. The educator must embody presence, not merely advocate it.
- *Knowledge in its right place*: Knowledge is essential for practical functioning, yet when it becomes the basis of identity, it generates division. As Krishnamurti observed, 'A man who knows how to split the atom but has no love in his heart becomes a monster'. Knowledge without compassion corrupts, producing weapons and systems of domination rather than relationships and care.
- *Freedom from the known*: Right education liberates the mind from memory, ideology and comparison. This does not discard knowledge but recognises its limits. Only when the mind is free from the compulsion of becoming can perception be fresh and intelligence act without distortion.
- *Learning how to think, not what to think*: Dewey (1938) argued for reflective inquiry, but reflection itself becomes bound if tied to ideology. Right education cultivates observation without bias, questioning without preconception and direct encounter with life.
- *Wholeness, not specialisation*: Fragmentation into subjects, careers and identities produces partial human beings. Right education nurtures the harmony of science, art, relationship and inward awareness so that knowledge is always guided by intelligence and love.

Movements Towards Right Education

Around the world, glimpses of this approach can be found. Krishnamurti Foundation schools in India, the United Kingdom and the United States place inquiry at the centre, fostering classrooms where teacher and student explore questions together rather than enforce answers. Finland's equity-based model emphasises trust, dialogue and holistic well-being over competition. Community-based experiments in Latin America, inspired by Freire, invite students to learn through dialogue with lived reality rather than rote repetition.

We do not claim authority to certify or idealise these spaces. They are acknowledged here for their vision and the spirit in which they began. If lived with awareness and perception, they embody the essence of right education. But when the living presence of awareness is absent, they risk collapsing into rules, methods or ideals. The capacity to see when a form is useful and when it has become redundant requires a responsible and clear mind.

Such spaces of transformation also emerge beyond formal programmes. They are present in the quiet, unstructured encounters between teacher and student:

when a child pauses to look at a flower or the moon, when a teacher responds without authority but with presence, when awareness itself becomes the field of learning. Compassion, too, cannot be modelled or imposed; it arises naturally when fear, comparison and ambition no longer dominate relationships.

Institutions such as Emerge, The Dais and the Centre for Complexity, Applied Spirituality and Public Policy (CEASP) have also sought to reimagine education as awakening. Emerge's *Awakening Intelligence in Education* workshops emphasise that transformation in classrooms comes not from tools or practices but from teachers observing their own conditioning and creating space for inquiry (Gupta, 2024). The Dais works globally with young people to foster dialogue and civic responsibility, showing how students can move beyond identity-bound nationalism towards shared humanity. CEASP integrates spirituality and complexity into higher education and policy, insisting that knowledge itself must be grounded in deeper consciousness. These initiatives are not perfect or complete, but they exemplify attempts to move beyond conditioning towards inquiry.

Right Education and Transformation

When education is rightly oriented, students learn not only mathematics, literature or science but also the art of living. They discover what it means to cooperate without competition, to see when not to conform, when not to obey and when not to participate in violence or exploitation. A rightly educated mind knows when to refuse and when to act, not out of ideology, but from clarity. In such action, there is no choice, for choice belongs to calculation; there is only perception. This is choiceless action: intelligence responding to what is.

The implications are profound. A generation educated in this way would not design weapons, exploit ecosystems or perpetuate inherited conflict. They would not separate knowledge from compassion or ambition from responsibility. Their intelligence would express itself in governance, science and art as coherence rather than fragmentation.

Education, rightly understood, is not preparation for life but life itself. It is the flowering of awareness in relationship, in learning and in action. Such education does not produce better citizens of fragmented societies but whole human beings. In their wholeness lies the ground for new governance. As Gupta (2025) argues in *Governance Without the Self*, institutions cannot substitute for awareness but can mirror it; their legitimacy depends on the quality of perception embodied by those who animate them. When knowledge is guided by intelligence, and intelligence is inseparable from love, education becomes the seed of coherence for society as a whole.

11.4 Systems Thinking and Complexity: Integration or Perception

In recent decades, systems thinking and complexity science have been advanced as tools to overcome fragmentation. Their language of interdependence, feedback loops and adaptive resilience promises a way to see wholes and understand the behaviour of the whole. Yet while these approaches are valuable in naming relations, they often stop short of dissolving the root division: the divided consciousness of the observer. They remain frameworks of thought, and as such, risk becoming reform loops rather than pathways to transformation.

The Appeal of Integration

Systems frameworks emerged to counter the mechanistic worldview that dominated industrial modernity. Thinkers such as Fritjof Capra (1996) emphasised the web of life, Gregory Bateson (1972) described the 'pattern that connects' and Donella Meadows (2008) offered tools for leverage in systems change. Today, organisations deploy 'complexity-informed' models in governance, policy and education.

The Inner Development Goals (IDGs), for example, claim to complement the SDGs by cultivating competencies such as empathy, critical thinking and collaboration. Similarly, Daniel Schmachtenberger and others in the meta-crisis discourse advocate for sense-making frameworks to integrate diverse fields into a coherent picture.

The appeal of these approaches lies in their capacity to gather fragments into patterns. Climate dashboards integrate emissions data with economic projections. Systems models of pandemics simulate interactions of health, mobility and policy. Complexity-informed governance proposes adaptive strategies for 'wicked problems'. At their best, these tools highlight the insufficiency of linear thinking and the need for multi-layered understanding.

Yet the moment we say 'frameworks illuminate relations', we must pause. Illumination is not transformation. At most, frameworks rearrange fragments into patterns; they cannot dissolve fragmentation itself. Naming the parts is not the same as perceiving the whole. Without insight, even integration becomes continuity.

The Limits of Frameworks

Integration of fragments is not the same as perceiving wholeness. Ken Wilber's integral theory organises domains of experience into quadrants and levels, but this structural map risks becoming another hierarchy to manage rather than a ground of insight. Jeremy Lent (2021) emphasises patterns of meaning, but the framework can still function as interpretation rather than direct perception. Nora Bateson (2016) reminds us that 'the system is not the whole', pointing to the need for relational seeing beyond diagrams.

The IDGs illustrate this danger clearly. By framing inner qualities as 'skills' or 'competencies' to be developed, they risk commodifying what is essentially immeasurable and emergent. Empathy or compassion cannot be cultivated as tools without turning into performance. Reading about empathy cannot dissolve violence in Benjamin Netanyahu, nor can 'resilience training' end greed in corporate boardrooms.

As Gupta (2025) notes, 'The self can endlessly practice mindfulness, resilience, or empathy while remaining the same self that creates disorder. To mistake practice for transformation is to strengthen continuity under a new vocabulary' (p. 214).

The danger is that systems thinking, while acknowledging interdependence, often reproduces the very logic of frameworks critiqued in §11.2. Dashboards and indicators can be more complex, but they still translate life into categories, metrics and targets. The SDGs themselves, often justified as an integrated framework, exemplify this paradox: their very design institutionalises fragmentation by dividing life into 17 goals and 169 targets. The IDGs then mirror this by dividing inner life into five dimensions and 23 skills, which is integration without perception.

The promise of integration becomes another architecture of delay (recall §11.2: when policy becomes dashboards and trade-offs, it stabilises delay; so too when complexity becomes integration without perception).

As Gupta (2025) further observes: 'The logic of frameworks is not to end disorder but to stabilize it by managing fragments. This creates the appearance of coherence while ensuring that incoherence continues under a new vocabulary' (p. 142). Systems approaches risk falling precisely into this trap: elaborate vocabularies of interconnection, yet little movement beyond continuity.

Morin's Dialogic Logic: Holding Opposites or Ending Them?

Edgar Morin's (2008) complex thought advances three core moves: *dialogic* (holding together order/disorder/organisation without reduction), *recursive causality* (effects feed back into causes in an 'ecology of action') and the *hologrammatic principle* (the part contains the whole, the whole in the part). These correct the linearity of reductionism and reintroduce the subject into knowledge.

Yet they remain movements within thought. Dialogic logic *holds* opposites; perception *ends* them. Recursivity maps unintended consequences; perception ends the controller who seeks control. Hologrammatics suggests the whole is reflected in the part; perception *is* the whole without a centre. Morin brings us closer to the edge of transformation by reintegrating the subject into knowledge, but without questioning the continuity of the self, his vision remains bound within knowledge. Complexity becomes another way of managing fragments, not ending fragmentation.

The Word and the Observer

Here lies the greater difficulty. Words and frameworks are necessary for communication, but they are not illumination. As Krishnamurti (1969/2009) insisted, 'the word is not the thing, the description is not the described'. A dashboard can name 'climate', 'poverty', or 'peace', but naming is not perceiving.

'The day you teach the child the name of the bird, he never looks at it again.' Words create the illusion of clarity, but they can blind the mind to seeing directly.

The problem is not the linearity of communication itself, but the mind that meets linearity with fragmentation. When the listener carries fear, comparison or ideology into each word, what is heard is filtered through the past. But when there is observation and listening without fragmentation, then knowledge or insight, whatever is spoken, written or read, finds its right place. In such communion, communication is no longer merely the transmission of ideas but the sharing of perception itself.

This was the simple truth the Buddha pointed to: see directly, end division. Yet instead of ending division, we created Buddhism. Likewise, Krishnamurti pointed to wholeness, yet institutions made 'Krishnamurtian' frameworks. The observer seeks comfort in a system rather than ending fragmentation.

Until the actor changes, no framework can bring change.

The Role of Thinkers: Insight, Misuse and Communion

Across governance, systems and consciousness, many thinkers have pointed towards wholeness. Edgar Morin's (2008) complex thought challenged reductionism, Fritjof Capra's (1996) *web of life* reframed science in relational terms, and Donella Meadows (2008) highlighted leverage points for systemic change. Elinor Ostrom (1990) revealed how commons could be managed without state or market domination. Ken Wilber (2000), Jeremy Lent (2021) and Daniel Schmachtenberger tried to provide integrative maps of experience and meaning. Nora Bateson (2016) reminded us that 'the system is not the whole', and Iain McGilchrist (2019) warned against the dominance of abstraction over lived perception. Philosophers of mind like Antonio Damasio (1999) and Varela et al. (1991) explored embodied cognition, while critics such as Shoshana Zuboff (2019), Tristan Harris (2019) and Jaron Lanier (2010) showed how technology exploits attention. Living traditions, from the Buddha's Dhamma to Krishnamurti's dialogues with Bohm (1996), pointed simply to the ending of division.

Yet what unites these contributions is also their danger: each is at risk of being absorbed into the very logic of fragmentation they sought to expose. Morin's complex thought often becomes another 'methodology of complexity studies'. Capra's *web of life* is distilled into ecological diagrams for classrooms. Meadows's leverage points are checklist tools for policymakers optimising systems without touching the consciousness behind them. Ostrom's commons governance is copied mechanically into management manuals. Wilber's integral quadrants risk turning perception into hierarchies of status. Lent's cultural patterns can become interpretive grids, not direct perception. Schmachtenberger's sense-making vocabularies risk creating new elites of 'meta-thinkers' whose expertise substitutes for seeing. Nora Bateson's warm data becomes another consulting product. Even the deeper traditions are not immune: Marx's critique hardened into state ideology, Plato's allegories became justification for philosopher-kings, Krishnamurti's rejection of method hardened into 'Krishnamurti philosophy', Bohm's communion became a corporate dialogue technique, and the Buddha's simple invitation became institutional Buddhism.

The problem is not their intent but their reception. A fragmented observer takes living insights and turns them into structures, methods and ideologies. What was alive becomes mechanical; what was meant to dissolve boundaries becomes another boundary. Frameworks can describe, compare and even inspire, but they cannot liberate. At best, they are pointers; at worst, they are prisons.

The decisive issue, then, is not whether one chooses Morin, Capra, Wilber or Krishnamurti, but whether one listens in fragmentation or in awareness. When filtered through fear, ambition or belief, even profound truth is reduced to method and absorbed into the architecture of delay. But when fragmentation ends, communication itself becomes communion. In that state, words and frameworks fall into their right place, not as substitutes for perception, but as secondary expressions of awareness.

11.5 Case Studies: Coherence in Action

The challenge for education, policy and systems is not to manage complexity but to act from coherence. Across the world, glimpses of such coherence have appeared: fragile, partial and often temporary. They reveal that perception, when free from identity, can bring order even within complex systems. These examples are not models to be replicated, for coherence cannot be copied. They are reflections of moments when perception briefly became the ground of action.

Krishnamurti Foundation Schools (India): KFI schools emphasise inquiry rather than competition or conformity. Teachers and students engage in dialogue, observing thought and fear as they arise, creating classrooms as spaces of relationship rather than performance. This refusal to condition the mind into conformity points towards an education that awakens perception rather than reproduces identity (Krishnamurti, 1972).

Finland's Equity-Based Education: By avoiding early standardised testing and focusing on trust, curiosity and equity, Finland demonstrates that education can privilege relationships over anxiety. The system's strength lies less in its results than in the perception that learning is not a race (Sahlberg, 2011).

Costa Rica's Abolition of the Military: In 1948, Costa Rica redirected defence budgets into education, health and the environment. The decision reflected the perception that true security lies in human flourishing, not militarisation. The coherence of this act was not in strategy but in seeing wholeness (Booth, 1998).

Kerala's Decentralised Planning (India): The 1990s People's Plan devolved authority to local assemblies where villagers, teachers and health workers set priorities. Though imperfect, it embodied governance rooted in dialogue and proximity, showing that coherence arises when communities deliberate together (Heller, 2001).

Mondragón Cooperatives (Spain): Worker-owned enterprises integrate education, work and governance through cooperative structures. Profits are reinvested locally, and schools teach both skill and solidarity. Though not without contradictions, Mondragón demonstrates how economies can be organised around interdependence rather than competition (Whyte & Whyte, 1991).

Zapatista Autonomous Schools (Mexico): In Chiapas, Indigenous communities design curricula that reflect ecology, history and collective governance. These schools resist assimilation into state or market systems, aligning learning with dignity and ecological relationship (Baronnet, 2008).

Rwanda's Reconciliation Villages: After the 1994 genocide, Rwanda created villages where survivors and perpetrators live and farm together. Despite criticisms, they represent an attempt to live reconciliation, seeing one another beyond victim and perpetrator, and align with SDG 16's vision of peace (Clark, 2010).

Revisiting Earlier Examples: South Africa's Truth and Reconciliation Commission, European Citizens' Assemblies, and New Zealand's legal personhood for rivers can also be seen as forms of education. Each turned governance into a living curriculum of awareness: societies learning to perceive wholeness through dialogue, listening and relationship. These moments were never pure, but they revealed that perception itself educates.

Implications: These diverse experiments from classrooms and cooperatives to constitutions and communities illustrate that coherence can emerge when systems act from perception rather than identity. Yet it is essential to remember that most of these examples no longer exist in their original form or may not retain the coherence they once expressed. Their validity today is secondary. What matters is not whether these models still function coherently, but whether we, in seeing them, remain coherent ourselves.

If we turn them into models, we reduce living intelligence into imitation. Coherence is not inherited from past examples but arises freshly in perception now. When attention is whole, frameworks and institutions may reflect that wholeness; when attention fragments, even the best-designed systems decay into repetition. What these cases reveal, therefore, is not a method but a mirror: coherence in action is possible only when there is coherence in the observer.

11.6 Acting From Coherence: Practical Principles

The failures of frameworks and reform show that fragmentation cannot be managed into wholeness. Acting from coherence is not about designing better trade-offs or inventing more sophisticated systems. It is about a radical shift of ground: from identity and knowledge to perception and intelligence. What does this look like in practice? The following principles illustrate how institutions, policies and education can embody coherence without reducing it to method.

Perception Before Prescription

Modern governance, education and institutions almost always begin with *prescription*. The logic is straightforward: identify a problem, design a solution, set measurable targets and implement. Constitutions prescribe rights and duties. Policy frameworks prescribe goals and indicators. Education systems prescribe curricula and standards. At first glance, this appears rational and orderly. Yet, beneath the order, something essential is missed. Prescription without perception inevitably reproduces disorder.

Prescription without perception means acting on categories, abstractions or projections of what should be, rather than responding directly to what is. Krishnamurti (1969/2009) emphasised that 'the moment you cultivate awareness, it ceases to be awareness'. By the same token, the moment institutions cultivate frameworks, they cease to perceive life as it unfolds. What they manage are categories, not realities.

Fragmentation Through Prescription

Policy dashboards are full of examples. The *SDGs* provide a comprehensive list of global prescriptions: end poverty, ensure education, reduce inequality and achieve climate action. Yet these prescriptions are pursued as silos. SDG 8 prescribes 'sustained economic growth', while SDG 13 prescribes 'urgent climate action' (United Nations, 2015). One is pursued through industrial expansion; the other requires contraction of the same industries. Dashboards light up green, but forests fall and emissions rise. Prescription produces the appearance of order while concealing systemic collapse.

Education follows the same trajectory. Prescribed curricula define what every child must know, when and how. Standardised tests prescribe how success is measured. Yet prescription blinds educators to the child before them. A student struggling with grief, hunger or curiosity is reduced to a score. As Paulo Freire (1970) warned in his critique of the 'banking model of education', prescription turns learning into deposits of information rather than awakening of understanding. Students learn to perform to prescriptions, not to see themselves, others or the world with clarity.

Health systems illustrate the danger even more starkly. During the COVID-19 pandemic, many governments prescribed blanket lockdowns or vaccination targets without perceiving the lived conditions of marginalised communities. Migrant workers in India walked hundreds of kilometres home because prescriptions ignored their realities (The Lancet, 2020). Vaccine stockpiles in high-income countries went unused while prescriptions of equity under COVAX collapsed into hoarding (WHO, 2022). Prescription without perception created both suffering and mistrust.

Perception as Ground

What, then, does it mean to put *perception before prescription*? It means governance begins not with abstract goals but with *seeing clearly what is*. Perception is not analysis alone; it is an unfragmented attentiveness to life. It recognises that ecological systems, human dignity and social relationships cannot be reduced to metrics or managed as separate silos.

In *education*, this shift is visible in schools that give autonomy to teachers to respond to students rather than merely deliver curriculum. Finland, often cited as a global exemplar, does not rank schools or rely heavily on standardised testing. Instead, it trusts teachers to perceive and respond, leading to some of the highest literacy rates and lowest achievement gaps globally (Sahlberg, 2015). Perception allows teaching to become a relationship, not conditioning.

In *ecology*, perception transforms policy from management to relationship. In the Amazon, dashboards prescribe conservation targets, yet forests continue to be razed. A perceptive approach would not calculate 'forest versus growth' but recognise the inseparability of human survival and forest vitality. Indigenous stewardship practices in the Amazon and Himalayas demonstrate this: conservation emerges not from targets but from a living perception of interdependence (Berkes, 2018).

In *justice*, perception allows reconciliation rather than punishment. South Africa's Truth and Reconciliation Commission showed glimpses of this when victims and perpetrators encountered each other beyond the categories of guilt and innocence. Where prescription would have confined them to roles, perception created the possibility of human encounter (Tutu, 1999).

Broader Dimensions

Seen across domains of life, the implications of perception-first governance are clear:

- *Political*: Prescription-driven politics stabilise polarisation. Parties prescribe manifestos to appeal to identities, not to perceive collective needs. Perception invites dialogue that transcends party lines.
- *Economic*: Economies prescribe growth as success, yet growth often deepens inequality and ecological harm. Perception redefines economy as embedded in ecology, dissolving the false trade-off.
- *Social*: Prescriptions of inclusion often remain tokenistic, managing categories like gender or race. Perception recognises the human being before the category, allowing genuine inclusion.
- *Technological*: Prescriptive faith in technology treats innovation as automatic progress. Perception places technology in its rightful place, guided by human and ecological well-being.
- *Legal*: Laws prescribe universal rights but often collapse into selective enforcement. Perception sees suffering directly, ensuring law does not become a tool of identity.

- *Environmental*: Prescriptions such as carbon credits commodify life. Perception reveals ecology as inseparable from survival, demanding care, not calculation.

Implications for Governance and Leadership

The principle of *perception before prescription* demands a radical humility in leadership. It means suspending the rush to design, legislate or reform, and beginning instead by *seeing without distortion*. Frameworks may still exist, but they follow perception, not precede it.

- For *policymakers*, this means sitting in communities before writing policy and perceiving realities before prescribing solutions.
- For *educators*, it means seeing the child before the curriculum, the learner before the score.
- For *leaders*, it means responding from clarity, not from identity or ideology.

As one of the insights in *Governance Without the Self* notes, governance fails when it begins from 'better continuity' rather than from perception (Gupta, 2025). Prescription manages fragments; perception sees the whole. Coherence arises only when the latter leads.

Dialogue as Foundation, Not Accessory

Modern governance often treats dialogue as an accessory rather than a foundation. Policies are drafted, targets agreed upon and frameworks negotiated long before dialogue occurs. Citizens are invited to consultations only after decisions are framed. Dialogue becomes a tool to legitimise prescription rather than a condition for perception. When dialogue is reduced to an instrument, it reproduces fragmentation.

Fragmentation Through Misuse of Dialogue

Global frameworks such as the *SDGs* explicitly emphasise participation and partnerships. SDG 16 (Peace, Justice and Strong Institutions) prescribes 'inclusive and participatory decision-making', and SDG 17 (Partnerships for the Goals) celebrates collaboration. Yet, in practice, these ideals often collapse into performative gestures. Public consultations are orchestrated to validate pre-decided plans. Partnerships between governments, corporations and NGOs are frequently driven by funding, visibility or identity-based interests rather than mutual understanding. Dialogue is simulated but not lived.

The result is dissonance. Migration summits bring leaders together for statements of solidarity while borders harden. Climate conferences create negotiation spaces but rarely spaces for true dialogue. In these forums, dialogue is not a foundation but a bargaining tool within the logic of identity and interest.

Illustrations of Genuine Dialogue

By contrast, when dialogue is genuine, coherence emerges. *Citizens' assemblies* in Ireland on abortion or in France on climate policy reveal how ordinary people, unbound by party or corporate identity, can deliberate together. They did not begin from prescription but from the openness of inquiry. Their recommendations carried legitimacy not because they represented interest groups but because they emerged from shared perception.

Truth and Reconciliation processes also highlight both the potential and fragility of dialogue. In South Africa, reconciliation was most alive when victims and perpetrators encountered one another directly, not as roles but as human beings. Where dialogue ossified into bureaucracy, the process faltered. But in moments of living dialogue, perception created the possibility of forgiveness and a new relationship.

Indigenous councils across Latin America, New Zealand and North America often embody this principle more consistently. Dialogue here is not a negotiation among interests but a shared listening to land, ancestors and community. Coherence arises not from agreement but from relationship.

Broader Dimensions

Looking across domains, the implications of dialogue as foundation become evident:

- *Political:* Prescriptive politics deepen polarisation, with parties speaking past one another. Dialogue as a foundation opens the possibility of seeing beyond identity and ideology.
- *Economic:* Trade negotiations framed as zero-sum deals perpetuate inequality. Dialogue that perceives interdependence could dissolve false dilemmas between growth and justice.
- *Social:* Inclusion programmes often reduce dialogue to token representation. Genuine dialogue perceives the human being before the category, enabling deeper equality.
- *Technological:* AI ethics frameworks prescribe principles, but dialogue with those most affected (workers and marginalised groups) grounds technology in lived reality.
- *Legal:* Courts often operate through adversarial prescription. Restorative justice shows that dialogue between victim and offender can reveal wholeness beyond guilt and punishment.
- *Environmental:* Climate negotiations prescribe emission targets, yet forests continue to fall. Dialogue with communities proximate to ecosystems allows action that reflects ecological rhythms rather than abstract numbers.

Implications for Governance and Leadership

If perception is to precede prescription, dialogue must precede governance. This is not dialogue as consultation or strategy, but dialogue as the ground of relationship itself. As Bohm (1996) described, dialogue is a suspension of fragmentation, a movement of 'thinking together'.

For policymakers, this means listening before legislating, entering spaces of relationship before drafting goals. For educators, it means dialogue with students as human beings, not merely transmitting curricula. For leaders, it means recognising that coherence is not secured by control but emerges from shared perception.

The SDGs gesture towards this through their language of participation and partnership, but they remain trapped in the prescriptive logic of frameworks. Conscious governance demands more. Dialogue must not be an afterthought to prescription but the ground from which any coherent prescription arises.

Proximity to Life

One of the greatest failures of modern governance and institutions is abstraction. Decisions are made far from the realities they govern. Policies are written in capital cities, education curricula in administrative offices and climate treaties in international conferences. The result is that governance becomes detached from the texture of life. Abstraction turns living realities into numbers, dashboards or categories, often with devastating consequences.

Fragmentation Through Abstraction

The *SDGs* reveal this paradox clearly. SDG 1 (No Poverty) is measured by income thresholds, yet these thresholds ignore local realities of deprivation and dignity. SDG 2 (Zero Hunger) counts calories and yields, yet food insecurity persists when supply chains collapse or cultural diets are ignored. Abstraction creates the illusion of order while concealing lived disorder.

The COVID-19 pandemic exposed this disconnect. Lockdowns were prescribed uniformly across nations without perceiving how millions of informal workers survive day-to-day. In India, migrant workers, unable to pay rent or access food, were forced to walk hundreds of kilometres back to their villages (The Lancet, 2020). Abstraction reduced 'flattening the curve' to numbers while obscuring human suffering.

Similarly, climate governance framed through global dashboards abstracts ecosystems into carbon credits. Forests are valued not as living beings but as carbon sinks. Programmes like REDD+ prescribe conservation targets, yet deforestation continues because proximate drivers – agribusiness, land rights and community needs – are ignored (Angelsen et al., 2018). Abstraction displaces responsibility from the immediacy of life into the safety of categories.

Illustrations of Proximity

In contrast, proximity allows coherence.

- *Community forestry in Nepal* demonstrates this vividly. Villagers directly responsible for forest stewardship have sustained ecosystems more effectively than state-led agencies. Their rules emerge not from distant policy but from living attentiveness to ecological rhythms and community needs (Agrawal & Gibson, 1999).
- *Disaster relief in Japan* shows a similar truth. Neighbourhood-level volunteer committees, rooted in direct relationships with affected communities, respond faster and with greater trust than distant bureaucracies. Their legitimacy comes not from frameworks but from proximity to suffering.
- *Indigenous ecological governance* in the Amazon, Himalayas and Arctic operates on proximity as a principle. Forests, rivers and animals are not abstractions but kin. Decisions emerge from relationships, not projections.

Broader Dimensions

Looking across various dimensions, the implications of proximity are profound:

- *Political:* Centralised governance risks losing touch with local realities. Decentralisation that maintains proximity, such as participatory budgeting in Porto Alegre, creates decisions grounded in lived needs rather than abstract categories.
- *Economic:* Growth models framed by GDP abstract away inequality and ecological destruction. Proximity reframes the economy around livelihoods, subsistence and reciprocity.
- *Social:* Welfare systems designed through databases exclude millions. Proximity recognises hunger or illness directly, as when community kitchens in Kerala responded to COVID-19 with immediacy rather than waiting for central aid.
- *Technological:* AI systems designed far from their users reproduce bias and exclusion. Proximity as participatory design with affected communities anchors technology in justice rather than efficiency.
- *Legal:* Abstraction in law often prioritises procedure over justice. Restorative justice programmes restore proximity, creating spaces where offender and victim encounter one another directly.
- *Environmental:* Carbon trading abstracts forests into credits, while proximity-based governance such as New Zealand's legal personhood for rivers, anchors care in relationship with ecosystems.

Implications for Governance and Leadership

Proximity dissolves the false distance between institutions and life. It reorients governance away from control through abstraction and towards coherence through relationship. For leaders, proximity means refusing the comfort of dashboards and statistics unless they are tethered to lived realities. For educators, it means teaching not only through prescribed curricula but also through dialogue with the student's actual condition. For policymakers, it means walking the streets, entering communities and listening before writing.

The SDGs aspire to universality, but without proximity they risk becoming exercises in abstraction. Conscious governance recognises that universality is not achieved through global dashboards but through countless proximate encounters where perception guides action.

Proximity is not anti-knowledge. Knowledge can inform, but perception rooted in proximity must lead. When institutions remain close to life, coherence emerges naturally. When they abstract away from life, disorder persists no matter how sophisticated the framework.

Accountability to Wholeness

Accountability is a defining feature of modern governance, education and institutional reform. Yet most accountability systems are built on categories and indicators rather than life itself. Dashboards measure progress towards the *SDGs*, governments track performance through metrics, schools rank students by scores and corporations publish ESG reports. These mechanisms create an appearance of rigour and transparency, but they often conceal deeper disorder. When accountability is reduced to indicators, the focus shifts from protecting life and dignity to satisfying categories.

Fragmentation Through Indicators

The SDGs exemplify this tension. Countries can report 'progress' on SDG 13 (climate action) by expanding renewable energy capacity, even as overall emissions rise due to continued fossil fuel dependence (IPCC, 2023). Indicators light up green, but atmospheric carbon continues to climb. Similarly, progress on SDG 4 (quality education) is often measured by enrolment rates, while the substance of learning, critical inquiry, freedom from conditioning or compassion is ignored.

The Aadhaar welfare system in India revealed the perils of indicator-driven governance. By streamlining beneficiary data, the programme delivered impressive numbers of 'authenticated transactions'. Yet millions were excluded from food rations and pensions due to authentication failures (Khera, 2019). On the dashboard, efficiency improved. On the ground, hunger deepened.

Global climate finance illustrates the same distortion. OECD (2020) data shows that much of what is reported as 'climate aid' is repackaged loans, saddling poorer nations with debt. Accountability to indicators allowed wealthy nations to claim billions in contributions while worsening the vulnerabilities they were meant to alleviate.

Illustrations of Wholeness-Based Accountability

Coherence arises when accountability is tethered to life rather than categories.

- *Health:* During COVID-19, some community health programmes in Africa and South Asia measured success not by vaccination counts alone but by reductions in mortality and improvements in trust between citizens and health workers. This alignment of accountability with lived well-being built resilience rather than mistrust.
- *Justice:* Restorative justice programmes hold individuals accountable not only through legal sanction but also through processes of recognition, repair and relationship. The goal is not compliance with law alone, but restoration of wholeness.
- *Ecology:* In New Zealand, granting legal personhood to the Whanganui River reframed accountability. Instead of monitoring pollutants alone, governance became accountable to the river as a living being. This shift anchored accountability in relationships rather than statistics.

Broader Dimensions

Across various, the implications of accountability to wholeness are clear:

- *Political:* Electoral politics often measure accountability through polling and electoral cycles. Accountability to wholeness would ask whether policies reduce suffering and division, not just whether they secure votes.
- *Economic:* Corporations measure success through quarterly profits or ESG scores. Accountability to wholeness demands alignment with ecological and social well-being, not just reputational management.
- *Social:* Social welfare programmes count beneficiaries served. Wholeness-based accountability asks whether dignity, security and inclusion have actually deepened.
- *Technological:* AI ethics boards often focus on compliance checklists. Accountability to wholeness means centring human dignity and ecological survival over performance benchmarks.
- *Legal:* Courts measure success by case clearance rates. Wholeness-based accountability asks whether justice was experienced by those involved.
- *Environmental:* Carbon markets measure offsets, but forests continue to die. Accountability to wholeness asks whether ecosystems are regenerating, not just whether credits are sold.

Implications for Governance and Leadership

The principle of accountability to wholeness reframes the very question of responsibility. Leaders are not accountable to numbers, dashboards or electoral cycles, but to life itself. For policymakers, this means rejecting metrics that allow suffering to persist behind green indicators. For educators, it means refusing to equate accountability with test scores and instead asking whether students are learning to see and live without division. For global institutions, it means holding themselves accountable not to treaty compliance but to the actual condition of people and planet.

When accountability is reduced to categories, institutions stabilise disorder under the guise of progress. When accountability is rooted in wholeness, coherence becomes visible. The measure of governance, education and systems is not whether indicators light up, but whether life flourishes.

Simplicity and Clarity

Modern governance and education often equate legitimacy with complexity. Elaborate bureaucratic procedures, intricate policy frameworks and detailed reporting mechanisms are assumed to guarantee fairness and rigour. Yet complexity, when rooted in fragmentation, obscures reality rather than clarifying it. True coherence arises not from multiplying layers of regulation, but from restoring simplicity and clarity.

Fragmentation Through Complexity

Governments frequently design programmes that appear sophisticated but alienate the very people they are meant to serve. Welfare systems often require multiple forms of identification, proof of residence, income verification and digital authentication. Each layer adds complexity in the name of accountability but creates barriers for the most vulnerable. India's Aadhaar welfare system again illustrates this: beneficiaries were required to authenticate with biometric scans linked to centralised databases. For millions of rural poor, connectivity failures or mismatched fingerprints meant exclusion from food and pensions (Khera, 2019). What was presented as administrative clarity became systemic opacity.

The *SDGs* themselves reflect this paradox. With 17 goals, 169 targets and more than 230 indicators, the framework aspires to comprehensiveness. Yet its complexity makes it unworkable for many governments, particularly in the Global South, where resources to track and report across dozens of indicators are limited (Fukuda-Parr & McNeill, 2019). The result is often selective reporting: countries highlight indicators that improve while ignoring those that worsen. Complexity allows disorder to hide behind dashboards.

Education systems reveal the same problem. In many countries, curricula are overloaded with information, leaving little room for exploration or inquiry. Students memorise facts across dozens of subjects to prepare for standardised tests, but rarely engage in deep learning. Ken Robinson (2006) noted that over-complex, fragmented curricula stifle creativity, producing conformity instead of clarity. The proliferation of benchmarks and testing regimes creates the illusion of accountability but undermines genuine learning.

Illustrations of Simplicity and Clarity

Coherence emerges when institutions remove unnecessary complexity and act with directness:

- *Welfare:* Studies show that unconditional cash transfers are often more effective than elaborate welfare schemes. Banerjee and Duflo (2019) demonstrated that when poor households receive cash directly, outcomes improve in health, nutrition and education. Simplicity restores dignity, avoids bureaucratic capture and enables people to respond to their own realities.
- *Ecology:* Recognising the Whanganui River in New Zealand as a legal person cut through decades of disputes about ownership, water rights and management. Instead of balancing competing claims, the law clarified a simple relationship: the river is a being to be cared for (Muru-Lanning, 2016). Simplicity revealed what complexity had obscured.
- *Education:* Finland's refusal to adopt standardised rankings and excessive testing simplified its education system. Teachers are trusted to perceive and respond to students directly, and the clarity of purpose, supporting learning rather than managing competition, has produced some of the world's most equitable outcomes (Sahlberg, 2015).

Broader Dimensions

Seen across domains, the implications of simplicity and clarity are significant:

- *Political:* Political systems often grow opaque through jargon-filled legislation and bureaucratic hurdles. Simplicity means laws written for citizens, not for technocrats.
- *Economic:* Economic governance thrives on complex financial instruments: derivatives, credit swaps and tax loopholes that obscure responsibility. Clarity would prioritise transparent systems where obligations and benefits are visible.
- *Social:* Social services frequently require individuals to navigate multiple agencies and documents. Simplification integrates services at the point of need, centring dignity over procedure.
- *Technological:* Technology regulation often becomes complex, with overlapping standards and competing jurisdictions. Simplicity would mean clear ethical boundaries, such as prohibiting lethal autonomous weapons outright, rather than regulating their 'responsible use'.
- *Legal:* Legal systems often reward those who can afford to navigate complexity. Simplicity would make justice accessible to all, not just those with resources.
- *Environmental:* Climate frameworks like carbon markets introduce layers of credits, offsets and trading schemes. Clarity would mean direct accountability for emissions reductions rather than shifting responsibility through complex accounting.

Implications for Governance and Leadership

The principle of simplicity and clarity requires leaders to resist the temptation to equate complexity with sophistication. It calls for stripping away procedures that obscure accountability and restoring directness of relationship. For policy-makers, this may mean prioritising straightforward welfare programmes that deliver directly to citizens rather than building systems of conditionality and control. For educators, it means focusing on the child's capacity to see and inquire rather than overloading them with fragmented curricula. For institutions, it means measuring success not by the number of frameworks or indicators created, but by whether dignity, justice and ecological vitality are preserved.

Complexity is often a symptom of fragmentation: when perception is absent, systems compensate with endless frameworks. Simplicity arises naturally when perception is whole. Clarity is not the absence of rigour but the presence of coherence. Institutions serve life best when they remove what obscures and protect what is essential.

Coherence as Present Action

Reform often negotiates within fragmentation; utopianism postpones transformation into a distant future. Both approaches evade the present. Yet coherence is not a vision for tomorrow; it arises now, when perception is whole, and action flows directly from seeing. Coherence as present action means dissolving the logic of delay and recognising that transformation is not a matter of method, but of immediacy.

Fragmentation Through Delay

Contemporary governance frequently institutionalises postponement. Climate policy illustrates this most clearly. Frameworks such as the Paris Agreement transform ecological urgency into distant timelines like 2030 and 2050, while emissions continue to rise (IPCC, 2023). The appearance of progress masks the reality of delay. What is deferred is not technical capacity but perception itself. As long as transformation is always 'by 2050', action today is absorbed into continuity.

Education embodies the same logic of postponement. Students are told that learning today is for success tomorrow: exams, careers and social mobility. The present moment, curiosity, relationship, the joy of discovery is subordinated to a future of becoming. Krishnamurti (1969/2009) warned that education obsessed with the future creates anxiety, comparison and conformity, leaving the mind incapable of perceiving what is. Reforming curricula does not dissolve this logic; it simply pushes the horizon further.

In justice, delay is often institutionalised through endless trials, backlogs or bureaucratic processes. Victims wait decades for redress while perpetrators exploit the system. Legal frameworks promise accountability but defer reconciliation. Coherence, however, is possible only when truth is encountered directly.

Archbishop Desmond Tutu (1999) emphasised that South Africa's Truth and Reconciliation Commission mattered not because it prescribed a final solution, but because it created spaces where people could face one another in the immediacy of encounter.

Illustrations of Coherence in the Present

When perception becomes primary, coherence appears spontaneously in action:

- *Leadership:* New Zealand's Jacinda Ardern resigned in 2023, acknowledging she no longer had the clarity or energy to serve. This decision was coherent not because it followed a framework, but because it arose from attentiveness to what is rather than clinging to role or ambition.
- *Education:* A teacher who sets aside rigid curricula to respond to a child's grief, hunger or creativity is acting coherently. The response arises from perception, not prescription.
- *Ecology:* Indigenous communities in the Amazon and the Himalayas practice conservation not by waiting for national targets but by living attentively to the rhythms of forest and land (Berkes, 2018). Their stewardship is not a reform but a present coherence with life.
- *Health:* During the COVID-19 pandemic, grassroots health workers in Kerala, India, organised immediate responses, door-to-door visits, food distribution and transparent communication while national systems focused on dashboards and targets. Their immediacy saved lives by responding directly to what was seen.

Broader Dimensions

Across domains, coherence as present action transforms delay into responsiveness:

- *Political:* Electoral politics often defers transformation into promises of the next term. Coherence requires responding to suffering without waiting for political convenience.
- *Economic:* Economic justice is often postponed through promises of 'growth first, redistribution later'. Coherence recognises that exploitation in the present cannot be justified by future prosperity.
- *Social:* Social change initiatives often delay transformation through 'awareness campaigns' while discrimination continues. Coherence arises when injustice is confronted directly in daily practice.
- *Technological:* AI ethics frameworks often promise safeguards in the future while systems of surveillance expand unchecked today. Coherence means acting on dignity now, not after regulation catches up.
- *Legal:* International law promises universality but defers enforcement. Coherence would act directly on violations, whether by allies or adversaries.

- *Environmental:* Climate action framed as 'net zero by 2050' defers coherence into a future that ecosystems may not survive to see. Coherence means reducing emissions now in recognition of inseparability with life.

Implications for Governance and Leadership

The principle of coherence as present action requires leaders and institutions to resist the temptation of delay. It does not deny planning or foresight, but it refuses to subordinate present perception to future targets. For governance, this means that climate, welfare or education policies must be evaluated not by distant projections but by their direct effects on dignity, ecology and justice today. For leadership, it means acting from clarity rather than role, ambition or identity. For education, it means awakening intelligence in the classroom now, not preparing students only for future achievement.

As noted in *Governance Without the Self*, frameworks that defer action institutionalise the self's tendency to postpone transformation. Awareness, however, acts immediately. Coherence is not achieved by managing fragments or waiting for conditions to align, it is present wherever perception is whole.

11.7 Conclusion: Moving Towards Institutions as Mirrors of Awareness

Institutions, policies and systems cannot create wholeness, but they can either obstruct or mirror it. Their true measure lies in whether they embody perception or perpetuate identity. Today, most institutions stabilise disorder by substituting prescription for perception, frameworks for relationship and negotiation for clarity. Dashboards light up while ecosystems collapse. Treaties multiply while wars continue. Schools expand while learning contracts. This is the continuity of fragmentation disguised as progress.

Yet glimpses reveal another possibility. Wherever perception is primary, coherence arises. When forests are seen as life itself rather than as carbon sinks or timber reserves, care flows without calculation. When a child is seen as a human being and not as a score or category, education becomes a relationship rather than conditioning. When the victim and perpetrator meet beyond their roles, reconciliation becomes possible. These moments show that governance, law and education can reflect awareness, even if they cannot manufacture it.

A critical distinction emerges here: the immediacy of awareness and the linearity of physical action. Awareness belongs to another order. The act of seeing another not as 'other' but as oneself is instantaneous. Compassion does not wait for procedure; clarity does not require negotiation. Yet in the physical world, actions unfold in time. To transport food from one region to another, to rebuild a collapsed bridge, to draft legislation, these require sequences, processes and coordination. Time is inevitable in the logistics of matter.

Time becomes an impediment when it is confused with the ground of action. When thought projects change into targets for 2030, when frameworks defer responsibility into distant horizons, when education postpones awakening into future achievement, time ceases to be a neutral sequence and becomes an alibi for continuity. The problem is not physical time but psychological time which is the movement of becoming, the attachment to methods, the identification with outcomes.

Awareness dissolves this false time. It meets disorder directly and acts without delay. Hunger does not demand calculation of charity but is seen as inseparable from one's own being. Violence does not require decades of negotiation but can end in the immediacy of encounter when enmity is dissolved. The delivery of food may take hours; the act of perceiving hunger as one's own is immediate. The rebuilding of a community may take years; the act of recognising human dignity is timeless.

This is what is meant by *change*: not reform stretched across decades, but the ending of disorder in the present. Transformation may unfold institutionally across time, but its source is the timeless perception of wholeness. Awareness acts now; time only carries its expression into the world.

Thus, the way forward is not to engineer perfect systems nor to dismantle all institutions overnight, but to see their rightful place. States, economies and organisations may continue for now, because their sudden dissolution would produce insecurity and chaos. But the very seeing of their limits, from a space of awareness, alters collective consciousness.

In that sense, institutions are no longer imagined as the ground of coherence, but as mirrors of it. They reflect the quality of perception held by those who operate within them. When that perception is fragmented, institutions enforce division. But when there is awareness, they may become spaces that do not obstruct intelligence. When there is the ground of awareness, what emerges is the space of coherence. In such a space, action is not governed by identity, method or scale, but by perception itself. Where awareness leads, wholeness informs action. In that movement, governance is transformed, not by reformist frameworks or activist strategies, but by intelligence in action, which moves without a centre, a plan or a method.

Coherence cannot be engineered or replicated. It is not born from design, but from the light of perception. Attempts to preserve it through tools, methods or best practices inevitably reduce it to structure. What is alive cannot be copied; it can only be seen. And when it is seen, action is immediate.

Yet the modern world does not stop trying. Even as glimpses of wholeness appear, most responses still emerge from the same consciousness that produced disorder. Activism surges. Institutions are reformed. New frameworks are introduced. But can these efforts, however well-intentioned, truly transform the root?

The next chapter explores this movement from fragmentation to coherence, asking whether institutions, once seen as mirrors and held as living spaces, can embody awareness rather than obstruct it.

References

Agrawal, A., & Gibson, C. C. (1999). Enchantment and disenchantment: The role of community in natural resource conservation. *World Development*, *27*(4), 629–649. https://doi.org/10.1016/S0305-750X(98)00161-2

Amnesty International. (2019). *The state of the world's human rights: Annual report 2018/19*. Amnesty International Ltd. https://www.amnesty.org/en/documents/pol10/6700/2019/en/

Angelsen, A., Hermansen, E. A., Rajão, R., & van der Hoff, R. (2018). *REDD+ results-based payments: Who should be paid, and for what?* https://www.cifor-icraf.org/publications/pdf_files/Books/BAngelsen180104.pdf

Arendt, H. (1958). *The human condition*. University of Chicago Press.

Austin, J. (1995). *The province of jurisprudence determined*. Hackett Publishing. (Original work published 1832).

Baker, L. (2023). South Africa's just energy transition partnership: Promise and contestation. *Energy Research & Social Science*, *101*, 103123. https://doi.org/10.1016/j.erss.2023.103123

Banerjee, A. V., & Duflo, E. (2019). *Good economics for hard times*. PublicAffairs.

Baronnet, B. (2008). Autonomy and education: The Zapatista experience. *Latin American Perspectives*, *35*(2), 23–40.

Bateson, G. (1972). *Steps to an ecology of mind*. Chandler.

Bateson, N. (2016). *Small arcs of larger circles: Framing through other patterns*. Triarchy Press.

Beck, U. (1992). *Risk society: Towards a new modernity* M. Ritter, Trans. Sage.

Berkes, F. (2018). *Sacred ecology* (4th ed.). Routledge.

Bohm, D. (1996). *On dialogue*. Routledge.

Booth, J. A. (1998). Costa Rica: Quest for democracy. In L. Diamond, J. J. Linz, & S. M. Lipset (Eds.), *Democracy in developing countries: Latin America* (pp. 115–157). Lynne Rienner Publishers.

Capra, F. (1996). *The web of life: A new scientific understanding of living systems*. Anchor Books.

Charpleix, L. (2018). The Whanganui River as *Te Awa Tupua*: Place-based law in a legally pluralistic society. *The Geographical Journal*, *184*(1), 19–30. https://doi.org/10.1111/geoj.12238

Clark, P. (2010). *The Gacaca courts, post-genocide justice and reconciliation in Rwanda: Justice without lawyers*. Cambridge University Press.

Courant, D. (2020). Citizens' assemblies for climate action: Lessons from France and the UK. *Environmental Politics*, *29*(4), 687–696.

Damasio, A. (1999). *The feeling of what happens: Body and emotion in the making of consciousness*. Harcourt Brace.

Dewey, J. (1938). *Experience and education*. Macmillan.

Escobar, A. (1995). *Encountering development: The making and unmaking of the third world*. Princeton University Press.

Freire, P. (1970). *Pedagogy of the oppressed*. Continuum.

Freire, P. (2005). *Pedagogy of the oppressed* (30th anniversary ed.). Continuum. (Original work published 1970).

Fromm, E. (1941). *Escape from freedom*. Farrar & Rinehart.

Fukuda-Parr, S., & McNeill, D. (2019). Knowledge and politics in setting and measuring the SDGs: Introduction to special issue. *Global Policy*, *10*(S1), 5–15. https://doi.org/10.1111/1758-5899.12604

Geneva Conventions. (1949, August 12). *International Committee of the Red Cross.* https://www.icrc.org/en/doc/assets/files/publications/icrc-002-0173.pdf

Gupta, D. (2024). *Eternal movement.* Emerge Publications. https://doi.org/10.65169/q7d2z6

Gupta, D. (2025). *Governance without the self: Consciousness and the end of fragmentation, systems, and complexity.* Emerge Publications. https://doi.org/10.65169/a9m3p4

Harris, T. (2019). *How a handful of tech companies control billions of minds every day.* TED Talk.

Hart, H. L. A. (2012). *The concept of law* (3rd ed.). Oxford University Press. (Original work published 1961).

Heller, P. (2001). Moving the state: The politics of democratic decentralization in Kerala, South Africa, and Porto Alegre. *Politics & Society*, *29*(1), 131–163. https://doi.org/10.1177/0032329201029001006

Human Rights Watch. (2021). A threshold crossed: Israeli authorities and the crimes of apartheid and persecution. *Human Rights Watch.* https://www.hrw.org/report/2021/04/27/threshold-crossed/israeli-authorities-and-crimes-apartheid-and-persecution

Intergovernmental Panel on Climate Change. (2023). *Climate change 2023: Synthesis report of the sixth assessment report of the intergovernmental panel on climate change.* IPCC. https://doi.org/10.59327/IPCC/AR6-9789291691647

International Committee of the Red Cross. (2023). *Annual report 2022.* https://www.icrc.org/en/document/annual-report-2022

Kant, I. (2012). *Groundwork of the metaphysics of morals* M. Gregor & J. Timmermann, Eds. & Trans. Cambridge University Press. (Original work published 1785).

Khera, R. (2019). Dissent on Aadhaar: Big data meets big brother. In R. Khera (Ed.), *Dissent on Aadhaar: Big data meets big brother* (pp. 1–30). Orient BlackSwan.

Kohn, A. (1992). *No contest: The case against competition.* Houghton Mifflin.

Krishnamurti, J. (1969/2009). *Freedom from the known.* Harper.

Krishnamurti, J. (1972). *The awakening of intelligence.* Victor Gollancz.

Krishnamurti, J. (1983/1992). *Krishnamurti to himself: His last journal.* Harper & Row.

Krishnamurti, J. (1992). *The wholeness of life.* HarperOne.

Krishnamurti, J. (1996). *Total freedom: The essential Krishnamurti.* HarperSan Francisco.

Krishnamurti, J. (2001). *The collected works of J. Krishnamurti (1952–1953)* (Vol. 7). Krishnamurti Foundation Trust.

Krishnamurti, J., & Bohm, D. (1985/2008). *The ending of time.* Harper.

Krishnamurti, J., & Bohm, D. (1986/2009). *The future of humanity.* Harper.

Lanier, J. (2010). *You are not a gadget: A manifesto.* Vintage.

Le, M. (2024, April 11). The bitter compromise of the EU migration pact. *Le Monde* (English edition). https://www.lemonde.fr/en/opinion/article/2024/04/11/the-bitter-compromise-of-the-eu-migration-pact_6668105_23.html

Lent, J. (2021). *The web of meaning: Integrating science and traditional wisdom to find our place in the universe.* New Society Publishers.

McGilchrist, I. (2019). *The master and his emissary: The divided brain and the making of the Western world* (2nd ed.). Yale University Press.
Meadows, D. H. (2008). *Thinking in systems: A primer*. Chelsea Green.
Morin, E. (2008). *On complexity*. Hampton Press.
Muru-Lanning, M. (2016). River co-governance and legal personhood: The Whanganui River settlement. *Journal of the Royal Society of New Zealand*, *46*(3–4), 207–214.
Office of the United Nations High Commissioner for Human Rights. (2025, September 9). *UN experts urge U.S. to grant visas to Palestinian officials and uphold diplomacy for peace*. https://www.ohchr.org/en/press-releases/2025/09/un-experts-urge-us-grant-visas-palestinian-officials-uphold-diplomacy-peace
Organisation for Economic Co-operation and Development. (2020). *Global competency for an inclusive world*. OECD.
Ostrom, E. (1990). *Governing the commons: The evolution of institutions for collective action*. Cambridge University Press.
Rawls, J. (1999). *A theory of justice* Rev. ed. Harvard University Press. (Original work published 1971).
Roberts, J. T., & Parks, B. C. (2007). *A climate of injustice: Global inequality, North–South politics, and climate policy*. MIT Press.
Robinson, K. (2006, February). *Do schools kill creativity?* Video. TED Conferences. https://www.ted.com/talks/ken_robinson_do_schools_kill_creativity
Sahlberg, P. (2011). *Finnish lessons: What can the world learn from educational change in Finland?*. Teachers College Press.
Sahlberg, P. (2015). *Finnish lessons 2.0: What can the world learn from educational change in Finland?*. Teachers College Press.
Stiglitz, J. E. (2002). *Globalization and its discontents*. W. W. Norton.
Stone, D. (2012). *Policy paradox: The art of political decision making* (3rd ed.). W. W. Norton.
The Civil Rights Project. (2024). *The case for a right to a racially just education*. University of California. https://civilrightsproject.ucla.edu/reports/the-case-for-a-right-to-a-racially-just-education/
The Lancet. (2020). *The Lancet COVID-19 commission statement*. The Lancet.
Tutu, D. (1999). *No future without forgiveness*. Rider.
UNICEF. (2023). *Annual report 2023*. https://www.unicef.org/reports/unicef-annual-report-2023
United Nations. (1945). *Charter of the United Nations*. https://www.un.org/en/about-us/un-charter
United Nations. (1947). *Agreement between the United Nations and the United States of America regarding the headquarters of the United Nations*. https://treaties.un.org/doc/Publication/UNTS/Volume%2011/v11.pdf
United Nations. (2015). *Transforming our world: The 2030 Agenda for sustainable development*. https://sdgs.un.org/2030agenda
Ura, K., Alkire, S., Zangmo, T., & Wangdi, K. (2012). *A short guide to gross national happiness index*. Centre for Bhutan Studies.
Varela, F. J., Thompson, E., & Rosch, E. (1991). *The embodied mind: Cognitive science and human experience*. MIT Press.
Vienna Convention on Diplomatic Relations. (1961). United Nations. https://legal.un.org/ilc/texts/instruments/english/conventions/9_1_1961.pdf

Whyte, W. F., & Whyte, K. K. (1991). *Making Mondragón: The growth and dynamics of the worker cooperative complex*. ILR Press.
Wilber, K. (2000). *A theory of everything: An integral vision for business, politics, science and spirituality*. Shambhala.
World Health Organization. (2022). *COVID-19 vaccine equity: One year of COVAX*. https://www.who.int/initiatives/act-accelerator/covax
Zuboff, S. (2019). *The age of surveillance capitalism: The fight for a human future at the new frontier of power*. PublicAffairs.

Chapter 12

Living Institutions: Mirrors, Spaces and Emergence

Abstract

Chapter 12 reimagines institutions as living mirrors rather than rigid systems. It explores how sovereignty, economy, law, education, technology and ecology can reflect awareness instead of perpetuating separation. Each domain is revisited as an expression of consciousness in relationship: sovereignty as coordination without domination, economy as care of life, law as restoration, education as awakening, technology as servant and ecology as the sacred foundation of existence. This chapter distinguishes immediate action, rooted in awareness, from continuity in time, proposing that institutions can embody awareness without solidifying it. When institutions function as mirrors of the whole, they support harmony; when they operate from identity, they sustain conflict. This chapter ends by affirming that the transformation of systems follows naturally from the transformation of perception.

Keywords: Living institutions; awareness in systems; sovereignty as coordination; economy of care; law as restoration; education as awakening; ecology as sacred; reflection vs division

12.1 Introduction: From Fragmentation to Living Institutions

The previous chapter revealed that institutions do not operate independently of human consciousness; they mirror it. When perception is fragmented, institutions enforce division; when there is awareness, they may reflect coherence. Chapter 11 showed that the difference lies in where systems begin: when governance starts from prescription, it produces dashboards, indicators and delay; when it begins from perception, coherence arises immediately and institutions may follow without obstruction.

From Fragmentation to Wholeness, 267–286

Published under exclusive licence by Emerald Publishing Limited
doi:10.1108/978-1-80686-677-920261012

To stop at critique, however, is to remain within the field of fragmentation. Chapter 12 begins from a different ground: the possibility that institutions are not fixed structures but living mirrors, spaces and emergent expressions of coherence.

Institutions are relationships crystallised. They are not separate from us; they are us, extended outward. When we look at a school, a parliament or a treaty, we are looking at a mirror of human consciousness. To see this clearly is already to loosen their grip as inevitabilities. Systems can then be re-seen not as permanent frameworks but as spaces that straighten, dissolve and adapt according to the depth of awareness in public life.

This chapter explores how that shift unfolds. Sovereignty, economy, education, law, technology and ecology are not isolated domains but interwoven reflections. Change in one reverberates through the others. Sovereignty reframed as care alters economic priorities. Education oriented towards awakening reshapes law and citizenship. Ecology recognised as sacred ground transforms every institution built upon it. Each fragment touches the whole, and the whole, once seen, reconfigures the fragments.

The distinction between immediate action and horizon is central here. Awareness acts now. Systems carry their expression in time. If institutions mistake continuity for change, they obstruct coherence. If they mirror awareness, they serve life provisionally, humbly and without permanence.

Chapter 12, therefore, steps beyond institutional critique. It invites us to re-see institutions as living: not fixed blueprints to be perfected but spaces that can hold coherence, straighten themselves when guided by perception and dissolve when they obstruct life. They are mirrors of consciousness, spaces for awakening and emergent movements that will always remain provisional. The task is not to reform them endlessly but to live from awareness, allowing institutions to follow.

From here, we turn first to the most direct insight: institutions as mirrors of consciousness, revealing not only our collective disorder but also our capacity, however brief, to hold coherence.

12.2 Institutions as Mirrors of Consciousness

The word *institutions* is used in many ways. Economists such as Douglas North (1990) describe them as the 'rules of the game' that shape incentives in political and economic life, while others use the term to mean organisations such as schools, parliaments or courts. In this chapter, the word is used in a deeper sense: institutions are crystallised relationships, the outer forms of inner patterns of consciousness. Rules, norms and organisations are only expressions of this deeper movement.

As Krishnamurti (1969/2009) observed, society is 'the projection of ourselves'. Institutions are not separate from human beings but are extensions of the same consciousness that creates them. Bohm (1992) similarly noted that thought generates structures, which then appear independent and controlling, even though they are products of the very fragmentation they mirror. To see institutions in this light is to recognise that they are not external arbiters of order but mirrors of the consciousness that sustains them.

This perspective directly continues from Chapter 11. Just as frameworks of systems thinking and complexity cannot dissolve fragmentation but only rearrange it, institutions too cannot generate coherence on their own. They stabilise disorder when the actor, the human being, remains fragmented. Frameworks and institutions share the same paradox: they can describe, organise and even inspire, but they cannot transform. Without perception, both become reform loops that preserve continuity under new names.

Peter Berger and Thomas Luckmann (1966) argued that institutions emerge from social processes, which, once routinised, acquire an air of objectivity. They are human constructs that paradoxically confront us as if they were external realities. Cornelius Castoriadis (1997) similarly described institutions as imaginary significations that are cultural inventions that structure experience while reflecting collective meaning. These perspectives echo what awareness reveals: institutions are mirrors, not origins.

History makes this plain. The League of Nations arose from a genuine impulse for peace after the First World War. Yet once translated into procedures and committees, it hardened under the weight of state interests. By the 1930s, when Japan invaded Manchuria and Italy attacked Ethiopia, it could only issue condemnations. The United Nations was created to correct these flaws, and while more robust in design, it remains caught in the same structural logic: veto powers ensure paralysis, peacekeeping is selective and humanitarian norms collapse into political bargaining. The International Committee of the Red Cross (ICRC, 2023) notes how fundamental humanitarian principles are routinely subordinated to Security Council politics. The mirror reveals a deeper truth: even noble aspirations, when institutionalised without awareness, can decay into continuity.

Grassroots movements display the same fragility. Occupy Wall Street in 2011 erupted as a raw cry against inequality, creating temporary spaces of solidarity and imagination beyond consumerist individualism. But as it fragmented into NGOs, campaigns and partisan projects, the coherence of perception gave way to the machinery of branding, funding and bureaucracy. Fridays for Future, though still vibrant, faces similar risks as its local chapters formalise. The lesson is not that movements must avoid structure, but that coherence cannot be stored. It can only be held momentarily as presence, and when presence returns, new forms of coherence may arise again.

Contemporary policy examples continue this pattern. The European Union's Green Deal was launched as a transformative sustainability initiative (European Commission, 2020). Yet under the pressure of economic competitiveness and energy security, member states have continued to subsidise fossil fuels while showcasing progress dashboards. Despite its ambition, EU fossil fuel support remains high (Reuters, 2022), undercutting transition goals. As Keohane and Victor (2011) observed, structural inertia often coexists with visionary rhetoric. What we see, again, is the architecture of delay, the system mirroring a divided psyche that glimpses clarity but soon retreats into identity and interest.

If institutions are mirrors, two implications follow. First, they expose the disorder of consciousness. Corruption, paralysis and selectivity are not institutional failures alone but symptoms of fragmentation within human beings. Second, they can also, even if briefly, hold the conditions for coherence. South Africa's Truth and Reconciliation Commission, fragile as it was, created moments where victim and perpetrator encountered each other beyond identity. In such encounters, institutions became spaces rather than structures, containers of perception, however temporary.

The task, then, is neither to idealise institutions nor to imagine that reform can redeem them. Their role is humbler: to refrain from obstructing intelligence and to hold a space where perception can act. Institutions cannot manufacture coherence; they mirror it. When awareness is absent, they ossify into machinery. When awareness is present, they can temporarily hold coherence before dissolving again into life. Their legitimacy lies not in permanence, but in their willingness to serve wholeness and to dissolve when they obstruct life.

12.3 Sovereignty, Borders and Collective Belonging

Sovereignty is treated as the cornerstone of the modern international order. States guard borders, defend authority and assert non-interference as a principle of legitimacy. Yet sovereignty is not natural. It is a construct of history and thought, a framework devised to stabilise relations among nations and later reified as if eternal. Borders were drawn through wars, treaties and colonisation, then taught as sacred lines of belonging. Awareness reveals their provisional character.

Benedict Anderson (1983) described nations as *imagined communities*, sustained by shared narratives rather than inherent essence. Ulrich Beck (2006) argued for *cosmopolitan realism*, warning that national frameworks are inadequate for global crises like pandemics or climate change. Achille Mbembe (2019) has gone further, describing borders as instruments of exclusion that perpetuate colonial hierarchies. These perspectives converge on a critical point: sovereignty and borders are contingent, not ultimate.

History illustrates both their utility and their danger. The 1648 Treaty of Westphalia institutionalised sovereignty to contain European wars, not only providing stability but also legitimising rivalry. In the 20th century, sovereignty provided a legal basis for decolonisation, yet the same principle shielded regimes committing atrocities within their borders. The UN's reports on Xinjiang, for example, highlight how sovereignty is invoked to block accountability (United Nations Human Rights Council, 2022). In Rwanda 1994, nearly a million people were killed as the UN failed to intervene, constrained by sovereignty and Security Council vetoes. By contrast, NATO's 1999 Kosovo intervention bypassed the UN, collapsing universality into selectivity. Ukraine further underscores the paradox: sovereignty invoked simultaneously to defend borders and justify aggression.

Borders, too, carry contradiction. They regulate trade and migration, offering stability, yet also exclude and divide. The Mediterranean refugee crisis reveals this vividly, with UNHCR data showing lives lost as Europe fortifies borders at the expense of survival. Climate-induced migration in the Sahel and South Asia is intensifying such dilemmas as rising seas and desertification push millions across lines that states insist must not be crossed.

Here, sovereignty can be seen as a mirror of consciousness. When identity and fear dominate, borders harden into weapons of exclusion. When awareness is present, sovereignty becomes provisional, a tool of coordination rather than an idol of belonging. In the short term, this recognition enables honesty: states can negotiate trade, climate or refugee agreements without pretending borders are sacred. In the longer horizon, belonging is reimagined as human and ecological rather than exclusively national.

This shift cannot happen as an idea or framework. It requires direct perception of sovereignty as the continuity of the self, seeking permanence, control and security. In seeing this movement directly, without the mediation of knowledge or ideology, the grip of sovereignty ends. Then sovereignty is reconfigured, no longer a fortress of identity but a space for care, coordination and the protection of life.

The implication is profound. Sovereignty, like knowledge, must find its rightful place. It is coherent when used to distribute food, regulate trade or reduce harm. It becomes incoherent when treated as absolute. Collective security, too, must be reimagined: instead of alliances that protect some while excluding others, coherence begins from the awareness that insecurity anywhere threatens security everywhere.

12.4 Economy in Wholeness

Modern economies mirror fragmentation. Growth is pursued as an unquestioned end, measured by GDP that counts weapons and oil spills as progress while ignoring the erosion of life. Finance and trade present themselves as neutral mechanisms but reproduce inequality and ecological harm. Even reform agendas reveal contradictions: the Sustainable Development Goals simultaneously prescribe sustained growth and urgent ecological restraint, as if perpetual expansion and planetary limits could coexist without tension (United Nations, 2015).

Joseph Stiglitz (2002) exposed how structural adjustment programs dismantled protections in developing nations, enriching creditors while deepening vulnerability. Arturo Escobar (2018) critiqued development itself as a paradigm rooted in Western notions of progress that fragments community and ecology. Kate Raworth (2017) proposed the doughnut economy, balancing social foundations with ecological ceilings. Elinor Ostrom (1990) demonstrated that communities can govern resources collectively, showing that relationships, not competition, sustain economy.

The contradictions remain stark. Cobalt mined in unsafe conditions in the Congo feeds global smartphone supply chains. COVID-19 revealed markets propped up by emergency interventions while millions faced hunger and unemployment; vaccine equity lagged under patent protections (WHO, 2022). Inequality continues to widen alongside climate stress (IMF, 2023; IPCC, 2023).

Yet glimpses of coherence emerge when perception interrupts fragmentation. Preston's community wealth-building redirected procurement towards local businesses, strengthening resilience (Manley et al., 2021). Basic income pilots in Finland, Kenya and the United States reduced stress and expanded participation by decoupling livelihood from competitive productivity (Government of Finland, 2020; Government of Kenya, 2021). Ecuador's constitutional embrace of *Buen Vivir*, rooted in Indigenous cosmologies, reframed the economy as stewardship of life, even if compromises limited its reach.

Re-seen from awareness, economy returns to its original meaning: *oikonomia*, the care of the household. Production is coherent when it sustains the conditions of its own possibility. Employment is coherent when it upholds dignity and ecology. Wealth is relevant only when it sustains well-being. Money becomes a medium of relationship, not an ultimate aim.

Short-term coherence demands redirecting priorities from growth to sufficiency. Fossil fuel subsidies still exceeded one trillion dollars in 2022–2023, undermining transition goals (IEA, 2023). Redirecting such funds towards renewable energy, local infrastructure and ecological restoration aligns economy with survival. Trade can shift from maximising efficiency at the cost of exploitation to strengthening local resilience. Financial accountability can move from rewarding quarterly profits to ensuring dignity across supply chains.

In the longer horizon, success is no longer measured by aggregate growth but by the vitality of life itself. A coherent economy does not externalise harm or sacrifice ecology for expansion. It sees human well-being and ecological health as one indivisible fact. Institutions may persist, but their legitimacy rests on whether they serve life.

The deeper implication is that economy, like sovereignty, is not separate from consciousness. When sovereignty is provisional, the economy loosens from rivalry. When education awakens perception, citizens resist being reduced to consumers. When awareness is lived, the economy ceases to be a machinery of interests and becomes the practice of care. Like all institutions, the economy is a mirror: fragmentation reflects exploitation and excess; awareness reflects sufficiency and relationship.

12.5 Education as Awakening, Not Conditioning

Education is one of the clearest mirrors of fragmentation. Instead of awakening intelligence, most systems train children to compete, conform and specialise. Success is measured by grades, rankings and credentials. Students are told to become something in the future rather than to understand themselves and the world in the present. The result is cleverness without wisdom. As noted by Krishnamurti (1954), the person who can split the atom without compassion becomes a monster. Knowledge without love destroys. This was already seen in

§7.4, where education was shown to institutionalise fear and desire, making schools, debating circuits and universities into theatres of performance and comparison rather than spaces of inquiry.

Paulo Freire described modern schooling as the banking model, where teachers deposit information into passive students (Freire, 1970). bell hooks called for an education as the practice of freedom, where learning is rooted in dialogue and presence rather than domination (hooks, 1994). Ivan Illich critiqued institutional schooling for reproducing inequality under the guise of neutrality, arguing for convivial forms of learning outside rigid systems (Illich, 1973). J. Krishnamurti emphasised throughout *Education and the Significance of Life* (1953/2009) that true education must free the mind from fear and conditioning, not prepare it for conformity. As he later observed in *A Wholly Different Way of Living* (Krishnamurti & Bohm, 1992), knowledge has its place, but when it dominates life, it produces disorder. This resonates directly with §7.4's sections on 'Curriculum as Commodity' and 'Universities, Knowledge, and False Equality'.

The consequences of fragmentation are stark. In Israel and Palestine, textbooks transmit narratives of conflict, embedding antagonism into the consciousness of children. In Afghanistan, girls are denied schooling under the Taliban, while boys are indoctrinated into rigid religious identities; UN reports detail the cascading harm of these bans. In the United States, curriculum battles over critical race theory and diversity mandates reveal how education becomes a proxy for political polarisation. Even in less extreme contexts, students worldwide are pressured into narrow definitions of success that disconnect knowledge from meaning. These mirror the distortions outlined in §7.4 'The Nation and the Insecurity of Structures' and 'Bias, Nationalism, and Instrumentalization'.

Higher education reflects the same cycle. Universities often equate success with specialisation, publication metrics and career pipelines, sidelining questions of ethics and responsibility. Students graduate trained to manage systems but rarely invited to ask whether those systems are coherent. As Martha Nussbaum (2010) has argued, the decline of the humanities signals a crisis of citizenship: education becomes training for profit rather than formation for democracy. This directly extends the critique in §7.4 'Universities, Knowledge, and False Equality'.

Yet glimpses of coherence exist. Krishnamurti Foundation schools in India, England and the United States emphasise inquiry, silence and relationship, encouraging students to question authority, tradition and even the teacher. Finland's system, though not explicitly spiritual, reduces standardised testing and emphasises equity, producing outcomes without anxiety. Indigenous education among Māori in New Zealand or Andean communities in Latin America ties learning directly to relationships with land and kinship, showing that knowledge and care can remain inseparable. These examples echo the case studies in §11.5, where coherence was glimpsed in practice.

Right education does not reject knowledge; it places it in its rightful place. Children must still learn language, mathematics, science and history. But knowledge cannot dominate. Right education begins in awareness. It invites inquiry into meaning, goodness, relationship and the possibility of living without division. These are not accessories to the curriculum; they are its ground. This responds directly to §7.4 'Performance in Schools and Debating Circuits', where performance displaced inquiry.

Immediate coherence means loosening the grip of exams and rankings, creating space for dialogue and reflection. Teachers can be companions in inquiry rather than managers of curricula. Ethical and ecological seeing can be integrated into daily lessons, not as separate subjects but as orientations. Technology, too, can be placed rightly: digital tools that expand access rather than platforms that addict or manipulate. This addresses the distortions in §7.4 'Curriculum as Commodity'.

Long-term coherence requires a more radical horizon. Schools would cease to be factories of identity or pipelines to labour markets. They would become spaces where children learn not only to calculate and analyse but also to perceive themselves and the world without distortion. Students educated in this way would not blindly cooperate with harmful systems, nor would they rebel destructively. They would act from choiceless intelligence, aware of when cooperation sustains life and when resistance is necessary. This is the counter-movement to §7.4 'Military Training', which mechanises obedience through fear and desire.

The implication for governance is decisive. Citizens educated in wholeness cannot be manipulated by propaganda, fear or identity politics. They will resist policies that perpetuate disorder and act from responsibility, dissolving the false division between private life and public duty. Education in wholeness therefore reshapes every other domain: it aligns economy with sufficiency (§12.3), sovereignty with care (§12.2) and law with dignity (§12.5).

Education as awakening is not a framework to be rolled out but a recognition of learning as life itself. The teacher, the student and the lesson are not separate but one movement of inquiry. When this is lived, education ceases to reproduce fragmentation and becomes the ground of coherence. This completes the through-line from §7.4 'Teachers as Students of Fear and Desire' and 'Inequality in Admissions and Teacher Distribution'. The child educated in this way is not merely a student but a whole human being, not trained for conformity but awakened into freedom.

12.6 Law and Justice in Wholeness

Law promises universality, yet in practice it often enforces division. Constitutions proclaim equality, but systemic exclusion persists. International law speaks of human rights, yet interventions are selective and inconsistent. Courts and treaties frequently become arenas where identity, interest and power dictate outcomes. Justice reduced to retribution freezes people into roles of victim and perpetrator, missing the possibility of restoration.

Michel Foucault showed how modern law and penal systems are bound to surveillance and discipline, more concerned with social control than justice

(Foucault, 1977). René Girard argued that legal systems arose historically as substitutes for sacrificial violence, channelling conflict into formal procedures while leaving the underlying rivalries unresolved (Girard, 1977). Martha Minow highlighted the importance of restorative justice, suggesting that acknowledgement, dialogue and repair can open spaces that punishment cannot (Minow, 1998). Together, these insights reveal that law is not a neutral guardian of order but a mirror of the consciousness that creates it. As explored in §7.4, institutions themselves embody fear and desire, and legal systems crystallise this movement when punishment becomes a projection of retribution rather than an opening to healing.

Examples abound. The United States proclaims equal justice under law, yet systemic racism continues to shape policing and incarceration. The International Criminal Court prosecutes leaders from Africa while powerful states evade accountability, reflecting a hierarchy of sovereignty. In India, laws against caste discrimination exist, but social exclusion and violence against Dalits persist in daily life. Israel cites self-defence in Gaza, while international humanitarian law condemns disproportionate civilian deaths; critiques of ICC selectivity and the politicisation of enforcement remain salient in Global South legal scholarship and civil society monitoring.

Yet glimpses of coherence show another possibility. South Africa's Truth and Reconciliation Commission, despite its limits, created conditions where victim and perpetrator could meet beyond categories, allowing acknowledgement and forgiveness to arise. New Zealand's Treaty of Waitangi settlements, though incomplete, represent efforts to restore dignity to Māori communities through land restitution and recognition. In Rwanda, community-based gacaca courts sought not only punishment but also restoration, enabling millions to return to village life after the genocide; the official final report records both community reintegration and due process concerns, underscoring the fragility of restorative mechanisms at the national scale (Government of Rwanda, 2019). These efforts remain fragile, but they reveal that justice is not only about retribution but also about restoring relationships. Relationship without image is the ground of right action; when law reflects this, it ceases to be an external coercion and becomes an expression of coherence.

Short-term coherence means strengthening accountability where the law already exists. Independent courts, protection for minorities and ecological rights recognised in constitutions, as in Ecuador and Bolivia, expand the scope of dignity. Restorative justice programmes in schools and communities reduce cycles of violence by inviting responsibility rather than imposing stigma. Internationally, mechanisms like universal jurisdiction, allowing states to prosecute crimes against humanity regardless of where they occurred, help prevent the worst abuses of selectivity.

Long-term coherence requires a more radical re-seeing. Justice in wholeness is not primarily about punishment or deterrence but about healing relationship. Law finds its right place not as an instrument of fear but as a mirror of dignity already perceived. When awareness dissolves division, the need for coercion diminishes. Communities organised around care require less policing; citizens who see others as themselves do not demand laws to regulate every encounter. Law does not disappear, but it humbly protects the space where awareness acts.

The deeper shift is psychological. As long as individuals are fragmented, law will be used selectively to protect identity and power. As perception deepens, law is re-understood as a safeguard, not a saviour. Justice is no longer an external mechanism imposed on disorder but the outer reflection of an inner seeing. As argued in §11.4, coherence is not achieved through frameworks or procedures but through perception itself. Justice in wholeness flows not from codes and procedures but from the consciousness that animates them.

12.7 Science, Technology and Innovation in Wholeness

Science and technology are among the most powerful mirrors of human consciousness. Knowledge, once freed from myth, opened extraordinary possibilities: curing diseases, exploring space and connecting people across continents. Yet the same knowledge, divorced from awareness, has enabled nuclear annihilation, mass surveillance and ecological devastation. The problem is not science or technology as such but the consciousness that wields them. Without love, knowledge becomes destructive. As discussed in §8.2, knowledge without awareness leads to fragmentation; without perception, information becomes power divorced from responsibility.

Martin Heidegger warned that modern technology reduces the world to a standing reserve, resources to be ordered and exploited (Heidegger, 1977). Lewis Mumford distinguished between authoritarian and democratic technics, showing how tools can either centralise power or distribute it (Mumford, 1934). Shoshana Zuboff revealed how digital platforms now operate as surveillance capitalism, extracting human experience for profit (Zuboff, 2019). Jaron Lanier argued that digital systems amplify division when reduced to algorithms that commodify attention (Lanier, 2018). These critiques converge: science and technology mirror the priorities of the psyche and the systems it creates. Bohm noted that thought fragments reality and then treats those fragments as independent; artificial intelligence (AI) and geoengineering are contemporary reflections of this same pattern.

The 20th century illustrates both extremes. Nuclear physics gave humanity the power to split the atom. It provided medical imaging, energy production and scientific insight, but also Hiroshima and Nagasaki, where cities were incinerated in seconds. Biotechnology has brought not only cures for genetic diseases but also patents that restrict access to lifesaving medicines in poorer countries. Artificial intelligence today offers potential for medical diagnosis, personalised education and ecological monitoring, while simultaneously enabling disinformation campaigns, predictive policing and algorithmic bias that deepen inequality.

Recent developments make the stakes urgent. Generative AI tools, including large language models, have entered education, law and healthcare, raising questions of trust, creativity and manipulation; education and labour markets are already reorganising around these tools, with uneven benefits and risks (The New York Times, 2023). Facial recognition systems are deployed in China for surveillance and in Europe for border control, embedding technological power into politics of exclusion. Technical reverse engineering shows how some of these systems operationalise mass profiling, as in Xinjiang (Human Rights Watch, 2019). Elon Musk's Starlink satellites not only provide internet access to rural communities but also enable militarised communication in Ukraine; the same network that connects rural clinics functions as a wartime communications backbone (Al Jazeera, 2022). Climate geoengineering proposals, from aerosol injection to ocean fertilisation, exemplify humanity's drive to treat planetary life as an object of control, with unknown risks.

Yet glimpses of coherence appear. Open-source software communities create tools as commons rather than commodities. Community-owned renewable energy cooperatives in Denmark and Germany distribute both power and decision-making. The push for open science and equitable access to COVID-19 research briefly showed that knowledge can be shared as a relationship, not hoarded as property. Citizen assemblies on AI in France and Canada demonstrate how democratic dialogue can shape technology rather than surrendering to corporate monopoly.

Short-term coherence requires embedding ethics, transparency and accountability into design and deployment. This means data rights for individuals, public-interest infrastructure and regulations that prevent the concentration of power. Scientific research must prioritise ecological and human well-being over profit, redirecting innovation towards sufficiency, not endless expansion.

Long-term coherence requires re-seeing science itself. True science is inquiry, not certainty. It is the humility to observe reality without imposing preconceptions. When guided by awareness, scientific exploration becomes a dialogue with nature rather than domination of it. Technology becomes a servant of care, extending the relationship rather than fragmenting it.

The shift is subtle but decisive. A society that sees the Earth as living will not design systems that profit from its destruction. A community that perceives others as itself will not code algorithms that manipulate, exclude or exploit. Innovation in wholeness is not a race for advantage but a movement of perception seeking coherence with life. As emphasised in §12.4 on education, knowledge placed in its right place becomes liberating rather than destructive. In that sense, the future of science and technology depends less on technical breakthroughs than on the consciousness from which they arise.

12.8 Ecology and the Sacred Ground of Life

Ecology is not one issue among many. It is the ground of all life. To treat it as a negotiable sector, balanced against growth targets or national security, is to mistake the table for just another chair. Institutions that measure success by dashboards of carbon offsets or GDP growth while forests fall and oceans acidify reflect the same fragmentation of consciousness that created the crisis. As we saw in §7.4, when fear and desire structure institutions, ecology becomes either a resource to exploit or a crisis to manage, never a life to live in relationship with.

Fritjof Capra argued that living systems are networks of relationships, not machines of separate parts (Capra, 1996). Arne Næss introduced the idea of deep ecology, calling for recognition of the intrinsic value of all beings, not just their utility to humans (Næss, 1989). Vandana Shiva has shown how industrial agriculture, driven by profit and monocultures, erodes both biodiversity and cultural knowledge, while communities rooted in ecological care sustain resilience (Shiva, 2016). Bruno Latour reminded us that climate change has dissolved the modern separation between nature and society; we are entangled in one living Earth (Latour, 2017). Krishnamurti (1992) put it simply: 'When man loses relationship with nature, he loses relationship with man'.

History illustrates the consequences of ignoring this entanglement. Industrial revolutions brought wealth to some but destabilised the climate, depleted soils and poisoned rivers. Colonial empires extracted resources from Asia, Africa and the Americas, leaving ecological and social scars that persist. Today, the Amazon rainforest faces deforestation at a scale that risks tipping the global climate into irreversible feedback loops, while Himalayan glacier loss threatens the water security of nearly two billion people in South Asia. The IPCC warns that multiple systems are nearing or have crossed thresholds beyond which recovery becomes nonlinear (IPCC, 2023). These are not isolated events but signs that fragmentation at the level of consciousness manifests as ecological collapse.

Recent cases underline the urgency. Governments are considering deep-sea mining licencing amid warnings of irreversible ecological damage. The European Union promotes renewable energy but continues to import gas and oil, externalising ecological harm (Reuters, 2022). Meanwhile, in Ecuador, a historic 2023 referendum saw citizens vote to halt oil drilling in Yasuni National Park, affirming Indigenous leadership in defending biodiversity (The Guardian, 2023). In New Zealand, the Whanganui River has been recognised in law as a living entity with rights, reflecting Māori cosmologies that see rivers as ancestors rather than resources. These examples reveal the spectrum: from treating nature as a standing reserve to glimpses of coherence where relationship is primary. Carbon offset schemes, often marketed as solutions, have also been shown to commodify ecosystems in ways that perpetuate fragmentation (Carbon Market Watch, 2021).

Short-term coherence requires immediate steps: ending subsidies for destructive industries, enforcing ecological rights and strengthening community stewardship. Community forestry in Nepal, where local groups manage forests sustainably, has shown remarkable success in restoring biodiversity and livelihoods. Policies that place Indigenous communities at the centre of land governance, as in parts of Canada and Australia, offer further glimpses of alignment between ecology and justice.

Long-term coherence is more radical. It requires ceasing to pretend that ecology is negotiable. The Earth is not a stakeholder at a table; it is the table. Governance in wholeness means re-seeing land, water, air and species not as categories of management but as inseparable from human survival. Guardianship replaces ownership. Relationship replaces extraction. In this horizon, national interest and ecological care no longer conflict because belonging itself is redefined: to be human is to be of the Earth.

The deeper shift is not policy but perception. When forests are seen as life itself rather than as carbon sinks, care flows without calculation. When rivers are kin, law and economy orient differently. When ecology is sacred ground, not a negotiable resource, institutions are humbled into serving life rather than managing it. This echoes §9 on relationship without image and §12.4 on education, where learning tied to land and community reveals that awareness itself sustains life.

12.9 Immediate Action and the Horizon Beyond Time

In the preceding sections of this chapter, we have spoken of short-term coherence and long-term coherence in each domain: sovereignty re-seen as provisional (§12.2), economy aligned with sufficiency (§12.3), education as awakening rather than conditioning (§12.4), law as restoration rather than retribution (§12.5), science and technology guided by humility (§12.6) and ecology recognised as sacred ground (§12.7). These distinctions were not frameworks imposed from outside but reflections of one movement of perception: coherence acting in the immediacy of the present and coherence continuing as horizon when sustained in time. This section now deepens that distinction by asking how action and vision, immediacy and horizon, can remain aligned without collapsing into the continuity of psychological time.

Every institution, system and reform wrestles with the problem of time. Food must be transported, laws drafted, bridges rebuilt and policies negotiated. These actions unfold in physical time, sequences of cause and effect necessary for life. Yet when institutions mistake physical time for the ground of change, coherence decays into continuity. Psychological time, the projection of becoming, infiltrates governance, turning transformation into endless targets and roadmaps. The 2030 Agenda for Sustainable Development is a clear example: urgent change is deferred into the comfort of future milestones, while present action continues in contradiction (United Nations, 2015). This pattern of future-heavy pledges with present-tense contradictions recurs across climate and development reporting (IPCC, 2023).

J. Krishnamurti insisted that truth is a pathless land, that it cannot be reached through gradual reform or temporal projection (Krishnamurti, 1981). David Bohm described thought as a movement that fragments reality when it carries the past into the present (Bohm, 1980). Hannah Arendt reminded us that action is rooted in natality, the capacity to begin anew, not in the accumulation of plans (Arendt, 1958). Together, these insights clarify the difference between the immediacy of awareness and the continuity of time.

Throughout this chapter, we have described short-term coherence and long-term coherence as two aspects of the same movement. Short-term coherence refers to immediate alignment of action with life: feeding the hungry, ending destructive subsidies, creating spaces of inquiry in classrooms, protecting rivers as legal persons. These are not reforms projected into the future but present responsibilities, grounded in awareness. Long-term coherence refers to the horizon that opens when action continues to unfold in wholeness: economies re-oriented around sufficiency (§12.3), sovereignty held provisionally (§12.2), law re-seen as restoration (§12.5), science guided by humility (§12.6) and ecology recognised as sacred ground (§12.7). Long-term coherence is not deferral into psychological time but the gradual straightening of institutions when they remain aligned with perception.

History shows the cost of mistaking time for change. Climate negotiations have produced decades of conferences and treaties, yet carbon emissions continue to rise. Peace processes in Israel–Palestine have been framed in roadmaps, while occupation and violence persist. Education reform promises transformation for the next generation, while millions of children remain trapped in systems of fear and competition today. In each case, projection into the future becomes an alibi for avoiding the ending of disorder in the present.

Yet glimpses of immediacy exist. When the Berlin Wall fell in 1989, change was not the outcome of a carefully sequenced roadmap but of perception reaching a breaking point, expressed in action. When South Africa ended apartheid, the immediate recognition of injustice enabled structures to shift more rapidly than decades of debate had suggested possible. When communities in Kerala, India, responded to devastating floods in 2018, neighbours acted without waiting for formal procedures, feeding, housing and rescuing one another in coherence born of necessity; contemporaneous reporting documented extensive mutual aid that outpaced formal logistics (BBC News, 2018).

Immediate action does not deny sequences; it aligns them with the whole. The delivery of food may take hours, but the perception of hunger as one's own is instant. Rebuilding a community may take years, but the recognition of dignity is timeless. The question is whether the sequences of time are animated by awareness or by continuity. If the means are fragmented, if they rely on competition, propaganda or fear, the ends will reproduce fragmentation. If the means are coherent, rooted in awareness and goodness, the ends carry the same quality.

Short-term orientation, then, is not a compromise with fragmentation but the expression of short-term coherence: aligning present action with wholeness here and now. Long-term orientation is not deferral but horizon: allowing institutions to adapt, straighten and evolve in response to coherence as it arises. The distinction is subtle: time is needed for action, but not for perception.

The danger is ossification. Once clarity is seen, institutions often try to preserve it through procedures. Yesterday's insight becomes today's duty, repeated mechanically. Awareness dissolves this habit by always returning to the present. Each moment is new. Each situation requires fresh seeing. The horizon is not a fixed model but a living movement that keeps adjusting to life.

The task, then, is not to choose between immediate action and long-term vision. It is to align both in wholeness. Awareness acts now; time carries its expression. The vision and the means are not separate but one movement. When governance learns this, it ceases to be the management of continuity and becomes intelligence in action. But if short-term coherence and long-term coherence are reduced to frameworks, prescriptions or ideas, they become another form of continuity. To adopt them as models without ending the self is to reproduce fragmentation under new language.

Chapter 3 already warned that knowledge, when abstracted from direct perception, becomes dependency, a substitute for seeing. Chapter 5 clarified that right action flows only from wholeness, not from methods or deferred ideals. Chapter 7 showed how fear and desire, projected through time, sustain the very fragmentation that institutions attempt to overcome. These earlier insights converge here: coherence cannot be stored as a model, practiced as a method or projected into the future. It can only be lived, immediately, in perception.

12.10 Living the Sacred in Public Life

The sacred is not confined to temples, scriptures or rituals. It is the quality of treating life as intrinsically worthy. To live the sacred in public life is not to impose doctrine but to meet the world with reverence. When institutions forget this, they reduce beings to cases, forests to carbon sinks, rivers to resources and communities to statistics. When they remember, governance itself becomes an act of care.

Max Weber warned that modern bureaucracy risks reducing life to rationalised procedures, stripping meaning from the iron cage of efficiency (Weber, 2004). By contrast, Mircea Eliade emphasised that every culture recognises the sacred as that which breaks through the profane order of calculation (Eliade, 1996). Charles Taylor traced how secular modernity has flattened moral horizons, yet noted that longing for fullness persists (Taylor, 2007). Together, these perspectives remind us that governance without reverence collapses into machinery.

Examples reveal the difference. In New Zealand, the Whanganui River is recognised in law as a living being, reflecting Māori cosmology and shifting governance from ownership to guardianship; legal personhood operationalises guardianship in statute and practice. In Ecuador, the constitutional principle of Pachamama affirms that nature has rights, not just uses; its articles remain a touchstone for jurisprudence on ecological personhood. In Kerala, India, the decentralised response to floods in 2018 reflected not only efficiency but reverence; neighbours acted as if each life mattered beyond statistics (BBC News, 2018). In contrast, climate dashboards that celebrate offset markets while deforestation accelerates show governance reduced to performance, with the sacred absent (Carbon Market Watch, 2021).

Living the sacred in public life means institutions holding space for presence, not performance. In education, it is the teacher who meets the child as a being, not a score. In healthcare, it is the doctor who refuses to reduce the patient to a case file. In law, it is the judge who listens not only to evidence but also to dignity. In ecology, it is the guardian who treats the river as kin, not a commodity. The sacred is not mystical excess but coherence in action.

Yet the sacred also demands dying to the known. Reverence is not the accumulation of rituals or practices but the ending of the 'me' that clings to memory, identity and control. Krishnamurti (1981) observed that to encounter life freshly, one must 'die every day to everything that is known', for otherwise perception is conditioned by the past. Death is not merely biological; it is the ending of psychological continuity: the attachments, hurts and ambitions that sustain division (Krishnamurti & Anderson, 1991). Each moment is new only when the past does not dictate perception.

To live the sacred in governance, then, is not to preserve the old in refined forms but to allow institutions and oneself to die to yesterday's knowledge, fears and ambitions. Institutions, like individuals, must dissolve when they harden so that new forms can arise in coherence. This is not a collapse but a renewal. An institution that clings to its history, prestige or procedures becomes machinery. An institution that can 'die to the known' becomes living: flexible enough to dissolve when it obstructs, emergent enough to reappear when awareness demands it. This dying to the known is the essence of meditation: the emptying of consciousness of its content so that awareness is free to act without distortion (Krishnamurti, 1996). Without this ending, even noble institutions fossilise; with it, action is reborn in coherence each moment.

Short-term coherence requires small but decisive shifts. Parliaments can create spaces for silence and reflection before deliberation. Hospitals can embed practices of dignity alongside treatment. Schools can reduce rankings and open space for inquiry and stillness. These are not symbolic gestures but reminders that presence is the ground of care.

Long-term coherence cannot be planned as a framework. Reverence is not installed like a policy. It appears wherever division ends. When citizens awaken to the sacredness of life, institutions that reflect them inevitably change. Governance does not need to advertise virtue; it embodies it. In that movement, systems straighten themselves, not by reform alone but by the emergence of awareness in public life.

The sacred is not an added dimension of governance; it is its essence. Without reverence for life, institutions may endure but will be hollow, producing order without meaning. With reverence, governance becomes the practice of honouring life. To live this sacredness is to live in meditation: dying to the 'me', ending the past at every moment, so that relationship, perception and action arise from wholeness. In this sense, the sacred is not outside politics but its deepest possibility: a public life that reflects the wholeness of existence.

Dying to the known, then, is not only the essence of meditation but the essence of living institutions. When individuals let go of the 'me', perception is freed to act without distortion. When institutions let go of their past forms, prestige, and hardened identities, they too become capable of reflecting wholeness rather than fragmentation. This is the thread that runs through Chapter 12: institutions are mirrors, but only when they die to continuity can they reflect the living movement of awareness.

12.11 Conclusion: Institutions, Awareness and the Movement of Wholeness

Institutions are not the source of coherence. They are mirrors of consciousness, crystallised expressions of human relationship. When fear, ambition and identity dominate, institutions reproduce division, reinforcing disorder with dashboards, treaties and reforms that mask continuity. When awareness is present, institutions can become spaces of coherence, holding conditions for perception to act, briefly, provisionally, but with transformative effect.

Across sovereignty, economy, education, law, technology and ecology, the same pattern recurs. Fragmentation produces systems that manage continuity rather than enable transformation. Yet glimpses of coherence emerge wherever perception precedes identity. A river recognised as living kin, a school that prizes inquiry over ranking, a community that responds to disaster without waiting for procedure, these moments reveal that institutions can host wholeness, even if only temporarily.

The distinction between immediacy and time clarifies the path. Awareness acts now; time carries its expression through logistics and systems. When psychological time, the projection of becoming, infiltrates governance, it turns transformation into distant targets. But when institutions align their sequences with the timeless perception of goodness, the means and the ends converge. Feeding the hungry is immediate. Building structures to sustain that care may take years. What matters is whether both arise from awareness, not from continuity.

The horizon of coherence is emergent. It cannot be designed once and for all. Sovereignty may persist, but loosened of absolutism, it becomes coordination for life rather than a fortress of identity. Economies may continue, but redefined as care rather than growth, they serve sufficiency over accumulation. Education, law and technology remain, but only as long as they mirror wholeness and dissolve when they obstruct it. Institutions in this sense are living: adapting, straightening, dissolving and re-emerging as consciousness shifts.

The sacred is not an addition to governance; it is its essence. Without reverence for life, institutions decay into machinery. With reverence, they become practices of honouring life. This is not about imposing doctrine but about seeing the world as intrinsically worthy. Governance that reflects this perception does not need to proclaim virtue; it embodies it in action.

To live this way is to die to the known. Just as meditation is the ending of the 'me' and the past at every moment, institutions too must release their attachments to procedure, prestige and permanence. Only in this ending can they mirror the living movement of awareness. This is the thread of Chapter 12: the recognition that coherence cannot be manufactured by frameworks but arises when both individuals and institutions allow the old to end so that the whole may act.

Final Reflection: The Whole Beyond the Mirrors

With this chapter, the inquiry into institutions comes full circle. We have seen that they are not engines of coherence but mirrors of consciousness, spaces that can either reflect division or hold moments of wholeness. To demand permanence from them is to miss their nature. To reject them entirely is to deny the need for sequences and structures in time. Their true place is humbler: to serve life without obstructing it, to dissolve when hardened and to re-emerge when awareness calls for new forms.

This recognition returns us to the core insight of the book: that transformation is not born of frameworks or reforms but of perception. When awareness acts, institutions straighten themselves. When wholeness is seen, systems adapt. The fragments of sovereignty, economy, education, law, technology and ecology are not separate battlegrounds but reflections of the same consciousness. To change one is to touch the whole.

To live this way is to die to the known. Institutions, like the individuals who create them, must allow yesterday's habits, rituals and securities to end. When the 'me' dissolves, so too do the structures that mirror its continuity. The fragments loosen, and new forms can emerge. Without this ending, governance becomes repetition dressed as reform. With it, governance becomes life: fresh, unburdened and aligned with the whole.

What follows, then, is not a roadmap but a horizon. Governance beyond the self is not a final model but a living movement: immediate in its seeing, emergent in its expression, timeless in its essence. The next chapter will deepen this arc, turning from the outer mirrors of institutions to the ground of action itself, where intelligence moves without division.

References

Al Jazeera. (2022, November 17). *Starlink becomes lifeline for Ukraine in war with Russia.* https://www.aljazeera.com/news/2022/11/17/starlink-becomes-lifeline-for-ukraine-in-war-with-russia

Anderson, B. (1983). *Imagined communities: Reflections on the origin and spread of nationalism.* Verso.

Arendt, H. (1958). *The human condition.* University of Chicago Press.

BBC News. (2018, August 21). Why the Kerala floods proved so deadly. https://www.bbc.com/news/world-asia-india-45243868

Beck, U. (2006). *The cosmopolitan vision.* Polity.

Berger, P. L., & Luckmann, T. (1966). *The social construction of reality: A treatise in the sociology of knowledge.* Anchor Books.

Bohm, D. (1980). *Wholeness and the implicate order.* Routledge.

Bohm, D. (1992). *Thought as a system.* Routledge.

Capra, F. (1996). *The web of life: A new scientific understanding of living systems.* Anchor Books.

Carbon Market Watch. (2021). *Carbon offsetting: Out of control, out of time.* Carbon Market Watch https://carbonmarketwatch.org/publications/carbon-offsetting-out-of-control-out-of-time/

Castoriadis, C. (1997). *The imaginary institution of society.* MIT Press.

Eliade, M. (1996). *The sacred and the profane* W. R. Trask, Trans. Harcourt. (Original work published 1957).

Escobar, A. (2018). *Designs for the pluriverse: Radical interdependence, autonomy, and the making of worlds*. Duke University Press.

European Commission. (2020). *The European Green Deal*. https://commission.europa.eu/strategy-and-policy/priorities-2019-2024/european-green-deal_en

Foucault, M. (1977). *Discipline and punish* A. Sheridan, Trans. Pantheon.

Freire, P. (1970). *Pedagogy of the oppressed*. Continuum.

Girard, R. (1977). *Violence and the sacred* P. Gregory, Trans. Johns Hopkins University Press.

Government of Finland, Ministry of Social Affairs and Health. (2020). *Results of Finland's basic income experiment 2017–2018*. https://julkaisut.valtioneuvosto.fi/handle/10024/162219

Government of Kenya. (2021). *Universal Basic Income Pilot: Interim evaluation report*. (public summary link varies; if you want, I can attach the PDF you're using.).

Government of Rwanda. (2019). *Gacaca courts: Contributions, limitations, and lessons learned*. Overview. https://www.gov.rw/

Heidegger, M. (1977). *The question concerning technology and other essays* W. Lovitt, Trans. Harper & Row.

hooks, b. (1994). *Teaching to transgress: Education as the practice of freedom*. Routledge.

Human Rights Watch. (2019). *China's algorithms of repression*.... https://www.hrw.org/report/2019/05/01/chinas-algorithms-repression/reverse-engineering-xinjiang-police-mass-surveillance

Illich, I. (1973). *Deschooling society* (p. 46). Harmondsworth, Middlesex.

Intergovernmental Panel on Climate Change. (2023). *AR6 Synthesis Report: Climate change 2023*. https://www.ipcc.ch/report/sixth-assessment-report-cycle/

International Committee of the Red Cross. (2023). *International humanitarian law and the challenges of contemporary armed conflicts*. https://shop.icrc.org/international-humanitarian-law-and-the-challenges-of-contemporary-armed-conflicts-2023.html

International Energy Agency. (2023). *World Energy Outlook 2023*. https://www.iea.org/reports/world-energy-outlook-2023

International Monetary Fund. (2023). *World economic outlook*. https://www.imf.org/en/Publications/WEO

Keohane, R. O., & Victor, D. G. (2011). The regime complex for climate change. *Perspectives on Politics*, *9*(1), 7–23. https://doi.org/10.1017/S1537592710004068

Krishnamurti, J. (1954). *The first and last freedom*. Victor Gollancz.

Krishnamurti, J. (1969/2009). *Freedom from the known*. Harper.

Krishnamurti, J. (1981). *The network of thought*. Harper & Row.

Krishnamurti, J. (1992). *The wholeness of life*. HarperOne.

Krishnamurti, J. (1996). *Total freedom: The essential Krishnamurti*. HarperSanFrancisco.

Krishnamurti, J., & Anderson, A. W. (1991). *A wholly different way of living*. Krishnamurti Foundation Trust Ltd.

Krishnamurti, J., & Bohm, D. (1992). *A wholly different way of living*. HarperSanFrancisco.

Lanier, J. (2018). *Ten arguments for deleting your social media accounts right now*. Henry Holt.

Latour, B. (2017). *Facing Gaia: Eight lectures on the new climatic regime*. Polity.

Manley, J., Bailey, D., & Foster, E. (2021). Community wealth building and the inclusive economy.... *Local Economy*, *36*(6), 493–512.
Mbembe, A. (2019). *Necropolitics* S. Corcoran, Trans. Duke University Press.
Minow, M. (1998). *Between vengeance and forgiveness: Facing history after genocide and mass violence*. Beacon Press.
Mumford, L. (1934). *Technics and civilization*. Harcourt, Brace & Company.
Næss, A. (1989). *Ecology, community and lifestyle* D. Rothenberg, Trans. Cambridge University Press.
National Electoral Council of Ecuador. (2023). *Referendum results: Yasuni National Park oil drilling*. https://resultados2023.cne.gob.ec/
North, D. C. (1990). *Institutions, institutional change and economic performance*. Cambridge University Press.
Nussbaum, M. C. (2010). *Not for profit: Why democracy needs the humanities*. Princeton University Press.
Ostrom, E. (1990). *Governing the commons*. Cambridge University Press.
Raworth, K. (2017). *Doughnut economics*. Chelsea Green.
Reuters. (2022, October 7). *EU fossil fuel subsidies remain high despite climate goals*. https://www.reuters.com/world/europe/eu-fossil-fuel-subsidies-remain-high-despite-climate-goals-2022-10-07/
Shiva, V. (2016). *Who really feeds the world?*. North Atlantic Books.
Stiglitz, J. E. (2002). *Globalization and its discontents*. W. W. Norton.
Taylor, C. (2007). *A secular age*. Harvard University Press.
The Guardian. (2023, August 21). Ecuador votes to halt oil drilling in Amazon's Yasuni National Park. *The Gaurdian*. https://www.theguardian.com/world/2023/aug/21/ecuador-votes-to-halt-oil-drilling-in-amazonian-biodiversity-hotspot
The New York Times. (2023, March 23). *Generative A.I. is reshaping education and work*. https://www.nytimes.com/2023/03/23/technology/ai-chatgpt-work-education.html
United Nations Human Rights Council/OHCHR. (2022). *Assessment of human rights concerns in the Xinjiang*. https://www.ohchr.org/en/documents/country-reports/ohchr-assessment-human-rights-concerns-xinjiang-uyghur-autonomous-region
United Nations. (2015). *Transforming our world: The 2030 Agenda for sustainable development*. https://sdgs.un.org/2030agenda
Weber, M. (2004). *The vocation lectures* D. Owen & T. B. Strong, Eds.; R. Livingstone, Trans. Hackett.
World Health Organization. (2022). *Global report on COVID-19 vaccine equity*. https://iris.who.int/server/api/core/bitstreams/1029b524-cc7a-4d5d-93ee-264a1b5d3604/content
Zuboff, S. (2019). *The age of surveillance capitalism*. PublicAffairs.

Chapter 13

Grand Conclusions and Summary: Awareness Is the Change

Abstract

This chapter gathers the movement of the entire book, from outer crisis to inner perception and from perception to action grounded in wholeness. It affirms that awareness itself is transformation; no external reform can substitute for seeing. This chapter revisits key relationships, self and other, human and nature, teacher and student, citizen and state, showing how they renew when freed from image. It restates the role of institutions as provisional mirrors of consciousness, not structures of control. Sovereignty, economy, education, law and ecology are reinterpreted as expressions of awareness, revealing that sustainability begins in the mind that sees. The closing reflections emphasise that awareness is love, and love is order. Transformation is not achieved through effort or belief but revealed when thought ends and perception is whole. The book concludes in stillness, where awareness, intelligence and life are one.

Keywords: Awareness as transformation; wholeness in action; relational renewal; institutions as mirrors; sovereignty revisited; economy as compassion; education as insight; law as awareness; the sacred in governance

13.1 The Whole Thread So Far

This book has followed a single thread through a mosaic of interlocking textures and patches. We began with diagnosis: how fragmentation in the mind becomes fragmentation in the world. We examined the self as a movement of memory and image, how psychological time projects becoming into the future, and how knowledge, when treated as the ground, turns cleverness into confusion. We saw that the same structure repeats at every scale. The divided mind manufactures divided systems. War, ecological collapse and inequality are not separate crises but the outward shape of inward disorder.

From Fragmentation to Wholeness, 287–300

doi:10.1108/978-1-80686-677-920261013

From that ground we questioned a cherished assumption: that more choice, more tools, more frameworks will deliver transformation. In Chapter 9, we discovered that choice often hides conditioning. Decisions arise from memory, identity, fear and desire, then claim the name of freedom. What breaks the loop is not better choosing but choiceless awareness: direct perception without the chooser. In such seeing, action is coherent because it is not filtered through image.

Chapter 10 carried this into public life. We contrasted leadership and governance performed as identity with leadership that arises from perception. Identity-driven leadership stabilises division while speaking the language of responsibility. Frameworks and reform loops manage symptoms while deferring endings into timelines and targets. By contrast, conscious governance acts from relationship with the whole. It does not discard institutions or knowledge; it places them in their right place beneath intelligence. Where perception is primary, action is immediate, clear and unforced.

Chapter 11 asked what this means for policy, education and systems. We showed how policy dashboards and grand agreements become the architecture of delay when they replace seeing with prescription. We contrasted education as conditioning with education as awakening, where knowledge serves but does not rule, and the learner is invited to inquire rather than conform. We examined systems thinking and complexity: useful for connecting fragments, yet still limited if they cannot dissolve the self that created the fragments. We then gathered practical principles for acting from coherence: perception before prescription, dialogue as foundation, proximity to life, accountability to wholeness, simplicity and clarity. These were not methods to be installed but orientations that keep perception alive inside institutions.

Chapter 12 turned the lens on institutions themselves. It offered a humbler view: institutions cannot manufacture wholeness; they can only mirror it or obstruct it. We traced how sovereignty, economy, education, law, technology and ecology are crystallised relationships, reflections of the psyche given form. Where awareness is absent, procedures ossify and dashboards glow while disorder deepens. Where awareness is present, institutions can briefly host coherence: a citizens' assembly that truly listens, a river recognised as kin, a classroom that prizes presence over performance, a community that responds to disaster before instructions arrive. We drew a crucial distinction that runs through the whole book: awareness acts now; time only carries its expression. Physical action needs sequences, logistics and coordination. Psychological time is the projection that postpones endings and protects identity. The task is to let the immediacy of seeing animate the sequences of action without being swallowed by them.

Across these chapters, certain recognitions have settled. Knowledge is indispensable, but it is not the ground. Frameworks can coordinate, but they cannot end disorder. Methods can be helpful, but the moment awareness is turned into a method, it collapses into performance. Identity can organise, but it inevitably divides. The only ground that does not fragment is awareness itself: the state in which the mind is still enough to see without distortion. In that seeing, love and intelligence are not added qualities. They are the very movement of perception.

From there, action is choiceless because there is nothing to choose between. The false dilemmas dissolve. Economy and ecology are one fact. Security and compassion are not opposites. Responsibility is not a burden carried by a role but the responsiveness of perception to what is.

This is the arc the book has traced: from seeing fragmentation to understanding the machinery of the self, to discovering choiceless awareness, to reimagining leadership and governance as perception in action, to placing institutions in their rightful place as mirrors and spaces rather than engines of salvation. The movement has been outward and inward at once because the division between inner and outer is itself part of the problem.

Chapter 13 now makes the simplest and most demanding statement of all. Awareness is the change. Not a precursor to action, not a competency to cultivate, not a private refuge from public life. It is the ending of disorder now, which then flows as coherent action through whatever forms are needed. In the pages that follow, we will not offer a new framework or a set of practices. We will clarify what it means for awareness to be both the ground and the act: how it relates to psychological time, how it transforms relationship, how it appears in public life without becoming ideology, and how it carries the sacred as everyday reverence. If earlier chapters showed why reform alone cannot deliver transformation, this chapter shows how transformation happens when the mind sees clearly.

Everything needed has already been encountered. What remains is to let the whole thread come together: to see directly that the crises of the world are mirrors and that the ending of division is not in some future arrangement but here, in the quality of seeing from which we live.

13.2 Awareness and the Ending of Time

To understand why awareness is change, we must distinguish between two orders of time. The first is chronological time, the neutral unfolding of physical processes: a seed grows into a tree, a river carries sediment to the sea and food is transported from one country to another. Human institutions also operate in this order: policies take months to draft, infrastructure takes years to build and learning unfolds through developmental stages. Chronological time is unavoidable; it is the rhythm of matter and nature.

The second is psychological time, the movement of thought projecting continuity into the future. It is the idea of 'becoming', of being something now and striving to become something else later. Psychological time sustains the self, which survives by imagining itself in progress: becoming successful, enlightened, secure or transformed. Yet this movement never ends because the self is continuity itself.

Reform belongs to psychological time. It sets distant goals like carbon neutrality by 2050, poverty eradication by 2030, and projects change into the future. While some coordination requires chronology, these projections often become alibis for postponement. Instead of transformation now, there is continuity disguised as progress. McGilchrist (2021) critiques this tendency of the left hemisphere to impose abstraction and control, noting how it blinds us to the living present.

Awareness is of a different order. It is choiceless seeing in the present moment. It does not project into becoming, nor is it cultivated as a practice or competency. Damasio (2010) distinguishes between memory-based cognition and the felt immediacy of awareness, showing that consciousness cannot be reduced to accumulated knowledge. Awareness is direct perception, free from the burden of past and the projection of future.

When awareness is present, psychological time ends. The illusion of becoming dissolves, and with it the continuity of the self. Bohm (1980) described such insight as a 'flash' in which disorder ends, not gradually, but in a movement outside time. This is not mystical abstraction but a practical recognition: violence, for instance, can end not after decades of negotiation but in the immediacy of seeing the other as oneself. Hunger can be met not as a statistic requiring future aid but as a present fact demanding response.

Thus, awareness is the ending of time in the psychological sense. It allows chronological time to play its necessary role of building, transporting, organising, etc., while freeing action from the distortions of projection and becoming. In that ending is true change.

13.3 Awareness and Relationship

Human life is relationship. Every thought, action and institution is woven through the web of how we relate: to ourselves, to one another, to nature, to knowledge and to the state. When relationship is mediated by images of who I am, who you are and what society should be, it fragments. Identity, comparison and projection distort perception, turning relationship into negotiation or conflict. Awareness, by contrast, dissolves the images that stand between human beings. It restores communion.

Self–Other

At the personal level, relationship is often governed by the 'self', a bundle of memory, fear and desire. We approach others not as they are but as reflections of our needs: as allies, threats, partners and competitors. This is why even intimate relationships collapse into conflict; we do not meet the person but the image we have of them. Awareness ends this cycle by revealing the image as an image. To see another directly, without projection, is to encounter them anew. As Levinas (1969) emphasised, the face of the other calls us to responsibility beyond categories.

Human–Nature

The ecological crisis is perhaps the starkest demonstration of image-based relationship. Nature is perceived as 'resource', a storehouse of timber, carbon or minerals. Policy debates reduce forests to carbon sinks, rivers to economic assets and species to biodiversity indexes. This utilitarian image allows extraction and destruction even when clothed in the language of sustainability. Awareness dissolves this division. When nature is perceived as inseparable from human survival, as life itself, the relationship shifts from management to care. Indigenous stewardship traditions around the world embody this awareness, living from a perception of interdependence (Berkes, 2018).

Teacher–Student

Education also mirrors the crisis of relationship. Standardised curricula turn the teacher into a transmitter and the student into a receptacle (Freire, 1970, 2000). Relationship is lost in the transaction of information. Awareness restores relationship by dissolving the image of the student as a 'score' or 'category'. When a teacher perceives the child directly, as a human being, not an abstraction, learning becomes communion. Finland's educational model, where teachers are trusted to respond to students without excessive standardisation, shows how attentiveness to the learner transforms outcomes (Sahlberg, 2015).

Citizen–State

Politics, too, is a question of relationship. Citizens are categorised into voters, taxpayers or members of identity blocs. States relate to citizens through policies designed for statistical groups rather than living beings. The result is alienation, mistrust and polarisation. Awareness invites a different relationship: the citizen not as a client of the state but as inseparable from it, the state not as power over but as responsibility with. Citizens' assemblies, such as Ireland's deliberations on abortion, demonstrated how ordinary people, when seen beyond identity, can deliberate responsibly and influence constitutional change (Farrell et al., 2019).

Case Illustrations

Moments of awareness in relationship have reshaped history. South Africa's Truth and Reconciliation Commission, when alive, allowed victims and perpetrators to meet beyond categories of guilt and innocence. For brief moments, enmity dissolved, and reconciliation became possible (Tutu, 1999). Grassroots peace movements, such as the Liberian women's movement that helped end civil war through persistent presence and refusal to be enemies, reveal the power of relationship rooted in awareness (Gbowee, 2011). These are not methods to replicate but glimpses of what arises when images are set aside.

Implications

Awareness transforms relationship because it dissolves separation. In personal life, it ends the cycle of projection and conflict. In ecology, it restores care by perceiving inseparability. In education, it allows dialogue rather than conditioning. In governance, it reframes responsibility beyond identity. The Sustainable Development Goals (SDGs) speak of 'partnerships for the goals' (SDG 17), but partnership without awareness becomes a negotiation of interests. With awareness, relationship is communion, not transaction.

In this sense, relationship itself is the test of awareness. If we see the other as other, awareness is absent. If we see directly, without image, awareness is alive. And when awareness is alive, relationship is wholeness in action.

13.4 Awareness in Action

Awareness is often mistaken for withdrawal, a private, inward state detached from the urgencies of the world. But true awareness is not escape; it is action. When awareness is present, it informs how food is distributed, how laws are interpreted, how classrooms are shaped and how forests are protected. Action without awareness reproduces fragmentation, no matter how well-intentioned. Awareness in action dissolves the roots of fragmentation and allows coherence to express itself in public life.

Beyond Thought as Controller

Most action today is driven by thought: calculation, strategy and policy design. Thought has its place, to plan logistics, construct a bridge or draft a treaty. But when thought assumes primacy, it projects images of the future and pursues them mechanically, blind to contradictions in the present. Climate policy illustrates this clearly: nations commit to 'net zero by 2050' while approving new oil and gas projects today (IPCC, 2023). This is thought in charge, promising coherence in the future while perpetuating incoherence in the present. Awareness, by contrast, begins with perception now. Action rooted in psychological time is continuity; action rooted in awareness is immediate.

Awareness as Direct Perception

Awareness in action is perception without distortion. When a hungry child is seen as one's own, the impulse to act is immediate; bureaucracy or ideology does not intervene. Neuroscience has shown that empathic perception activates brain regions linked to care and social bonding, bypassing purely cognitive evaluation (Decety & Cowell, 2015). Yet awareness is deeper than empathy; it is the dissolution of the separation between 'me' and 'the other'. Action that flows from this perception does not calculate cost–benefit; it expresses necessity.

Short-Term and Long-Term Dimensions

Awareness in action does not deny sequences. Feeding the hungry requires cooking and distribution; restoring a forest requires planting and decades of care. But the animating quality matters. If the short term is guided by fear, propaganda or identity, the long term inherits the same disorder. If the short term arises from awareness, the long term expresses coherence. Awareness ensures that every beginning is fresh, not a repetition of past patterns.

Implications for the SDGs

The SDGs are often framed as targets to be achieved by 2030. Yet progress reports reveal a gap between vision and action, with most goals 'off track' or regressing (United Nations, 2023). The reason is not only lack of resources or political will but also the absence of awareness in action. Goals become projections, while present action continues to contradict them. Awareness reorients the SDGs from a checklist of targets to a horizon of coherence: act rightly now, and structures will follow.

Towards Choiceless Action

Awareness in action is choiceless. It is not a matter of weighing alternatives but of seeing what must be done. When perception is clear, the question of 'should I act?' dissolves. As Bohm (1980) observed, when thought fragments reality, action fragments too; when perception is whole, action is whole. Such action may be small, listening without judgement, refusing corruption or protecting a river, but its quality reshapes the field.

13.5 Awareness and Institutions

If institutions are mirrors of consciousness, then the quality of public life depends not on their design alone but on the awareness that animates them. Constitutions, treaties and global frameworks can reflect either division or wholeness. Reform, by itself, cannot guarantee transformation; it merely rearranges continuity. Awareness, when alive, reconfigures institutions not by blueprint but by dissolving the disorder they mirror.

From Frameworks to Mirrors

Modern governance is dominated by frameworks: the SDGs, climate agreements and human rights charters. These structures provide direction but often collapse into dashboards and reports, with action lagging far behind commitments (United Nations, 2023). Institutions mistake measurement for meaning, assuming that reform in structure guarantees transformation in life. Awareness shows the opposite: without perception, frameworks become hollow. Krishnamurti

(1981) insisted that no framework or system can bring about radical change; only direct seeing can. Institutions, therefore, cannot be the origin of coherence; they can only mirror it.

Sovereignty as Provisional Coordination

Awareness dissolves the absolutism of sovereignty. States remain necessary to organise logistics and protect communities, but when sovereignty becomes identity, it produces rivalry, exclusion and paralysis. Ulrich Beck (2006) argued for cosmopolitan realism: only by loosening the grip of the nation-state can humanity respond to global risks like climate change and pandemics. Awareness aligns with sovereignty as provisional coordination for life, not as fortress identity. Leaders who perceive this shift act differently: they negotiate not for national advantage alone but for the protection of the whole.

Economy as Care of Life

As Chapter 12 showed, economies become incoherent when growth is pursued as an end in itself, externalising harm and measuring success in GDP. Kate Raworth's (2017) doughnut model and Elinor Ostrom's (1990) commons governance offer glimpses of reorientation, but awareness pushes further: economy as *oikonomia*, the care of the household of life. From awareness, production serves sufficiency, wealth serves dignity, and finance becomes a medium of relationship, not accumulation. Policy designed from this perception, such as redirecting fossil fuel subsidies towards ecological restoration, is not just efficient but coherent.

Education as Awakening

Education institutionalises either conditioning or inquiry. Paulo Freire (1970) critiqued the 'banking model' of schooling as depositing information without freedom. Krishnamurti (1974) went deeper, urging that true education must free the mind from fear and conditioning. Awareness in education means classrooms as spaces of inquiry, not pipelines to labour markets. When awareness is present, exams and rankings lose their dominance, and learning becomes the living exploration of relationship, ethics and meaning. Citizens educated in this way resist manipulation, weakening the grip of identity-based politics.

Law as Restoration, Not Retribution

Law mirrors the consciousness of fear and control when it reduces justice to punishment. Michel Foucault (2008) exposed how legal systems often reproduce discipline rather than protect freedom. Yet when awareness enters law, justice becomes restorative. South Africa's Truth and Reconciliation Commission, fragile though it was, revealed how acknowledgement and forgiveness can emerge when roles of victim and perpetrator dissolve (Minow, 1998). Awareness reframes law from external enforcement to the humble protection of spaces where dignity can be restored.

Technology as Servant, Not Master

Technological innovation amplifies the psyche that wields it. Without awareness, it deepens exploitation through surveillance capitalism (Zuboff, 2019), algorithmic bias (Lanier, 2018) and ecological domination (Heidegger, 1977). With awareness, technology becomes an extension of care: renewable energy cooperatives, open-source knowledge and artificial intelligence (AI) designed for ecological monitoring. True science, Bohm (1980) reminded us, is inquiry without preconception. Awareness restores science to humility and technology to service, ensuring they extend wholeness rather than fragment it.

Ecology as Sacred Ground

Above all, awareness reconfigures the human relationship with ecology. Awareness dissolves the illusion of separateness, revealing that forests, rivers and species are not 'resources' but inseparable life. Legal personhood for rivers in New Zealand reflects this perception (Charpleix, 2018). Ecuador's recognition of *Pachamama* in its constitution attempts to embody the same (Government of Ecuador, 2008, Art. 71). Awareness places ecology not as a stakeholder in negotiation but as the living ground of all action, the table itself, not just another seat.

Implications for the SDGs and Global Governance

The SDGs aspire to coherence but often reproduce fragmentation: growth and climate action side by side, equity promised while inequality widens. Without awareness, the SDGs remain dashboards; with awareness, they can become provisional tools aligned with life. This means governments and institutions must act from perception first, metrics second. As Damasio (2010) noted, attention and feeling guide action before calculation. Awareness grounds the SDGs not in targets but in the immediacy of seeing hunger, inequality or ecological collapse as intolerable now.

In sum, awareness does not abolish institutions but reconfigures them. Sovereignty becomes coordination, economy becomes care, education becomes awakening, law becomes restoration, technology becomes servant and ecology becomes sacred ground. This is not a reformist agenda but a perceptual shift. Institutions are not perfected but placed rightly as humble mirrors of awareness, provisional expressions of coherence, dissolving when they obstruct life.

13.6 Awareness, Love and the Sacred

If awareness is the ground of transformation, then its essence is not abstract neutrality but love. Awareness is inseparable from love, intelligence and sacredness. To see without distortion is to care without condition. This movement cannot be reduced to sentiment, doctrine or belief. It is the living quality of attention in which division ends.

Awareness as Love Without Image

In Chapter 8, we saw that love is distorted when mediated by images of self and other. Awareness dissolves these images, allowing relationship without projection. Love without image is not selective attachment but the perception of inseparability. Krishnamurti (1975) insisted that love cannot be cultivated; it arises only when fear, desire and memory no longer dominate perception. Similarly, Martin Buber (1970) described the 'I–Thou' encounter as meeting the other directly, not as object but as presence. Awareness is this unmediated seeing, and love is its natural expression.

Intelligence as Responsiveness to What Is

Awareness is also inseparable from intelligence. Not the intelligence of accumulation or cleverness, but the responsiveness of perception to reality. Bohm (1980) argued that insight is intelligence itself, a movement of wholeness beyond fragmented thought. Damasio (1999) demonstrated that feeling and attention guide human reasoning, suggesting that perception is prior to calculation. Intelligence in this sense is not stored knowledge but the immediate alignment of action with life. Awareness is the ground from which this intelligence arises.

The Sacred as Wholeness Beyond the Self

Awareness reveals not only love and intelligence but also the sacred. The sacred here is not confined to religion, ritual or belief. It is the quality of perceiving life as intrinsically worthy, beyond measure and utility. Eliade (1959) described the sacred as that which breaks through the profane order of calculation. Levinas (1969) spoke of infinite responsibility to the other as an encounter with the sacred. Awareness uncovers this quality whenever division ends. To meet another human being without image is already to touch the sacred. To see a forest as life itself rather than resource is to live the sacred.

The Sacred in Public Life

When awareness infuses public life, institutions shift. Politics becomes service, not power. Economics becomes care, not growth. Education becomes inquiry, not conditioning. Law becomes restoration, not punishment. Ecology becomes guardianship, not extraction. Dag Hammarskjöld, the former UN Secretary-General, embodied this orientation when he described service as 'the road to holiness' in daily work (Hammarskjöld, 1964/1993). Indigenous traditions also reflect this: the Māori recognition of rivers as ancestors, or Andean cosmologies of Pachamama, show that governance can embody reverence. The sacred is not outside public life but its deepest possibility.

Beyond Identity and Belief

Awareness as sacred is not bound to identity. Religious affiliation can provide symbolic grounding, but awareness is prior to any doctrine. As Taylor (2007) argued, modernity has flattened moral horizons into secular frameworks, yet the longing for fullness persists. Awareness responds to this longing not with new dogma but with seeing. The sacred is revealed whenever division ends, whether in silence, in relationship or in collective action that arises without self-interest.

In sum, awareness, love and the sacred are not separate. To live in awareness is to live in love (Gupta, 2024). To live in love is to live intelligently, responding to what is. To live intelligently is to act with reverence for life. The sacred, then, is not an external addition to governance or sustainability but the essence of awareness itself. It is this inseparability of awareness as love, love as intelligence and intelligence as sacredness that reveals the possibility of coherence in a fragmented world.

13.7 Conclusion: The Whole Beyond Frameworks

The journey of this book has traced a movement through fragmentation and into wholeness. Part I diagnosed the disorder of the world as the mirror of disorder within the self: systems fractured, policies contradictory and institutions hollow because consciousness itself was divided. Part II turned inward, exploring the nature of self, thought, time and knowledge, revealing that continuity and becoming are the roots of conflict. Part III turned outward again, examining reform, governance, education, law, technology, ecology and institutions as reflections of the same fragmentation, while opening to glimpses of coherence wherever awareness was alive.

The central discovery across this arc is simple yet radical: awareness itself is the change. Not reform, not new models, not frameworks of improvement, but direct perception beyond the self. Where there is awareness, division ends. Where division ends, love and intelligence act. And where love and intelligence act, coherence manifests, whether in individual life, community or institutions.

Beyond Time and Frameworks

Frameworks, whether in governance or spirituality, promise transformation by organising steps in time. The SDGs (United Nations, 2015), the Inner Development Goals (IDG Initiative, 2021) and countless national roadmaps exemplify this: change projected into 2030 or 2050, mapped through indicators and targets. Yet as history shows, emissions rise, wars continue, and inequality deepens despite these frameworks (IMF, 2023; IPCC, 2023). Time becomes an alibi for continuity. Awareness, by contrast, is not in time. The ending of division, fear or conflict happens only in the present moment of seeing (Krishnamurti, 1975).

The Role of Institutions and Systems

Institutions are necessary in the physical world. Food must be delivered, laws drafted, schools built and bridges maintained. These require sequences, processes and coordination. Yet institutions cannot generate coherence; they can only mirror it (Berger & Luckmann, 2016; Castoriadis, 1987). When awareness is absent, institutions become machinery of continuity. When awareness is present, they can briefly hold coherence, enabling life to flow. Their rightful role is provisional, to serve without obstructing.

Awareness as the Ground of Sustainability

Sustainability is not achieved through policy frameworks alone. Dashboards may glow green while forests fall. Conferences may proclaim progress while inequality widens. True sustainability emerges only when life is seen as inseparable from oneself. Awareness dissolves the illusion of separateness, making exploitation impossible. As Capra (1996) argued, living systems thrive through relationship, not competition. Awareness is the perception of this relationship in real time. Sustainability, then, is not a strategy but the living expression of awareness.

The Sacred Dimension of Awareness

Awareness is not merely practical; it is sacred. To see life as intrinsically worthy is to live in reverence. The sacred is not belief or ritual but the quality of treating every being as inseparable from oneself (Eliade, 1959; Levinas, 1969). Public life without reverence collapses into machinery, but when reverence is alive, governance becomes service, economy becomes care, education becomes awakening, and ecology becomes guardianship. Awareness is the doorway to this sacredness.

Closing the Arc

The crises of the world with war, climate breakdown, inequality and technological exploitation are mirrors of the crisis within. Reform and frameworks, however well-intentioned, remain movements of continuity when rooted in the self. The only real change is the ending of the self, the ending of becoming and the ending of psychological time. This ending is awareness itself.

In this light, governance without the self is not a model to be designed but a way of living. Institutions straighten when awareness is alive. Communities cohere when perception replaces identity. Global cooperation becomes possible when separateness is no longer believed. The whole is not reached through fragments; it is seen directly, and in that seeing, action flows.

The book's movement closes here, yet it does not conclude. Awareness is not an endpoint but the ground of every moment. To live in awareness is to live in change, in intelligence and in love. It is to live in the sacred inseparability of life. That is the only foundation for coherence, the only ground for true sustainability, and the only horizon beyond fragmentation.

References

Beck, U. (2006). *Cosmopolitan vision*. Polity.

Berger, P., & Luckmann, T. (2016). The social construction of reality. In *Social theory re-wired* (pp. 110–122). Routledge.

Berkes, F. (2018). *Sacred ecology* (4th ed.). Routledge.

Bohm, D. (1980). *Wholeness and the implicate order*. Routledge.

Buber, M. (1970). *I and thou* (W. Kaufmann, Trans., Vol. 57). Charles Scribner's Sons.

Capra, F. (1996). *The web of life: A new scientific understanding of living systems*. Anchor Books.

Castoriadis, C. (1987). *The imaginary institution of society*. MIT Press.

Charpleix, L. (2018). The Whanganui River as *Te Awa Tupua*: Place-based law in a legally pluralistic society. *The Geographical Journal*, *184*(1), 19–30. https://doi.org/10.1111/geoj.12238

Damasio, A. (2010). *Self comes to mind: Constructing the conscious brain*. Pantheon Books.

Damasio, A. (1999). *The feeling of what happens: Body and emotion in the making of consciousness*. Harcourt Brace.

Decety, J., & Cowell, J. M. (2015). Empathy, justice, and moral behavior. *AJOB Neuroscience*, *6*(3), 3–14. https://doi.org/10.1080/21507740.2015.1047055

Eliade, M. (1959). *The sacred and the profane: The nature of religion* (Vol. 81). Houghton Mifflin Harcourt.

Farrell, D. M., Suiter, J., Harris, C., & Cunningham, K. (2019). The effects of mixed membership in a deliberative forum: The Irish Constitutional Convention on marriage equality. *Political Studies*, *67*(1), 184–206. https://doi.org/10.1177/0032321719830936

Foucault, M. (2008). "Panopticism" from discipline & punish: The birth of the prison. *Race/Ethnicity: Multidisciplinary Global Contexts*, *2*(1), 1–12.

Freire, P. (1970). *Pedagogia do oprimido* [*Pedagogy of the oppressed*]. Paz e Terra.

Freire, P. (2000). *Pedagogy of the oppressed* 30th anniversary ed.; M. B. Ramos, Trans. Continuum. (Original work published 1970).

Gbowee, L. (2011). *Mighty be our powers: How sisterhood, prayer, and sex changed a nation at war*. Beast Books.

Government of Ecuador. (2008). *Constitution of the Republic of Ecuador*. https://www.constituteproject.org/constitution/Ecuador_2021.pdf?lang=en

Gupta, D. (2024). *Eternal movement*. Emerge Publications. https://doi.org/10.65169/q7d2z6

Hammarskjöld, D. (1993). *Markings* L. Sjöberg & W. H. Auden, Trans. Vintage. (Original work published 1964).
Heidegger, M. (1977). *The question concerning technology, and other essays* W. Lovitt, Trans. Harper & Row. (Original essay published 1954).
Inner Development Goals. (2021). *Inner development goals*. https://innerdevelopmentgoals.org/
Intergovernmental Panel on Climate Change. (2023). *AR6 Synthesis Report: Climate change 2023*. IPCC.
International Monetary Fund. (2023). *World economic outlook: Navigating global divergences*. IMF.
Krishnamurti, J. (1974). *Education and the significance of life*. Harper & Row.
Krishnamurti, J. (1975). *The first and last freedom*. Harper & Row. (Original work published 1954).
Krishnamurti, J. (1981). *The network of thought*. Harper & Row.
Lanier, J. (2018). *Ten arguments for deleting your social media accounts right now*. Henry Holt.
Levinas, E. (1969). *Totality and infinity: An essay on exteriority* A. Lingis, Trans. Duquesne University Press.
McGilchrist, I. (2021). *The matter with things: Our brains, our delusions, and the unmaking of the world*. Perspectiva Press.
Minow, M. (1998). *Between vengeance and forgiveness: Facing history after genocide and mass violence*. Beacon Press.
Ostrom, E. (1990). *Governing the commons: The evolution of institutions for collective action*. Cambridge University Press.
Raworth, K. (2017). *Doughnut economics: Seven ways to think like a 21st-century economist*. Chelsea Green.
Sahlberg, P. (2015). *Finnish lessons 2.0: What can the world learn from educational change in Finland?*. Teachers College Press.
Taylor, C. (2007). *A secular age*. Harvard University Press.
Tutu, D. (1999). *No future without forgiveness*. Doubleday.
United Nations. (2023). *The sustainable development goals report 2023: Special edition*. United Nations.
United Nations. (2015). *Transforming our world: The 2030 Agenda for sustainable development*. United Nations.
Zuboff, S. (2019). *The age of surveillance capitalism: The fight for a human future at the new frontier of power*. PublicAffairs.

Final Message: The Mirror of the Whole

This book has not been a call to action, reform or integration. It has been a mirror. A mirror in which the self, its fears, desires, images and ideas may be seen clearly. Not to correct it, not to perfect it, but to understand its movement entirely. And when the movement of the self is fully seen, it comes to an end. In that ending, a different quality of action begins.

We have explored how every crisis in the world, whether ecological, social or psychological, is not separate from human consciousness. The nation, the market, the institution, the classroom and the battlefield, the policy are all projections of the same fragmented mind. The frameworks we build to fix them often carry the same division they seek to heal. Sustainability collapses into measurement. Equity becomes competition. Reform becomes repetition. We do not need new models. We need a new mind.

Such a mind does not begin from knowledge. It begins from awareness, a silent observation that holds the whole, without interference or judgement. This awareness is not cultivated through method or time. It emerges when fear and desire are understood, not obeyed. When the observer is no longer separate from the observed, what remains is attention. And from that attention arises intelligence, not stored in books or systems, but living in the present.

Intelligence is not what acts from an idea of justice. It is what sees clearly, and in that seeing, acts without division. Such action has no motive, no reward and no fear. It is not driven by success or failure. It does not wait for consensus or authority. It does not belong to a person or a nation. It is the flowering of love, and love is the only ground on which a sustainable world can exist.

We have tried policies, programmes, principles and philosophies. We have measured outcomes and optimised systems. But the root of disorder remains untouched as long as we seek solutions outside. The transformation begins not in society, but in relationship, in the mirror of daily life, where thought, image and reaction can be observed freshly.

When this seeing becomes whole, when the mind no longer escapes, no longer becomes, no longer clings, then there is silence. And in that silence, the sacred. Not as a belief, but as the stillness in which life and death are one. A wholeness where the individual is not separated from the world, and action flows from the total.

From Fragmentation to Wholeness, 301–305

doi:10.1108/978-1-80686-677-920261014

This book ends with the next sections. But what it points to has no end. It is not a path, because there is no distance to travel. It is not a method, because it begins now, not later. It is not a hope, because it is not born of fear. It is this very moment, when the mind is quiet, and the heart is awake.

And in that moment, the whole is.

This Is Not a Resolution (Because Awareness Does Not Resolve. It Sees.)

Introduction

This document is not a resolution in the conventional sense. It does not propose operative clauses, implementation plans or targets. It does not speak from authority, consensus or demand. It arises, instead, from the final movement of this book, a movement that sees clearly that every crisis in the world is a mirror of human consciousness. And that true change begins not from ideas, but from the ending of the self.

What follows is not framed in the language of the United Nations nor in the machinery of reform. It is written in the quiet clarity of observation. We do not use 'We are resolved' or 'We commit' or 'we urge', for awareness does not operate in time. It does not resolve; it sees. And in seeing, action flows.

This text may end up as just another file on someone's drive. But it lives only when seen. Until this seeing happens, there is no real going forward. There is only repetition.

This is not a draft. This is not a resolution. This is a mirror.

Articles

Seeing 1: Article 1 – On the Nature of the Crisis

The crisis is not only environmental, social or institutional. It is perceptual. It arises from the fragmentation of human consciousness, its separation between the observer and the observed, the self and the other, the idea and the actual.

Seeing 2: Article 2 – On the Failure of Frameworks

All solutions born of division carry division within them. Sustainability becomes measurement, equity becomes comparison, and reform becomes repetition. No framework, however sophisticated, can substitute for intelligence born of direct perception.

Seeing 3: Article 3 – On the Illusion of Time

Psychological time – the movement of becoming, improvement and ideal is illusion but has been made real by thought. Transformation is not a matter of future planning. It begins now, or not at all.

Seeing 4: Article 4 – On the Insecurity of Structures

Nations, markets, institutions and ideologies are built on fear and desire. As long as education, policy and governance operate within these structures without observing their psychological roots, they perpetuate the very disorder they seek to resolve.

Seeing 5: Article 5 – On the Role of the Individual

The individual is not separate from the world. Systems reflect selves. A new world does not arise from mass mobilisation, but from the ending of psychological division in relationship, in daily life, in attention.

Seeing 6: Article 6 – On Awareness as Ground

Awareness is not a practice or a method. It is the silent observation of what is. When fear and desire are seen without escape, they end. This ending is the beginning of intelligence.

Seeing 7: Article 7 – On Action Without Image

Right action is not driven by ideology, identity or reward. It arises when the self is not. Such action is not perfecting the world; it is acting from wholeness. It does not belong to a person, a nation or a belief. It is love in motion.

Seeing 8: Article 8 – On Love and the Sacred

Love is not cultivated. It is not born of attachment or effort. It arises when division ends. Where there is love, there is order. And in that order, the sacred, the unnameable totality of life reveals itself.

Seeing 9: Article 9 – On the End of Seeking

There is no path to truth. All becoming is delay. When the mind stops escaping through belief, effort or knowledge, then there is silence. In that silence, life and death are not two.

Seeing 10: Article 10 – On the Responsibility of Seeing

This document does not ask you to agree. It invites you to see. If it is seen completely, it ends. And in that ending, a different action begins, not later, but now.

Let this not be remembered as an idea, a goal or a set of targets. Do not turn it into a belief to follow, a method to apply or a future to pursue. See what you are, here and now. Not through analysis, comparison or effort, but through direct observation. In that seeing, there is no becoming, no path, no process. There is only the fact, and the ending of illusion. This is not a message to preserve. It is something to see. And once it is seen, it no longer needs to be said. And may awareness, choicelessly, remain seized of the matter.

Closing Note: The Illusion of Reform, the Simplicity of Seeing

There is a strange paradox that marks the end of this book. To the one who has read with intellect and stored what was said, there will be urgency to act, to fix, to spread. The instinct is familiar: take the insight and apply it, convert it into practice, build systems, scale change, design new movements. But to do that is to miss the point entirely.

The world is full of reformers. Some speak of justice, others of compassion. Some build institutions, others break them. All are convinced they are doing good. But conviction is not clarity. It is the hardened form of confusion. A mind that does not understand itself becomes a missionary for change, blindly extending its disorder into the very world it seeks to transform.

Most action is a flight from inner disorder. One feels disturbed by inequality, war and environmental destruction and moves urgently to intervene. But the mind that intervenes is rarely quiet. It carries its own fragmentation: the desire to help, the fear of passivity, the image of being a saviour, the identity of the reformer. So action continues, but it is fragmented activity. One joins movements, launches initiatives, writes policies, teaches values. And yet the world burns on. Each act, no matter how noble, is often driven by the same field of the known: memory, identity, fear, desire. There is no stillness in it. No clarity. No freedom. This is not to say one should do nothing. It is to say that action must emerge from the ground of total perception, not from the momentum of thought. Only a mind that has come upon its own disorder and ended it can respond to the world with intelligence. Today's world demands urgency; we do not have time to waste. But urgency does not justify confusion. If the problem is important, our approach must be clear. We have learnt that punishment, reward, rule or idea cannot change the other; only love can open that door. If it is urgent, then clarity is essential, for what has been clearly seen and understood does not repeat itself. That is what change truly is.

The chapters in this book have attempted to mirror that journey: not by offering a method, but by helping the reader observe. Not to solve global problems through plans, but to see that the crisis out there is inseparable from the crisis within. And if that has been seen, even in part, then something profound has already begun.

For when the mind sees itself in totality, something ends. The seeker disappears. The reformer vanishes. The impulse to become dissolves. And what remains is a choiceless awareness, a stillness without centre, from which action flows, not as reaction, not as identity, but as love and intelligence. This is not heroic. It is not loud. It may not be seen by anyone. But it is the only action that transforms.

To Those Who Arrive Here

This book is not a movement. It is not a blueprint for the future. It is not meant to be followed. It is a mirror, for those who arrive here in silence, without expectation, ready to look. Many will not arrive. Some will skim, others will quote. A few will organise discussions. But only the one who comes without the burden of becoming may find something else here. And what is found is not knowledge. It is not a new idea. It is not another system. It is the ending of psychological time, the dropping of the known, the flowering of clarity. Then, there is nothing to do, and that very nothing is the ground of right action. From that stillness, love moves. Intelligence acts. And the world, though still in flames, no longer frightens or seduces. It is seen clearly, wholly, without separation. That is enough. Let that seeing be the final word.

Index

Printed and bound by CPI Group (UK) Ltd, Croydon, CR0 4YY

15/09/2026

14962484-0003